PCR
STRATEGIES

PCR
STRATEGIES

Edited by

Michael A. Innis
Chiron Corporation, Emeryville, California

David H. Gelfand and John J. Sninsky
Roche Molecular Systems, Inc., Alameda, California

ACADEMIC PRESS
San Diego   New York   Boston   London   Sydney   Tokyo   Toronto

This book is printed on acid-free paper.

Copyright © 1995 by ACADEMIC PRESS, INC.

All Rights Reserved.
No part of this publication may be reproduced or transmitted in any form or by any means, electronic or mechanical, including photocopy, recording, or any information storage and retrieval system, without permission in writing from the publisher.

Academic Press, Inc.
A Division of Harcourt Brace & Company
525 B Street, Suite 1900, San Diego, California 92101-4495

United Kingdom Edition published by
Academic Press Limited
24-28 Oval Road, London NW1 7DX

Library of Congress Cataloging-in-Publication Data

PCR strategies / edited by Michael A. Innis, David H. Gelfand, John J. Sninsky
    p.   cm.
   Includes index.
   ISBN 0-12-372182-2 (case)   ISBN 0-12-372183-0 (paper)
   1. Polymerase chain reaction—Methodology.   I. Innis, Michael A.
II. Gelfand, David H.   III. Sninsky, John J.
QP606.D46P367   1995
   574.87'3282—dc20   94-42506
                                                              CIP

PRINTED IN THE UNITED STATES OF AMERICA
95 96 97 98 99 00 DO 9 8 7 6 5 4 3 2 1

CONTENTS

Contributors ix
Foreword xiii
Preface xv

Part One
KEY CONCEPTS FOR PCR

1. The Use of Cosolvents to Enhance Amplification by the Polymerase Chain Reaction 3
   P. A. Landre, D. H. Gelfand, and R. M. Watson

2. DNA Polymerase Fidelity: Misinsertions and Mismatched Extensions 17
   Myron F. Goodman

3. Extraction of Nucleic Acids: Sample Preparation from Paraffin-Embedded Tissues 32
   Hiroko Shimizu and Jane C. Burns

4. Thermostable DNA Polymerases 39
   Richard D. Abramson

5. Amplification of RNA: High-Temperature Reverse Transcription and DNA Amplification with *Thermus thermophilus* DNA Polymerase 58
   Thomas W. Myers and Christopher L. Sigua

6. Nucleic Acid Hybridization and Unconventional Bases 69
   James G. Wetmur and John J. Sninsky

7. Practical Considerations for the Design of Quantitative PCR Assays   84
   Robert Diaco

Part Two
ANALYSIS OF PCR PRODUCTS

8. Carrier Detection of Cystic Fibrosis Mutations Using PCR-Amplified DNA and a Mismatch-Binding Protein, MutS   111
   Alla Lishanski and Jasper Rine

9. Single-Stranded Conformational Polymorphisms   121
   Anne L. Bailey

10. Analysis of PCR Products by Covalent Reverse Dot Blot Hybridization   130
    Farid F. Chehab, Jeff Wall, and Shi-ping Cai

11. High-Performance Liquid Chromatography Analysis of PCR Products   140
    John M. Wages, Xumei Zhao, and Elena D. Katz

12. Heteroduplex Mobility Assays for Phylogenetic Analysis   154
    Eric L. Delwart, Eugene G. Shpaer, and James I. Mullins

13. PCR Amplification of VNTRs   161
    S. Scharf

Part Three
RESEARCH APPLICATIONS

14. Site-Specific Mutagenesis Using the Polymerase Chain Reaction   179
    Jonathan Silver, Teresa Limjoco, and Stephen Feinstone

15. Exact Quantification of DNA–RNA Copy Numbers by PCR–TGGE   189
    Jie Kang, Peter Schäfer, Joachim E. Kühn, Andreas Immelmann, and Karsten Henco

16. The *in Situ* PCR: Amplication and Detection of DNA in a Cellular Context   199
    Ernest Retzel, Katherine A. Staskus, Janet E. Embretson, and Ashley T. Haase

17. Y Chromosome-Specific PCR: Maternal Blood   213
    Diana W. Bianchi

18. Genomic Subtraction   220
    Don Straus

19. DNA Amplification-Restricted Transcription Translation (DARTT): Analysis of *in Vitro* and *in Situ* Protein Functions and Intermolecular Assembly   237
    Erich R. Mackow

20. DNA and RNA Fingerprinting Using Arbitrarily Primed PCR   249
    John Welsh, David Ralph, and Michael McClelland

21. PCR-Based Screening of Yeast Artificial Chromosome Libraries   277
    Eric D. Green

22. Oligonucleotide Ligands That Discriminate between Theophylline and Caffeine   289
    Rob Jenison, Stanley Gill, Jr., and Barry Polisky

23. Generation of Single-Chain Antibody Fragments by PCR   300
    Jeffrey R. Stinson, Vaughan Wittman, and Hing C. Wong

24. Longer PCR Amplifications   313
    Suzanne Cheng

25. Direct Analysis of Specific Bonds from Arbitrarily Primed PCR Reactions   325
D. A. Carter, A. Burt, and J. W. Taylor

Part Four
ALTERNATIVE AMPLIFICATION STRATEGIES

26. Detection of Leber's Hereditary Optic Neuropathy by Nonradioactive Ligase Chain Reaction   335
John A. Zebala and Francis Barany

27. Detection of *Listeria monocytogenes* by PCR-Coupled Ligase Chain Reaction   347
Martin Wiedmann, Francis Barany, and Carl A. Batt

Index   363

CONTRIBUTORS

Numbers in parentheses indicate the pages on which the authors' contributions begin.

Richard D. Abramson (39), Program in Core Research, Roche Molecular Systems, Inc., Alameda, California 94501

Anne L. Bailey (121), AT Biochem Inc., Division of Applied Technology Genetics Corporation, Malvern, Pennsylvania 19355

Francis Barany (335, 347), Department of Microbiology, Cornell University Medical College, New York, New York 10021

Carl A. Batt (347), Department of Food Science, Cornell University, Ithaca, New York 14853

Diana W. Bianchi (213), Department of Pediatrics, Obstetrics, and Gynecology, Tufts University School of Medicine, Boston, Massachusetts 02111

Jane C. Burns (32), Department of Pediatrics, Pediatric Immunology and Allergy Division, School of Medicine, University of California, San Diego, La Jolla, California 92093

A. Burt (325), Department of Plant Biology, University of California, Berkeley, Berkeley, California 94720

Shi-ping Cai (130), Department of Laboratory Medicine, University of California, San Francisco, San Francisco, California 94143

D. A. Carter (325), Department of Infectious Diseases, Roche Molecular Systems, Inc., Alameda, California 94501

Farid F. Chehab (130), Department of Laboratory Medicine, University of California, San Francisco, San Francisco, California 94143

Suzanne Cheng (313), Department of Human Genetics, Roche Molecular Systems, Inc., Alameda, California 94501

Eric L. Delwart (154), Aaron Diamond AIDS Research Center, New York University School of Medicine, New York, New York 10016

Robert Diaco (84), Diaco Communications, Underhill Center, Vermont 05490

Janet E. Embretson (199), Department of Microbiology, University of Minnesota, Minneapolis, Minnesota 55455

Stephen Feinstone (179), Laboratory of Hepatitis Research, Center for Biologics, Evaluation, and Research, Food and Drug Administration, Bethesda, Maryland 20892

D. H. Gelfand (3), Program in Core Research, Roche Molecular Systems, Inc., Alameda, California 94501

Stanley Gill, Jr. (289), Department of Ligand Analysis, NeXagen, Inc., Boulder, Colorado 80301

Myron F. Goodman (17), Department of Biological Sciences, University of Southern California, Los Angeles, California 90089

Eric D. Green (277), National Center for Human Genome Research, National Institutes of Health, Bethesda, Maryland 20892

Ashley T. Haase (199), Department of Microbiology, University of Minnesota, Minneapolis, Minnesota 55455

Karsten Henco (189), EVOTEC BioSystems GmbH, D-22529 Hamburg, Germany

Andreas Immelmann (189), Department of Analysis, Biomedizinische Test Gmbh, D-60596 Frankfurt, Germany

Rob Jenison (289), Department of Molecular Biology, NeXagen, Inc., Boulder, Colorado 80301

Jie Kang (189), QIAGEN GmbH, D-40724 Hilden, Germany

Elena D. Katz (140), Department of Biotechnology, Perkin-Elmer, Wilton, Connecticut 06897

Joachim E. Kühn (189), Institut für Virologie, Universität zu Köln, D-50935 Köln, Germany

P. A. Landre (3), Engineering Research Division, Lawrence Livermore National Laboratory, Livermore, California 94551

Teresa Limjoco (179), Laboratory of Molecular Microbiology, National Institute of Allergy and Infectious Disease, National Institutes of Health, Bethesda, Maryland 20892

Alla Lishanski (111), Research Department, Syva Co., Palo Alto, California 94304

Erich R. Mackow (237), Departments of Medicine and Microbiology, State University of New York at Stony Brook, and Northport VA Medical Center, Stony Brook, New York 11794

Michael McClelland (249), California Institute of Biological Research, La Jolla, California 92037

James I. Mullins (154), Department of Microbiology, University of Washington, Seattle, Washington 98195

Thomas W. Myers (58), Program in Core Research, Roche Molecular Systems, Inc., Alameda, California 94501

Barry Polisky (289), Department of Research, NeXagen, Inc., Boulder, Colorado 80301

David Ralph (249), California Institute of Biological Research, La Jolla, California 92037

Ernest Retzel (199), Department of Microbiology, University of Minnesota, Minneapolis, Minnesota 55455

Jasper Rine (111), Department of Molecular and Cell Biology, University of California, Berkeley, Berkeley, California 94720

Peter Schäfer (189), Institut für med. Mikrobiologie und Immunologie, Universitätskrankenhaus Eppendorf, D-20251 Hamburg, Germany

S. Scharf (161), Roche Molecular Systems, Inc., Alameda, California 94501

Hiroko Shimizu (32), Department of Pediatrics, Pediatric Immunology and Allergy Division, School of Medicine, University of California, San Diego, La Jolla, California 92093

Eugene G. Shpaer (154), Applied Biosystems, Foster City, California 94404

Christopher L. Sigua (58), Program in Core Research, Roche Molecular Systems, Inc., Alameda, California 94501

Jonathan Silver (179), Laboratory of Molecular Microbiology, National Institute of Allergy and Infectious Diseases, National Institutes of Health, Bethesda, Maryland 20892

John J. Sninsky (69), Program in Core Research, Roche Molecular Systems, Inc., Alameda, California 94501

Katherine A. Staskus (199), Department of Microbiology, University of Minnesota, Minneapolis, Minnesota 55455

Jeffrey R. Stinson (300), Department of Molecular Biology, Biology Skills Center, Dade International, Inc., Miami, Florida 33174

Don Straus (220), Department of Biology, Brandeis University, Waltham, Massachusetts 02254

J. W. Taylor (325), Department of Plant Biology, University of California, Berkeley, Berkeley, California 94720

John M. Wages (140), Department of HIV and Exploratory Research, Genelabs Technologies, Inc., Redwood City, California 94063

Jeff Wall (130), Department of Laboratory Medicine, University of California, San Francisco, San Francisco, California 94143

R. M. Watson (3), Program in Core Research, Roche Molecular Systems, Inc., Alameda, California 94501

John Welsh (249), California Institute of Biological Research, La Jolla, California 92037

James G. Wetmur (69), Department of Microbiology and Human Genetics, Mount Sinai School of Medicine, New York, New York 10029

Martin Wiedmann (347), Department of Food Science, Cornell University, Ithaca, New York 14853

Vaughan Wittman (300), Department of Molecular Biology, Biology Skills Center, Dade International, Inc., Miami, Florida 33174

Hing C. Wong (300), Department of Molecular Biology, Biology Skills Center, Dade International, Inc., Miami, Florida 33174

John A. Zebala (335), Department of Microbiology, Cornell University Medical College, New York, New York 10021

Xumei Zhao (140), Department of Cellular Pathology, Armed Forces Institute of Pathology, Washington, District of Columbia 20306

FOREWORD

During the past forty years there has been tremendous growth and development in molecular biology. Much of this revolution has been driven by breakthroughs that were the result of the introduction of new instruments and new techniques. Each of these innovations built on previous breakthroughs and have in sum led to an exponential increase in our understanding of biology. The rate of growth continues to accelerate. Perhaps there is no other single technique that has contributed as much as the polymerase chain reaction (PCR). The original concept with its impact on biological research has been recognized both in the scientific literature and by a variety of prizes and awards. However, one of the most remarkable properties of the PCR approach has been its malleability and the variety of applications, modifications, and added innovations that have grown from the initial insights. PCR has been applied directly to research in molecular biology, yielding large families of homologous genes and forcing us to address the complexity inherent in the parallel functions found in different cells and tissues. PCR has been used in a variety of assays for specific transcripts and may eventually provide the quantitative tools for the analyses of complex patterns of gene expression. PCR has provided new insight into the process of transcription and the control of cell growth, differentiation, and development. In addition, PCR analyses have been applied to such disparate fields as anthropology, forensics, genomic analysis, and the diagnosis of human disease. Thus, it is almost impossible for a single volume to encompass all of the protocols, recipes, and methods in an archival fashion; rather a series of "cookbooks" are required to keep up with the latest innovations and detailed applications. In a previous book, *PCR Protocols*, a number of authors collected and presented the basic approaches and variations that initially emerged from an exploration of the PCR technique. In this volume, *PCR Strategies*, the authors have tried to focus on strategies that extend PCR in new and different ways. Thus, *in situ* PCR, quantitative PCR, the analysis of PCR products, and long-range PCR are all introduced with an attempt to demonstrate the breadth of approaches that can be taken with this versatile tool. In addition to appropriate protocols, many of the authors have discussed the background, provided references, and described some of the questions and problems faced in using the methods that are described.

While this book does not represent the last word in the development of the polymerase chain reaction, it does represent a chapter in the ongoing conversation that has enriched this area and provided numerous, important tools.

Melvin I. Simon

PREFACE

Introduced just a decade ago, the use of the polymerase chain reaction (PCR) as a means to amplify DNA sequences has revolutionized biological research, from investigations of evolutionary relationships among extinct animals to localization of gene activity in single brain cells. It is impossible to underestimate the importance of PCR.

PCR Protocols (Academic Press, 1990) has served as a sourcebook for researchers using PCR or contemplating its use. Since its publication, the usefulness of PCR has been expanding exponentially, with new applications appearing constantly. *PCR Strategies* equips its readers with knowledge necessary to exploit the power of PCR fully by incorporating these newer techniques into their arsenal.

The book is divided into four parts. The first part gives readers theoretical background they need to understand why procedures work and to troubleshoot effectively. The second part describes various procedures used to analyze PCR products and includes detailed, easy-to-follow, step-by-step protocols for analyzing PCR products by covalent reverse dot blot hybridization and high-performance liquid chromatography. The third part gives examples of the application of PCR to a variety of interesting research areas. The last part describes two other techniques used to amplify DNA, the ligase chain reaction and the PCR-coupled ligase chain reaction.

We hope this book will be a practical tool for researchers in a wide range of areas in the life and clinical sciences and that it will make their busy lives easier.

This book would not have been completed without the expertise and dedication of the contributors and of the editors and other talented people at Academic Press, especially Mr. Craig Panner.

Part One

KEY CONCEPTS FOR PCR

1

THE USE OF COSOLVENTS TO ENHANCE AMPLIFICATION BY THE POLYMERASE CHAIN REACTION

P. A. Landre, D. H. Gelfand, and R. M. Watson

The polymerase chain reaction (PCR) (Saiki *et al.*, 1985) uses repeated cycles of template denaturation, primer annealing, and polymerase extension to amplify a specific sequence of DNA determined by oligonucleotide primers. The use of a thermostable DNA polymerase, such as *Taq* from *Thermus aquaticus,* permits repeated cycles of heating at high temperatures for reliable template denaturation without the addition of more enzyme at each cycle. It also allows higher temperatures to be used to denature templates that have a high G + C content or stable secondary structures. *Taq* DNA polymerase also has a relatively high temperature optimum for DNA synthesis (75° − 80°C). The use of higher annealing/extension and denaturation temperatures increases specificity, yield, and the sensitivity of the PCR reaction.

While the PCR is widely used with great success, certain templates, for example, those with high G + C content or stable secondary structures, may amplify inefficiently, resulting in little or no intended product and often nonspecific products. Thus new applications of the PCR may require some optimization. The PCR parameters most effective in optimizing the reaction are the concentration of enzyme, primers, deoxynucleotide 5'-triphosphates and magnesium; primer annealing/extension temperatures and times; template denaturation temperature and time; and cycle number (Innis and Gelfand, 1990; Saiki, 1989). Use of higher annealing/extension or denaturation temperatures may improve specificity in some cases (Wu et al., 1991). Some investigators have found that incorporation of the nucleotide analog 7-deaza-2'-deoxyguanosine triphosphate (c7dGTP) in addition to deoxyguanosine triphosphate (dGTP) helped destabilize secondary structures of DNA and reduced the formation of nonspecific product (McConlogue et al., 1988).

Various additions or cosolvents have been shown to improve amplification in many applications. Dimethyl sulfoxide (DMSO) increased the amplification efficiency of human leukocyte antigen DQ alpha sequence with the Klenow fragment of *Escherichia coli* DNA polymerase I (Scharf et al., 1986). DMSO and glycerol were found to improve amplification of G + C rich target DNA from herpesviruses and of long products from ribosomal DNAs containing secondary structure (Smith et al., 1990). Bookstein et al. (1990) found that DMSO was necessary to successfully amplify a region of the human retinoblastoma gene. Sarkar et al. (1990) reported that formamide improved specificity and efficiency of amplification of the G + C rich human dopamine D2 receptor gene, whereas DMSO and dc7GTP did not improve specificity in this application. Hung et al. (1990) found that tetramethylammonium chloride (TMAC), a reagent that improves stringency of hybridization reactions, improved the specificity of PCR, although at concentrations far below those effecting hybridization reactions. DMSO and nonionic detergents have been reported to improve DNA sequencing, presumably by reduction of secondary structures (Winship, 1989; Bachmann et al., 1990).

It is not fully known which PCR parameters are influenced by cosolvents. Some cosolvents, such as formamide, are known to reduce the melting temperature (T_m) of DNA, thus possibly affecting the T_m of the primers and template in PCR. Also, Gelfand and White (1990) reported that agents such as DMSO and formamide inhibit *Taq* DNA polymerase activity in incorporation assays. In order to understand how cosolvents influence the PCR, their effects on template melting properties and on the activity and thermostability of *Taq* DNA polymerase were determined.

Materials and Methods

Taq DNA Polymerase Activity Assay

Taq DNA polymerase activity was assayed by determining the level of incorporation of labeled nucleotide monophosphate into an activated salmon sperm DNA template (Lawyer et al., 1989). The assay was performed for 15 min at 74°C (the optimum temperature for *Taq* DNA polymerase activity) and at 60°C (to minimize melting effects on the template). Reaction mixes (50 μl) contained 25 mM TAPS–HCl (pH 9.5), 50 mM KCl, 2 mM MgCl$_2$, 1 mM β-mercaptoethanol, 200 μM each deoxyadenosine triphosphate (dATP), deoxythymidine triphosphate (dTTP), and deoxyguanosine triphosphate, 100 μM α-[^{32}P]deoxycytidine triphosphate (dCTP) at 100 cpm/picomole; 12.5 μg activated salmon sperm DNA, and co-solvents. Reactions were stopped with 10 μl 60 mM EDTA, precipitated with trichloroacetic acid (TCA), and filtered through Whatman GF/C filters. The filters were counted in a liquid scintillation counter and the picomoles of incorporated label calculated from the CPM and specific activity. Percent control was calculated from the number of picomoles incorporated in the presence of cosolvents divided by the picomoles incorporated in the absence of cosolvents.

Thermal Inactivation Assay

The thermostability of *Taq* DNA polymerase was determined by steady-state thermal inactivation performed in a constant-temperature waterbath at 95° or 97.5°C. The enzyme was prepared in a standard GeneAmp PCR (Perkin–Elmer) master mix containing: 10 pmoles each of lambda primers 1 and 2200 μM each dATP, dTTP, and dGTP, 1.25 units *Taq* DNA polymerase, 0.5 ng lambda DNA template, 1X PCR buffer, and cosolvents. The PCR buffer contained 10 mM Tris–HCl (pH 8.3), 50 mM KCl, and 2 mM MgCl$_2$. Fifty microliters were dispensed to 0.5-ml tubes and overlaid with 50 μl of light mineral oil. Following incubation at high temperature, 5 μl from each tube, in duplicate, were assayed for DNA polymerase activity by measuring the incorporation of labeled nucleoside monophosphate into activated salmon sperm DNA, as described previously. The percent of activity remaining for each time point was determined as the fraction of the initial activity for each enzyme. The half-life, $t_{1/2}$, was the time at which there was 50% remaining activity.

Determination of DNA Temperature Melting Profiles

DNAs with different G + C contents were denatured in PCR buffer [10 mM Tris–HCl (pH 8.3), 50 mM KCl, and 3 mM MgCl$_2$], and in PCR buffer containing cosolvents to measure their respective melting temperatures (T_m) and strand separation temperatures (T_{ss}). Samples overlaid with mineral oil were heated in a spectrophotometer in stoppered and thermojacketed cells. All samples were heated at an initial concentration of about 1 OD^{260}/ml and the OD^{260} and temperature were recorded as the temperature was increased. Optical density was plotted as the fractional change in OD^{260} with the T_m determined as the temperature at the midpoint of the hyperchromic transition, and the T_{ss} as the temperature at the end of the transition.

PCR Reactions

The effects of the cosolvents glycerol and formamide and the denaturation temperature on PCRs were measured in standard 50-μl GeneAmp PCR (Perkin–Elmer) reactions containing 1.25 units *Taq* DNA polymerase, 0.5 ng DNA, 10 pmole each lambda primers (500 bp product), 200 μM each dNTP, cosolvents, and PCR buffer. PCR buffer contained 10 mM Tris–HCl (pH 8.3), 50 mM KCl, and 2 mM MgCl$_2$. The reactions were thermocycled in a Perkin–Elmer 480 thermocycler programmed for 25 cycles and two-temperature PCR: denaturation temperatures of 90°, 95°, or 98°C for 1 min and an anneal/extend temperature of 60°C for 1 min were used. A 5 min extension at 75°C was done after thermocycling. The samples were analyzed on 5% polyacrylamide gels by ethidium bromide staining.

For the effect of glycerol and formamide on amplification from a high G + C template, the 50-/μl PCRs contained the following: 1.25 units *Taq* DNA polymerase (Perkin–Elmer), 0.01 μg plasmid DNA containing the *Thermus thermophilus* DNA polymerase gene, 50 pmoles each primer (amplified sequence: 2600 bp segment of the cloned gene with 67% G + C), 200 μM each dNTP, cosolvents, and PCR buffer. The thermocycler was programmed for two-temperature PCR: a denaturation temperature of 96°C for 1 min and an anneal/extend temperature of 60°C for 1 min. A 5 min extension of 75°C was done after thermocycling. Samples were analyzed on a 0.7% agarose gel by ethidium bromide staining.

Results

Effect of Cosolvents on *Taq* DNA Polymerase Activity

There were two types of effects seen on *Taq* DNA polymerase activity. A decrease in polymerase activity at both 74° and 60°C was exhibited in the presence of DMSO and glycerol, which is characteristic of enzyme inhibition (Fig. 1A). In contrast, 1-methyl-2-pyrolidone (NMP) and formamide were inhibitory at 74°C but stimulatory at 60°C, presumably as a result of increased template denaturation. *Taq* DNA polymerase has optimal activity at 74°C on activated salmon sperm DNA. Cosolvents inhibited activity with this template at this temperature. With 10% glycerol, 79% of the activity remained and with 10% NMP about 20% remained. Ten percent formamide or DMSO produced about 50% of the control activity.

When the activity assay was done at 60°C, NMP and formamide increased activity compared with the control (Fig. 1B). This is presumably due to melting effects on the activated salmon sperm DNA template used. This template is composed of undefined nicks and gaps, and an increased incorporation of nucleotide monophosphate into this template may occur if pieces of DNA are melted.

The effects of combinations of glycerol and formamide or glycerol and NMP on activity were determined at 74° and 60°C (data not shown in Fig. 1). Formamide (5 or 10%) and glycerol (10%) are somewhat more than additive in their inhibition of *Taq* DNA polymerase activity at 74°C. However at 60°C, the activity of the combinations is about the same as the activity with formamide alone. Glycerol and NMP in combination did not show an increase in activity at 60°C, as is seen with NMP alone.

Effect of Cosolvents on the Thermostability of *Taq* DNA Polymerase

The thermostability of *Taq* DNA polymerase was determined as described earlier (Fig. 2). At 97.5°C, *Taq* DNA polymerase has a half-life ($t_{1/2}$) after steady-state thermal inactivation of 11 min (Fig. 2A).

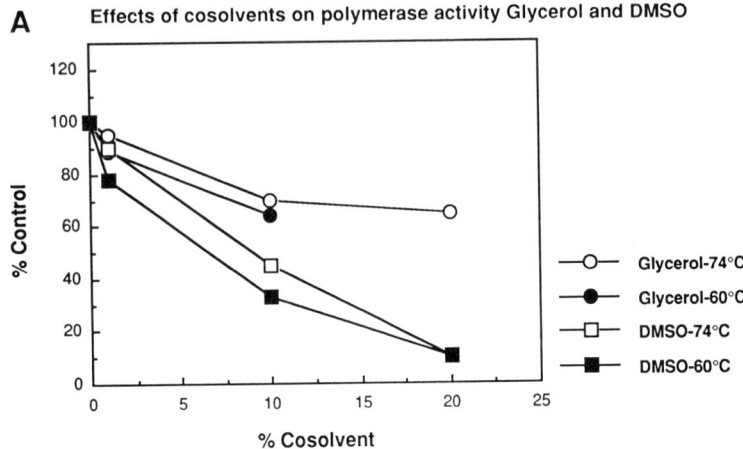

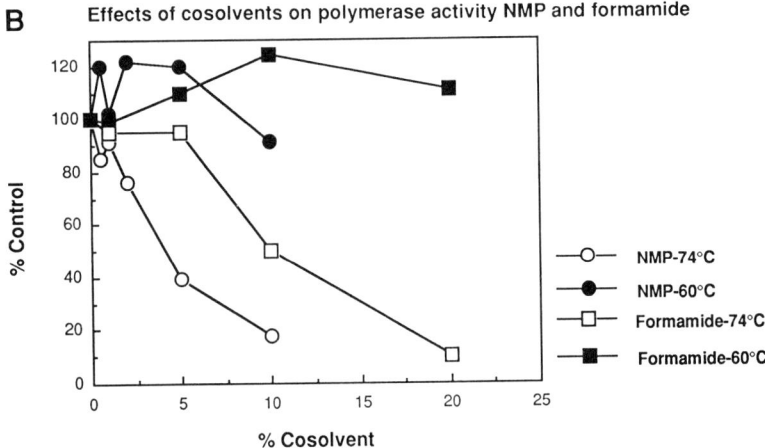

Figure 1 Effect of cosolvents on *taq* DNA polymerase activity. *Taq* DNA polymerase activity was assayed by determining the level of incorporation of labeled nucleotide monophosphate into activated salmon sperm DNA template. The assay was performed for 15 min at 74°C (the temperature optimum for *Taq* DNA polymerase activity) and at 60°C (to determine template melting effects). Reaction mixes (50 μl) contained: 25 mM TAPS-HCl (pH 9.5); 50 mM KCl; 2 mM MgCl$_2$; 1 mM β-mercaptoethanol; 200 μM each dATP, dTTP, and dGTP; 100 μM α-[^{32}P]dCTP (100 cpm/picomole); 12.5 μg activated salmon sperm DNA and cosolvents. Reactions were stopped with 10 μl 60 mM EDTA, TCA precipitated and filtered through Whatman GF/C filters. The number of picomoles of labeled nucleotide monophosphate were determined. Percent control was calculated from the number of picomoles incorporated in the presence of cosolvents divided by the picomoles incorporated in the absence of cosolvents. Panel A shows the effects of glycerol and dimethylsulfoxide (DMSO) on *Taq* DNA polymerase activity. Panel B shows the effects of formamide and *N*-methyl-2-pyrrolidone (NMP) on *Taq* DNA polymerase activity.

At 95°C the $t_{1/2}$ is 40 min (Fig. 2B). Glycerol improves the thermostability of *Taq* from 11 min to more than 60 min with 20% glycerol at 97.5°C. Formamide reduces the thermostability from 40 min at 95.0°C to 16 min with 5% formamide, and 2 min with 10% formamide. Combining glycerol and formamide improves the thermostability of *Taq* over formamide alone: 5% formamide and 10% glycerol had a $t_{1/2}$ of 24 min compared with 16 min for 5% formamide alone.

Table 1A summarizes the effects of single cosolvents on thermostability. Of the cosolvents tried, only glycerol improved the thermostability of *Taq*. The other cosolvents reduced thermostability, NMP most dramatically; at 5% NMP the enzyme was immediately inactivated at 97.5°C. Table 1B shows a summary of combinations of glycerol and NMP or formamide. In both cases, glycerol ameliorated the thermal inactivation of NMP or formamide.

Effect of Cosolvents on Melting Temperature of DNA

The effect of cosolvents on the melting temperature of DNA depended on template composition. Figure 2A shows that the T_ms of various DNAs were lowered using 5% formamide with 10% glycerol, but to different degrees on each template. The templates had different G + C contents and the measured T_ms were proportional to the G + C content, as expected. The change in T_m using cosolvents, or delta T_m, was also somewhat dependent on the G + C content of the template and ranged from $-6.5°$ to $-4.5°C$ for the various templates. Also shown in Table 2A is the effect of this cosolvent on the strand dissociation temperature, or T_{ss}. The T_{ss} is significantly lowered for all the DNAs examined, although the delta T_{ss} shows less-dependence on the G + C content of the template than does the delta T_m.

The effect of different cosolvents on λ DNA was determined and is shown in Table 2B. All the cosolvents studied lowered the T_m, ranging from a reduction of about 1.1° per percent for NMP to 0.3° per percent for glycerol. The effect of formamide on delta T_m was close to reported values of 0.7°C per percent. The effect of the combination of 5% formamide and 10% glycerol on T_m was additive.

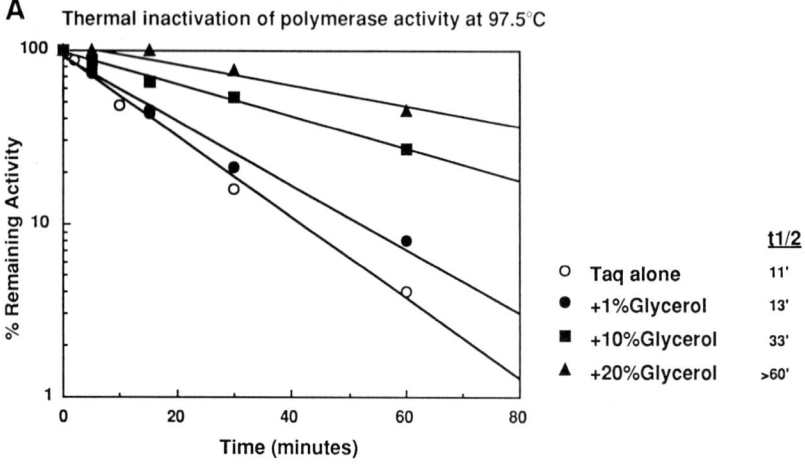

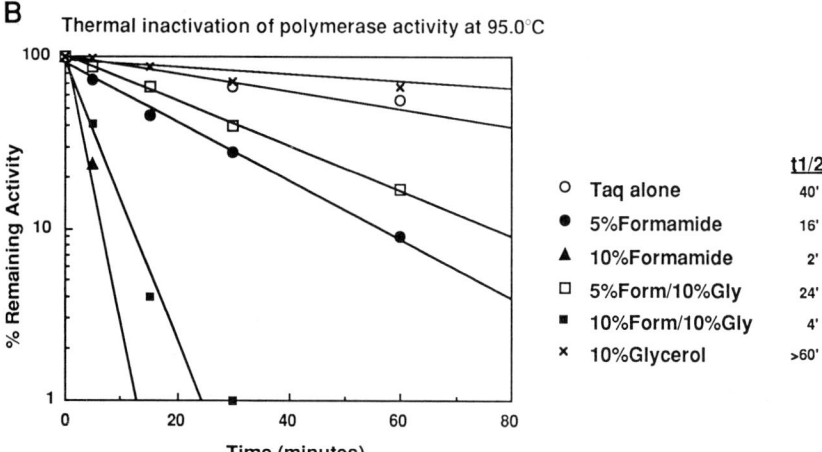

Figure 2 Thermostability of *Taq* DNA polymerase in the presence of cosolvents at 97.5° and 95°C. The thermostability of *Taq* DNA polymerase was determined by steady-state thermal inactivation. Thermal inactivation was performed in a constant temperature water-bath at 95° or 97.5°C. The temperature was monitored using a calibrated thermometer and fluctuated no more than 0.5°C during the experiment. Enzyme was prepared in a standard GeneAmp PCR master mix containing the following: 10 pmol lambda primers 1 and 2; 200 μM each dATP, dTTP, and dGTP; 1.25 U *Taq* DNA polymerase; 0.5 ng lambda template; 1× PCR buffer; and cosolvents. PCR buffer contained: 10 mM Tris–HCl, pH 8.3; 50 mM KCl; and 2 mM MgCl$_2$. Fifty μl were dispensed to 0.5 ml tubes and overlaid with 50 μl of light mineral oil. Five μl from each tube, in duplicate, were assayed for DNA polymerase activity by the incorporation of labeled nucleotide monophosphate into activated salmon sperm DNA as described previously. The percent remaining activity for each timepoint was determined as a fraction of the initial activity for each enzyme. The half-life, $t_{1/2}$, was the time at which there was 50% remaining activity. Panel A shows the thermal inactivation of *Taq* DNA polymerase at 97.5°C in the presence of glycerol. Panel B shows thermal inactivation of *Taq* at 95°C with the combination of glycerol and formamide.

Table 1A

Effects of Single Cosolvents on Thermostability

Cosolvent		$t_{1/2}$(min) 97.5°C
Taq	alone	9–11
Glycerol	1%	13
	10%	33
	20%	>60
DMSO	1%	10
	10%	6
	20%	4
Formamide	1%	9
	10%	1.5
	20%	3
NMP	0.5%	6
	1%	4
	5%	0
KCl	50mM	9–11
	25mM	7
	10mM	5
	0mM	3
K_2SO_4	100mM	>35
	50mM	35
	25mM	20
	5mM	7

Table 1B

Combinations of Cosolvents and Thermostability

Cosolvent		$t_{1/2}$ (min) 95°C	97.5°C
Taq	alone	40	11
Glycerol	10%	>60	23
Formamide	5%	16	3
	10%	2	<1
10% Glycerol–5% formamide		24	5
10% Glycerol–10% formamide		4	1.5
NMP	0.5%	26	6
Glycerol	5%	>60	20
	10%	>60	23
	15%	>60	32
0.5% NMP–5% glycerol		32	11
0.5% NMP–10% glycerol		42	14
0.5% NMP–15% glycerol		38	18

Table 2A

DNA Melting and Strand Dissociation in Cosolvents

DNA	%G + C	PCR buffer $T_m(°)$	PCR buffer $T_{ss}(°)$	PCR buffer 5% formamide 10% glycerol $T_m(°)$	$T_{ss}(°)$	$\Delta T_m(°)$	$\Delta T_{ss}(°)$
Lambda	50	90.2	95	84	90–91	−6.2	−4-5
Herring sperm	42	86.5	(98)*	82	86–92	−4.5	(−6)*
Clostridium perfringens	32	80.6	85	76	78–79	−4.6	−6
Micrococcus luteus	72	99.5	(102.5)*	93	94–97	−6.5	(−7)*
Human-HL60	42	86	96–98	80.6	87–92	−5.4	−6

* Estimated.

Cosolvents and PCR

After determining the effects of cosolvents on activity, thermostability, and template melting, a PCR was designed to look at effects of cosolvents on denaturation temperature in a PCR. A standard lambda system (as described earlier) was used with denaturation temperatures (T_{den}) set at 90°, 95°, and 98°C (Fig. 3). The reactions were expected not to work with a T_{den} of 90°C. At 95°C the reactions were expected to work and at 98°C were expected to work but not optimally because of enzyme inactivation. As expected, a T_{den} of 90°C did not allow amplification. The addition of 5% and 10% glycerol improved amplification. Glycerol (20%) has slightly less product,

Table 2B

Effects of Cosolvents on λ DNA Denaturation

DNA	$T_m(°)$	$\Delta T_m(°)$
PCR buffer Alone	90.2	—
+10% Glycerol	87.2	−3.0
+10% DMSO	84.4	−5.8
+10% NMP	79.6	−10.6
+10% Formamide	83.6	−6.6
+5% Formamide and 10% Glycerol	84.0	−6.2

1. Use of Cosolvents to Enhance Amplification   13

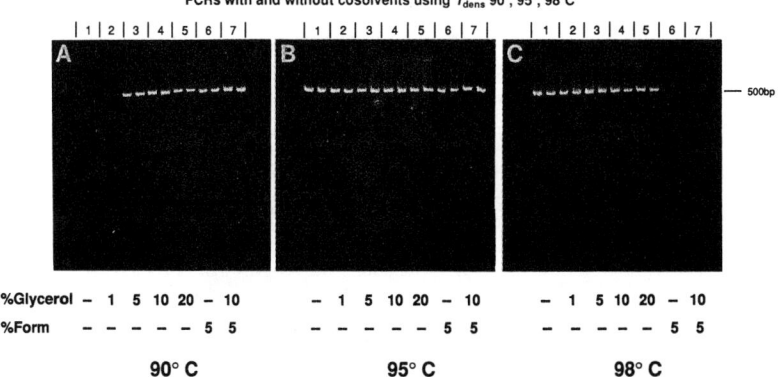

Figure 3 Effect of glycerol and formamide and denaturation temperature on PCR. PCR reactions (50 μl) contained the following: 1.25 U *Taq* DNA polymerase; 0.5 ng lambda DNA; 10 pmol each lambda primers (lambda 500 bp product); 200 μM each dNTP; cosolvents; and PCR buffer. PCR buffer contained: 2 mM MgCl$_2$, 50 mM KCl, and 10 mM Tris–HCl, pH 8.3. The reactions were thermocycled for 25 cycles in a Perkin–Elmer Cetus thermocycler programmed for two temperature PCR: denaturation temperatures of 90°, 95°, or 98°C for one min and an anneal/extend temperature of 60°C for one min. A five min extension at 75°C was done after thermocycling. The samples were analyzed on 5% polyacrylamide gels by ethidium bromide staining. PCR was performed in the absence of cosolvent (lane 1); 1, 5, 10, and 20% glycerol (lanes 2 through 5); 5% formamide (lane 6); or 5% formamide and 10% glycerol (lane 7). Reactions were done in duplicate. Panels A, B, and C show PCR products from reactions performed at denaturation temperatures of 90°, 95°, and 98°C, respectively.

probably owing to inhibition of *Taq* activity. The addition of 5% formamide or 5% formamide plus 10% glycerol also improved amplification at a T_{den} of 90°C (Fig. 3A). The effects of cosolvents here are presumably due to a lowering of the T_{ss} of the template and more reliable denaturation of the template. At a higher T_{den} of 98°C, the samples without cosolvents amplified well; however, reactions containing 1% or 5% glycerol looked somewhat better owing to their stabilizing effects on the enzyme (Fig. 3C). With glycerol added at 10% or 20%, there was less product because of its inhibition of enzyme activity. Reactions with 5% formamide and 5% formamide plus 10% glycerol had no product, presumably owing to the effect formamide has on thermostability and activity of *Taq* DNA polymerase.

The effect of cosolvents on amplification of a template with a high G + C content was explored. We attempted to amplify a region of *Thermus thermophilus* DNA contained on a plasmid (67% G + C

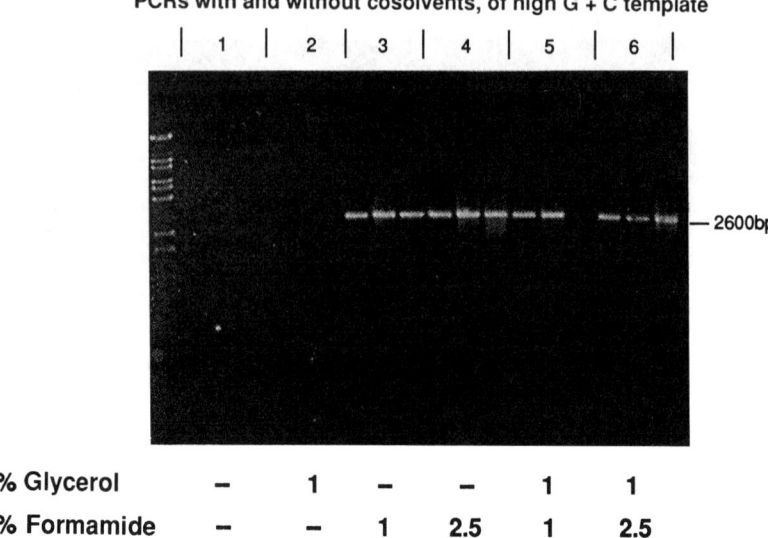

Figure 4 Use of glycerol and formamide to improve PCR amplification from a high G/C content template. PCR reactions (50 μl) contained the following: 1.25 U Taq DNA polymerase; 0.01 μg target DNA (plasmid DNA containing *Thermus thermophilus* DNA polymerase gene, 55% G/C); 50 pmol each primer (amplified sequence: 2600 bp, 67% G/C); 200 μM each dNTP; cosolvents; and PCR buffer. A Perkin–Elmer Cetus thermocycler was programmed for two temperature PCR: a denaturation temperature of 96°C for one min and an anneal/extend temperature of 60°C for two min. A five min extension at 75°C was done after thermocycling. Samples were analyzed on 0.7% agarose gels by ethidium bromide staining. The gel shows PCR products from 20, 25, and 30 cycles performed with: no cosolvent added (set 1); 1% glycerol (set 2); 1% formamide (set 3); 2.5% formamide (set 4); 1% glycerol and 1% formamide (set 5); 1% glycerol and 2.5% formamide (set 6).

in the amplified region) but were not successful (Fig. 4). The addition of small amounts of formamide improved amplification, as did the combinations with 1% glycerol. The absence of product in the 30-/cycle lane of 1% glycerol and 1% formamide was probably a "dropout" or otherwise unexplained failure to amplify.

Discussion

Many investigators have shown that the addition of cosolvents can improve PCR amplification efficiency or specificity in many cases.

We have shown that cosolvents can influence template melting properties, *Taq* DNA polymerase activity, and thermostability. It is likely that appropriate cosolvents for PCR could be predicted based on these effects, but this is complicated by their diverse influences on enzyme activity.

Our data show that the use of glycerol and formamide in a PCR facilitated both amplification at lower denaturation temperatures and amplification from templates with high G + C content. The ability to lower T_m by the use of cosolvents may allow the use of lower denaturation temperatures in a PCR, but may also affect the annealing temperature of the primers. Five percent formamide and 10% glycerol can lower the template melting point by 6.2°. As shown in Table 1, the thermostability of *Taq* with these cosolvents at 95°C is 24 min compared with 40 min alone. The thermostability of *Taq* at 97.5°C is 11 min. If one is operating at T_{den} of 95°C with both 5% formamide and 10% glycerol, the effective T_{den} is 6.2° higher and the thermostability of *Taq* is greater than if it were programmed for a higher T_{den} without cosolvents.

It is important to remember that in some cases the specificity of the PCR may only appear to be improved with cosolvents. If the enzyme is partially inhibited or other parameters of the PCR adversely affected, less overall amplification of both intended and unintended sequences way occur without any real improvement in reaction specificity.

Literature Cited

Bachmann, B., W. Luke, and G. Hunsmann. 1990. Improvement of PCR amplified DNA sequencing with the aid of detergents. *Nucleic Acids Res.* **18**:1309.

Bookstein, R., C. Lai, H. To, and W. Lee. 1990. PCR-based detection of a polymorphic Bam H1 site in intron 1 of the human retinoblastoma (RB) gene. *Nucleic Acids Res.* **18**:1666.

Gelfand, D. H., and T. J. White. 1990. Thermostable DNA polymerases. In: *PCR protocols: A guide to methods and applications* (eds. M. A. Innis, D. H. Gelfand, J. J. Sninsky, and T. J. White), pp. 129–141. Academic Press, San Diego.

Hung, T., K. Mak, and K. Fong. 1990. A specificity enhancer for polymerase chain reaction. *Nucleic Acids Res.* **18**:4953.

Innis, M. A., and D. H. Gelfand. 1990. Optimization of PCRs. In: *PCR protocols: A guide to methods and applications* (eds. M. A. Innis, D. H. Gelfand, J. J. Sninsky, and T. J. White), pp. 3–12. Academic Press, San Diego.

Lawyer, F. C., S. Stoffel, R. K. Saiki, K. Myambo, R. Drummond, and D. H. Gelfand. 1989. Isolation, characterization, and expression in *Escherichia coli* of the DNA polymerase gene from *Thermus aquaticus*. *J. Biol. Chem.* **264**:6427–6437.

McConlogue, L., M. D. Brow, and M. A. Innis, 1988. Structure-independent DNA amplification by PCR using 7-deaza-2'-deoxyguanosine. *Nucleic Acids Res.* **16**:9869.

Saiki, R. K. 1989. The design and optimization of the PCR. In: *PCR technology-principles and applications for DNA amplification* (ed. H. A. Erlich), pp. 7–16. Stockton Press, New York.

Saiki, R. K., S. Scharf, F. Faloona, K. B. Mullis, G. T. Horn, H. A. Erlich, and N. Arnheim. 1985. Enzymatic amplification of β-globin genomic sequences and restriction site analysis for diagnosis of sickle cell anemia. *Science* **230**:1350–1354.

Sarkar, G., S. Kapelner, and S. S. Sommer. 1990. Formamide can dramatically improve the specificity of PCR. *Nucleic Acids Res.* **18**:7465.

Scharf, S. J., G. T. Horn, and H. A. Erlich. 1986. Direct cloning and sequence analysis of enzymatically amplified genomic sequences. *Science* **233**:1076–1078.

Smith, K. T., C. M. Long, B. Bowman, and M. M. Manos. 1990. The use of cosolvents to enhance PCR amplifications. *Amplifications* **5**:16–17.

Winship, P. R. 1989. An improved method for directly sequencing PCR. *Nucleic Acids Res.* **17**:1266.

Wu, D. Y., L. Ugozzoli, B. K. Pal, J. Qian, and R. B. Wallace. 1991. The effect of temperature and oligonucleotide primer length on the specificity and efficiency of amplification by the polymerase chain reaction. *DNA Cell Biol.* **10**:233–238.

2

DNA POLYMERASE FIDELITY: MISINSERTIONS AND MISMATCHED EXTENSIONS

Myron F. Goodman

PCR has become a standard method for cloning genes. Since it is necessary to minimize replication errors to prevent mutations from being introduced into a DNA product during replication, amplification of selected genes using PCR requires that the target DNA sequences be copied with exceptionally high fidelity. A related PCR technique, "allele-selective amplification," is used to amplify rare mutant genes differing from a wild-type allele by a single base substitution. A primer molecule complementary to a mutant allele, but forming a mismatch at its 3' terminal with the wild-type sequence, is used to copy the mutant DNA. Selective amplification of a mutant DNA sequence in the presence of a large excess of wild-type sequences requires that the polymerase elongate and amplify primer-template DNA containing correctly matched terminals in preference to those with mismatched 3' terminal base pairs. This chapter discusses the properties of DNA polymerases, DNA, and deoxyribonucleoside triphosphate (dNTP) substrates that determine the fidelity of nucleotide incorporation and favor the elongation of correctly matched over mismatched primer terminals.

Discrimination during Insertions and Mismatched Extensions

DNA polymerases strongly favor incorporation of proper Watson–Crick base pairs. At each template position, there are one correct and three incorrect dNTPs competing for insertion. Several checkpoints in the enzymatic pathway for insertion can be used to discriminate between formation of matched and mismatched base pairs. These include differences in the binding stabilities of matched and mismatched substrate dNTPs to the polymerase–DNA complex, enzyme conformational changes that allow preferential escape of an incorrectly bound substrate, and differences in the rates of catalysis for correct and incorrect phosphodiester bond formation. Nucleotide misinsertion frequencies are, on average, between 10^{-3} and 10^{-5}; for a general review, see Echols and Goodman (1991).

At least three major factors can affect insertion error rates: (1) the type of mispair formed; G(primer)·T and T(primer)·G transition mispairs can generally be made more easily than say C·C or G·G transversion mispairs (Mendelman et al., 1989); (2) the sequence context where the mispair occurs (Mendelman et al., 1989); and (3) the identity of the polymerase (i.e., different polymerases appear to exhibit different mutational spectra; Kunkel and Bebenek, 1988). When *Taq* polymerase, which is devoid of 3'→5' exonucleolytic proofreading activity, is used to amplify a gene to be cloned, one can expect to find A·T ↔ G·C transitions which result from G·T or A·C mismatches about once every 10^3 to 10^4 base pairs copied. Transversions arising from Pur·Pur and Pyr·Pyr mismatches are likely to occur one to two orders of magnitude less frequently—about one per 10^5 to 10^6 base pairs.

Average nucleotide misinsertion rates for a wide variety of mispairs have been measured for several well-characterized polymerases. Mutational hot and cold spots, representing large positive and negative deviations in mutation rates, occur through mechanisms that are likely to involve base stacking interactions between incoming substrate dNTPs and primer 3' terminals (Mendelman et al., 1989). The random nature of mutations, and their dependence on the properties of the polymerases used and the sequences being copied, make it mandatory to sequence PCR products to ensure that errors are not introduced during amplification.

The presence of a 3'-exonuclease proofreading activity, either as part of a single polypeptide, for example, *Escherichia coli* DNA polymerase I or bacteriophage T4 polymerase, or as part of a complex involving separate polymerase and proofreading subunits (e.g., the E. coli pol III holoenzyme complex; Kornberg and Baker, 1992), can, on average, reduce insertion errors by about 5- to 500-fold (Schaaper, 1988). The same three factors that affect nucleotide misinsertion efficiencies—identity of the mispair, sequence context, and specific enzyme properties—also control proofreading efficiencies (Echols and Goodman, 1991). There are thermal stable polymerases available, for example, vent polymerase, that have proofreading exonucleases. Although DNAs amplified by polymerases with proofreading capability should contain fewer errors unless they happen to be less accurate at the insertion step, the product DNAs still must be sequenced to guarantee that errors are not introduced during amplification. Note that active proofreading exonucleases can excise from about 5 to 15% of correctly inserted nucleotides (Clayton et al., 1979; Fersht et al., 1982), a factor that might affect PCR cycling times and DNA product yields.

Polymerases used to carry out allele-selective amplification, such as Taq, must be devoid of proofreading activity. The basis for allele-selective amplification is the preferential extension of matched over mismatched primer 3' terminals. Thus, the absence of a proofreading exonuclease is necessary to eliminate the possibility of proofreading and thus destroying the mismatched allele. A key determinant of the fidelity required to effectively carry out allele-selective amplification is the specificity of excluding mismatched compared with correctly matched primer 3' terminals (Petruska et al., 1988; Perrino and Loeb, 1989; Mendelman et al., 1990; Creighton et al., 1992). As for the case of misinsertion spectra, which may differ significantly depending on the polymerase used to copy DNA, primer extension fidelity is also strongly dependent on polymerase identity (Mendelman et al., 1990; Huang et al., 1992). Taq polymerase, for example, exhibits high specificity and favors extension of matched over mismatched base pairs (Creighton et al., 1992), while avian myeloblastosis virus (AMV) and human immunodeficiency virus (HIV-1) reverse transcriptases, in contrast, show less discrimination in extension of terminally mismatched compared to matched primer-templates (Mendelman et al., 1990; Yu and Goodman, 1992).

Additional important factors that can alter the fidelity of insertion and extension are reaction pH and the presence of divalent cations other than Mg^{2+}. There is ample evidence based on NMR and enzyme kinetic studies that protonation and ionization of nucleotide bases can stablize base mispairs in duplex DNA (Lawley and Brooks, 1962; Sowers et al., 1986a,b, 1987, 1989) and stimulate misinsertions by polymerases (Driggers and Beattie, 1988; Yu et al., 1993). Replacement of Mg^{2+} by other divalent cations, for example, Mn^{2+}, has been observed to strongly degrade polymerase fidelity (Goodman et al., 1982; Fersht et al., 1983; Eger et al., 1991; Copeland et al., 1993).

Polymerase processivity is defined as the average number of nucleotides inserted per template binding event. Highly processive polymerases synthesize long stretches of DNA before dissociating from the primer-template. It is possible that increased processivity could result in either increased or decreased fidelity, depending on the properties of the polymerase–primer–template interaction. For example, increased mutations in homopolymeric regions of DNA could occur if the polymerase continued synthesis on distorted primer terminals rather than dissociating. Alternatively, it is possible that factors that increase processivity could reduce strand slippage and thereby enhance fidelity. Kunkel and Wilson and co-workers (Bebenek et al., 1993) have suggested that frameshift fidelity for HIV-1 reverse transcriptase in homopolymeric runs correlates with processivity—higher processivity leads to higher fidelity. The degree of processivity could also affect the relative efficiencies of extending mismatches compared with correct matches. This effect would depend on whether or not dissociation rates from mismatched compared to correctly matched primer terminals were influenced by polymerase processivity. Current data appears to suggest that polymerases tend to bind with roughly equal affinities to both matched and mismatched primer-template terminals (Mendelman et al., 1990; Wong et al., 1991; Creighton et al., 1992; Huang et al., 1992; Yu and Goodman, 1992; Copeland et al., 1993).

For proofreading proficient enzymes, the fraction of extended to excised mismatched terminals depends on a number of factors, including the concentration of dNTP substrates (Clayton et al., 1979; Fersht, 1979) and the ratio of exonuclease to polymerase activity (Muzyczka et al., 1972) which, in turn, depends on the efficiency of switching between polymerase and nuclease active sites (Sinha, 1987; Cowart et al., 1989; Joyce, 1989; Reddy et al., 1992). As is true for the case of exonuclease-deficient polymerases, it remains an open question whether processivity per se directly affects misincorporation frequencies by proofreading proficient polymerases.

A Simple Kinetic Picture for Nucleotide Insertion and Mismatched Extensions

It is useful to briefly focus on the similarities and differences in elements governing fidelities of nucleotide insertion and extension. Figure 1 depicts the competition events governing insertion (Fig. 1a) and extension fideltity (Fig. 1b). For the case of insertion, "right" (R) and "wrong" (W) dNTP substrates compete with each other to form a complex with the polymerase bound to a correctly matched primer 3' terminal (Fig. 1a). The nucleotide misinsertion efficiency f_{ins}, is defined as the ratio of wrong to right insertions, when dWTP and dRTP are present at equal concentrations in solution. A direct measurement of f_{ins} can be made in a double-label experiment, for example, using [^{32}P]dWTP and [^{3}H]dRTP, by determining the ratio of the two isotopes in the product DNA (Echols and Goodman, 1991). Insertion fidelity is defined as the reciprocal of the misinsertion efficiency, $F = 1/f_{ins}$.

a. INSERTION KINETICS

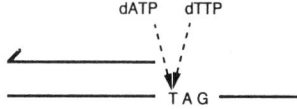

b. EXTENSION KINETICS

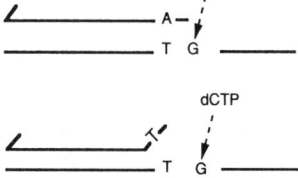

Figure 1 Competition events for nucleotide insertion and mismatch extension. (a) Insertion kinetics with right (dATP) and wrong (dTTP) substrates competing for addition onto the same 3'-primer terminal. (b) Extension kinetics for addition of the "next correct" (dCTP) nucleotide onto either matched or mismatched primer 3' terminals.

Although simple in concept, the direct competition measurement is limited in practice for several reasons. First, competition between normal dNTP substrates strongly favors correct insertions, usually by $> 10^4$, and thus a large excess of dWTP over dRTP is required to detect misincorporation. However, the presence of large nucleotide pool imbalances can introduce two major sources of error in the analysis: (1) significant errors in estimating f_{ins} can arise when multiplying dWMP/dRMP incorporation ratios by a large factor to correct for dWTP/dRTP pool bias, especially when nucleotide misincorporation levels are not much above background; (2) the presence of dWTP in a millimolar concentration range can inhibit polymerization, either through substrate inhibition or by "chelation" of Mg, the normal divalent cation required for polymerase activity.

In place of the competition assay for insertion fidelity, a convenient kinetic method can be used to measure f_{ins}, where insertion of right and wrong substrates is measured in separate reactions as a function of dNTP concentration (Boosalis et al., 1987; Randall et al., 1987). In the kinetic assay, the ratio of V_{max}/K_m is determined from the "linear" portion of the insertion velocity versus [dNTP] Michaelis–Menten curves for right and wrong substrates. The misinsertion efficiency is given by (Fersht, 1985),

$$f_{ins} = (V_{max,W}/K_{m,W})/(V_{max,R}/K_{m,R}). \tag{1}$$

For insertion fidelity (Fig. 1a), either a right or wrong nucleotide is added in the identical sequence context, that is, where the polymerase is bound to the same (correctly matched) primer terminal. As a consequence, f_{ins} contains no explicit polymerase–DNA binding term since the polymerase–DNA binding constant cancels in the ratio given in Eq. (1) (Fersht, 1985).

The analysis of extension fidelity is more complex than insertion fidelity, because instead of having right and wrong dNTPs competing for insertion by polymerase bound to a matched primer 3' terminal (Fig. 1a), the initial competition involves polymerase binding to different primer 3' terminals "right" (R) or "wrong" (W), followed by addition of the next correct nucleotide (Fig. 1b). Since a polymerase binding to the two different primer terminals, melted versus annealed, could occur with different affinities, polymerase–DNA binding terms, K_{DW} and K_{DR}, appear explicitly in the expression for mismatch extension efficiency, f_{ext}, [see Eq. (2); Mendelman et al., 1990; Creighton et al., 1992].

A second important factor that distinguishes f_{ins} from f_{ext} is the effect of dNTP concentration. The relative rate of insertion of right

2. DNA Polymerase Fidelity 23

and wrong nucleotides depends only on the ratio dWTP/dRTP, the pool bias ratio, but not on absolute [dNTP] (Fersht, 1985). The pool bias ratio does not appear in Eq. (1) because the f_{ins} is defined as the fraction of wrong to right insertions when dWTP and dRTP compete at equimolar concentrations. However, for extension, the concentration of the next correct dNTP clearly influences the relative rate of extending matched compared to mismatched primer 3' terminals (Mendelman et al., 1990). The key point is that the specificity of allele-selective PCR is reduced significantly as dNTP concentrations are increased (Ehlen and Dubeau, 1989; Mendelman et al., 1990).

In order to estimate specificities for extension of mismatched compared to correctly matched primer terminals taking place during the first PCR cycle, it is necessary to determine how polymerase binding, processivity, and dNTP substrate concentrations influence the relative rates of extending incorrectly versus correctly paired primer terminals.

Primer Extension Kinetics: Efficiency for Extension of Mismatched Compared with Correctly Matched Primer 3' Terminals

The use of PCR to amplify an allele preferentially depends primarily on the ability to elongate a more stable primer-template terminal compared to a less stable terminal, as shown in Fig. 1b, during the initial round of amplification. In the type of allele-selective amplification carried out by Ehlen and Dubeau (1989), the polymerase can choose between two types of primer template terminals, right (D_R) and wrong (D_W). We have analyzed a kinetic model (Fig. 2), consisting of a branched pathway that corresponds to elongation of each allele, to calculate the fraction of mismatched compared with correctly matched primer terminals extended during the first PCR round (Mendelman et al., 1990). In protocols in which an allele is extended in the absence of its counterpart (Wu et al., 1989; Ugozzoli and Wallace, 1991), the analysis is reduced to considering each branch separately.

In the first step (Fig. 2, left side), DNA polymerase encounters a DNA molecule containing either a correctly matched (D_R) or incorrectly matched (D_W) primer-template terminal, and associates and dissociates from the DNA with rate constant k_{0R} and k_{0W} respectively. The equilibrium dissociation constants to matched and mis-

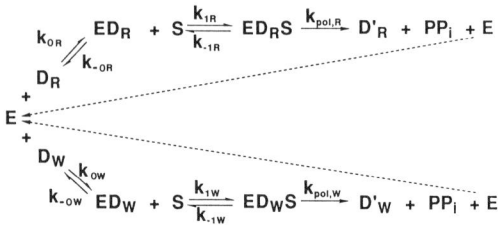

Figure 2 Kinetic model for polymerase extensions of matched or mismatched primer terminal bases. (Reproduced from Mendelman et al. 1990 with permission of The American Society for Biochemistry and Molecular Biology.) The upper branch illustrates kinetic steps and rate constants for extension of a correctly base-paired terminal (D_R). The lower branch is the corresponding pathway for extension of a mismatched terminal (D_W). The same dNTP substrate (S) is used to extend both matched and mismatched primer terminals, D_R' and D_W' are, respectively, matched and mismatched primer-template DNAs extended by a single next correct nucleotide accompanied by the release of inorganic pyrophosphate, PP_i. Both branches compete for free polymerase (E).

matched primer-templates are $K_{DR} = k_{-0R}/k_{0R}$ and $K_{DW} = k_{-0W}/k_{0W}$. The initial interaction favoring elongation, and hence amplification of matched over mismatched primer terminals, could in principle occur in the enzyme–DNA binding step, assuming the enzyme exhibits higher affinity for matched over mismatched primer terminals. However, polymerase–DNA binding measurements indicate that a variety of polymerases (Mendelman et al., 1990; Wong et al., 1991; Creighton et al., 1992; Huang et al., 1992; Yu and Goodman, 1992; Copeland et al., 1993), including *Taq* (Huang et al., 1992), exhibit similar K_D values for matched and mismatched primer terminals.

The model (Fig. 2) illustrates the kinetic steps for a polymerase (E) bound to a matched terminal proceeding along the upper branch and to a mismatched terminal along the lower branch, where incorporation of a next correct nucleotide, S, into each type of terminal occurs with catalytic rate constants $k_{pol,R}$ and $k_{pol,W}$, respectively. The expression for the mismatch extension efficiency, f_{ext}, in terms of dNTP concentration [S], polymerase–DNA equilibrium dissociation constants, K_D, and Michaelis constants, $K_m = (k_{-1} + k_{pol})/k_1$, is

$$f_{ext} = \frac{v_W}{v_R} = \frac{[D_W]}{[D_R]} \times \frac{V_{maxW}}{V_{maxR}} \times \frac{K_{DR}}{K_{DW}} \times \frac{\{K_{mR} + p_R[S]\}}{\{K_{mW} + p_W[S]\}} \qquad (2)$$

The parameter, $p = k_{pol}/k_{-0}$, defined by the ratio of the rate constants for nucleotide extension compared with polymerase–DNA dissociation, is a measure of enzyme processivity at a given template site, and $V_{max} = k_{pol}[E]$, where [E] is the concentration of active DNA polymerase present in the reaction.

A plot of f_{ext} as function of [S] shows that maximum discrimination, f^0_{min}, is achieved when the dNTP concentration → 0 (Fig. 3). (The superscript zero in f^0_{ext} is used to indicate that the concentration of wild-type and mutant alleles has been assumed, for purposes of illustration, to be constant, $[D_R] = [D_W]$.) Loss of specificity in allele-selective amplification has been observed experimentally, where selectivity decreased from about 10^5 for [dNTP] < K_m to approximately unity, when dNTP concentrations were saturating (Ehlen and Dubeau, 1989). Note that the numerator on the x axis contains the product of polymerase processivity, p_W, with the dNTP concentration [S]. Thus, as enzyme processivity increases, lower dNTP levels are required to extend a mismatch with the efficiency f_{ext} given by Eq. (2). Enzymes having excessively high processivity may therefore prove disadvantageous for achieving high allelic specificity. Polymer-

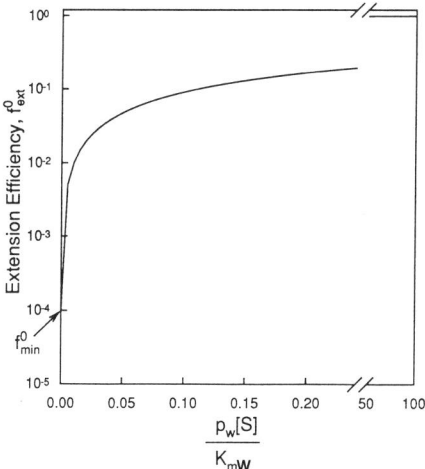

Figure 3 Predicted change in extension efficiency as a function of normalized substrate concentration; see Mendelman et al. (1990). In graphing f_{ext} [Eq. (2); see text], we have assumed for simplicity that the concentrations of DNA with matched and mismatched primer terminals are the same ($f^0_{ext} = f_{ext}$ when $[D_W] = [D_R]$), and that their equilibrium dissociation constants with the polymerase are the same. The maximum possible discrimination between extension of mismatched versus correctly matched primer terminals [f^0_{min}, Eq. (3); see text], is given by the minimum value of f^0_{ext} when [S] → 0. For this illustration, f^0_{min} has been set equal to 10^{-4}, corresponding roughly to extension of C(primer)·A mismatches by *Taq* polymerases (see Fig. 4). p_W is the ratio of the polymerase rate constant, $k_{pol,W}$, to the rate constant for dissociation, k_{-0W}, from a mismatched primer terminal (Fig. 2). Note that as the concentration of the next correct nucleotide, [S], increases, mismatched terminals are extended with efficiencies that approach those for correctly paired terminals, a prediction that was borne out in an experiment by Ehlen and Dubeau (1989).

ases having excessively low processivity may provide sufficient allelic selection, but there may be problems synthesizing adequate amounts of product DNA.

The "standard extension efficiency," f^0_{min}, in Eq. (3) gives the maximum relative efficiency (specificity) for extending mismatched versus correctly matched primer terminals (Mendelman et al., 1990).

$$f^0_{min} = [k_{pol,W}/K_{mW}]/(k_{pol,R}/K_{mR})] \times \{K_{DR}/[D_R]/(K_{DW}/[D_W]\} \quad (3)$$

We have shown that kinetic measurements to evaluate f^0_{min} can be carried out by measuring apparent V_{max}/K_m ratios in separate reactions with DNA containing either matched or mismatched primer terminals (Mendelman et al., 1990). The extension fidelity measurements are analogous to those measuring the fidelity of nucleotide insertion, provided that the extension efficiencies, given by k_{pol}/K_m or alternatively by V_{max}/K_m, are obtained at a DNA concentration in a range where primer extension velocities are linearly dependent on $[D]$ (Mendelman et al., 1990; Creighton et al., 1992). However, if $K_{DW} \sim K_{DR}$, then f^0_{min} is independent of $[D]$ (Mendelman et al., 1990; Creighton et al., 1992), and the kinetic parameters can be measured using any convenient DNA concentration.

Base Mispair Extension Efficiencies Using *Taq* Polymerase

Generally, $k_{pol,W}/K_{mW} << k_{pol,R}/K_{mR}$, in the expression for f^0_{min} [Eq. (3)]. The much smaller values for ratios of k_{pol}/K_m for extension of mismatched rather than correctly matched primer terminals reflect a strong kinetic barrier to extension of poorly annealed terminals, and are primarily responsible for determining the greatest possible discrimination between the two alleles. Measurements of f^0_{min} have been carried out using *Taq* polymerase for all configurations of matched and mismatched primer terminals using a single surrounding DNA sequence (Huang et al., 1992).

The data from Huang et al. (1992) have been reproduced as Fig. 4. The four transition mispairs, A·C, C·A, G·T, and T·G, are extended 10^{-3} to 10^{-4} less efficiently than the corresponding correct pairs. In contrast, AMV reverse transcriptase extends transition mispairs with efficiencies between 10^{-1} and 10^{-3} (Mendelman et al., 1990). Thus, *Taq* polymerase is roughly one to two orders of magnitude more selective than AMV-RT. Values of f^0_{min} for transversion mispairs are between 10^{-4} and 10^{-5} for C·T and T·T, about 10^{-6} for A·A, and below the level of detection in the assay, $<10^{-6}$, for G·A, A·G,

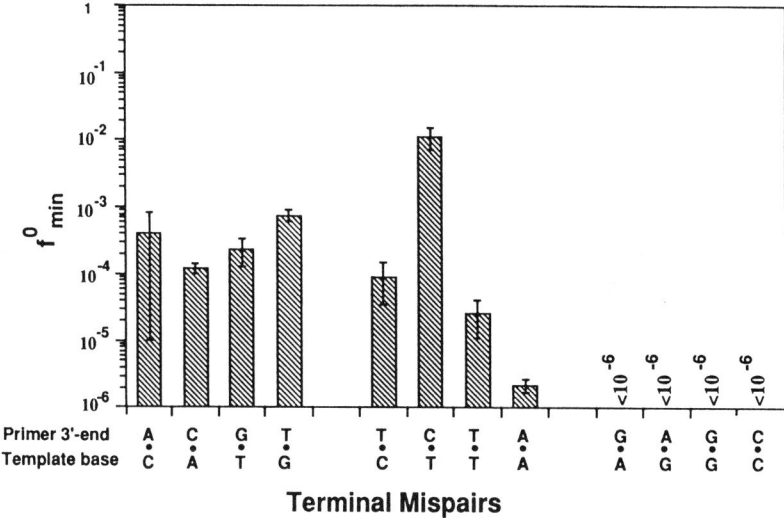

Figure 4 Standard extension efficiencies of terminal mispairs by *Taq* DNA polymerase. (Reproduced from Huang et al. 1992 with permission of Oxford University Press.) Extension efficiencies calculated according to Eq. (3) (see text) are grouped from left to right for Pur·Pyr and Pyr·Pur mismatches causing transitions and Pyr·Pyr and Pur·Pur mismatches causing transversions. Extension of the four transversion mispairs at the right-hand side of the figure was below the detection sensitivity of the assay ($f^0_{ext} < 10^{-6}$).

G·G, and C·C. *Taq* polymerase also exhibits considerably greater selectivity than AMV-RT against extension of the transversion mispairs. An unexpected result was the relative ease with which *Taq* polymerase was able to extend the C·T mispair, that is, with an efficiency of about 1% compared with an A·T pair. We suspect, however, that C·T mispairs will turn out to be considerably more difficult to extend when placed in other primer-template sequence contexts. Based on the data in Fig. 4, we offered a tentative "prognosis," reproduced in Table I, for the use of *Taq* polymerase in carrying out allele-selective amplification for each mispaired primer terminal.

Reaction temperature obviously affects the stability of primer-template terminals, but may not alter the prognosis given in Table I. We examined the effect of varying temperature between 45° and 70°C for extension of two base mispairs, A(primer)·C and T(primer)·C, and observed no measurable effect on f^0_{min} (Huang et al., 1992). As expected, based on the optimum temperature for synthesis by *Taq* polymerase of around 70°C, the V_{max}/K_m values for extension of the two incorrect and the single correct base pair increased. However, the synthesis rates for the correct pair and both mispairs were

Table 1

Summary of the Prognosis for Allele-Specific Amplification[a]

Alleles		Mismatches	Prognosis for allele-specific amplification
A·T	T·A	A·A AND T·T	Excellent for A·A, good for T·T
C·G	G·C	C·C AND G·G	Excellent for both
C·G	T·A	A·C AND G·T OR	Good
		T·G AND C·A	Good
G·C	T·A	A·G AND C·T OR	Excellent for A·G, poor for C·T
		T·C AND G·A	Excellent for G·A, good for T·C
A·T	G·C	C·A AND T·G OR	Good
		G·T AND A·C	Good
A·T	C·G	G·A AND T·C OR	Excellent for G·A, good for T·C
		C·T AND A·G	Excellent for A·G, poor for C·T

[a] All of the possible allelic differences are shown on the left side of the table. For some allelic differences there are two different possible sets of primers, depending upon which of the two strands of the DNA is used as template for extension of the allele-specific primers. A·T represents A(primer) · T(template); T·A represents T(primer) · A(template). Excellent, good, and poor refer to an f^0 of $<10^{-5}$, 10^{-5} to 10^{-3}, $> 10^{-3}$ respectively. Note that the prognosis for allele-specific amplification using C·T mismatches may be much more favorable when it is located at other primer-template positions. This summary is from Huang et al. (1992), with permission of Oxford University Press.

found to increase essentially in parallel with increasing temperature, so that the ratio of extension efficiencies given by Eq. (3) remained invariant with temperature.

The prognosis (Table I), which is based solely on measurements made within a single sequence context, can be influenced by surrounding sequences. Nearest-neighbor base stacking effects can affect the equilibrium between melted and annealed primer termini for both correctly matched and mismatched base pairs. Also, depending on the surrounding template sequence, a mispaired primer terminal might slip, by one or more bases, to become transiently misaligned but correctly paired with a complementary template base (Kunkel, 1986; Kunkel and Soni, 1988). A slippage mechanism may be responsible for the G·G mismatches that produce PCR products with the same efficiency as G·C or C·G matches in the experiment by Kwok et al. (1990), where the position of the mismatch was followed by a downstream C in the template strand. It is noteworthy that, in accordance with the predictions from Eq. (2) and Fig. 4, reduction of dNTP from 200 μM to 50 μM caused the G·G mismatch to be amplified poorly (Kwok et al., 1990). Melting temperature effects can be further enhanced by placing additional mismatches near the 3' end of the primer (Newton et al., 1989; Wu et al., 1989; Ugozzoli and Wallace, 1991; Cha et al., 1992). For example, the "MAMA" discrimination scheme carries out extension with two mismatches versus one mismatch (in place of one mismatch versus a correct match) (Cha et al., 1992). This scheme, which has kinetic blocks in both primer templates, effectively slows down the first PCR cycle, so that higher dNTP concentrations can be used and still maintain good allele selectivity. However, optimal selectivities obtained by either method are probably similar.

Acknowledgment

Research support from National Institutes of Health grants GM21422, GM 42554, and AG11398 is gratefully acknowledged.

Literature Cited

Bebenek, K., J. Abbots, S. H. Wilson, and T. A. Kunkel. 1993. Error-prone polymerization by HIV-1 reverse transcriptase. *J. Biol. Chem.* **268**:10324–10334.

Boosalis, M. S., J. Petruska, and M. F. Goodman. 1987. DNA polymerase insertion fidelity: Gel assay for site-specific kinetics. *J. Biol. Chem.* **262**:14689–14696.

Cha, R. S., H. Zarbl, P. Keohavong, and W. G. Thilly. 1992. Mismatch amplification mutation assay (MAMA): application to the c-H-*ras* gene. *PCR Methods and Applications* **2**:14–20.

Clayton, L. K., M. F. Goodman, E. W. Branscomb, and D. J. Galas. 1979. Error induction and correction by mutant and wild type T4 DNA polymerases: Kinetic error discrimination mechanisms. *J. Biol. Chem.* **254**:1902–1912.

Copeland, W. C., N. K. Lam, and T. S.-F. Wang. 1993. Fidelity studies of the human DNA polymerase α. *J. Biol. Chem.* **268**:11041–11049.

Cowart, M. C., K. J. Gibson, D. J. Allen, and S. J. Benkovic. 1989. DNA substrate structural requirements for the exonuclease and polymerase activities of procaryotic and phage DNA polymerases. *Biochemistry* **28**:1975–1983.

Creighton, S., M.-M. Huang, H. Cai, N. Arnheim, and M. F. Goodman. 1992. Base mispair extension kinetics: binding of avian myeloblastosis reverse transcriptase to matched and mismatched base pair termini. *J. Biol. Chem.* **267**:2633–2639.

Driggers, P. H., and K. L. Beattie. 1988. Effect of pH on base-mispairing properties of 5-bromouracil during DNA synthesis. *Biochemistry* **27**:1729–1735.

Echols, H., and M. F. Goodman. 1991. Fidelity mechanisms in DNA replication. *Annu. Rev. Biochem.* **60**:477–511.

Eger, B. T., R. D. Kuchta, S. S. Carroll, P. A. Benkovic, M. E. Dahlberg, C. M. Joyce, and S. J. Benkovic. 1991. Mechanism of DNA replication fidelity for three mutants of DNA polymerase I: Klenow fragment KF(exo$^+$), KF(polA5), and KF(exo$^-$). *Biochemistry* **30**:1441–1448.

Ehlen, T., and L. Dubeau. 1989. Detection of ras point mutations by polymerase chain reaction using mutation-specific, inosine-containing oligonucleotide primers. *Biochem. Biophys. Res. Comm.* **160**:441–447.

Fersht, A. R. 1979. Fidelity of replication of phage ϕX174 DNA by DNA polymerase III holoenzyme; spontaneous mutation by misincorporation. *Proc. Natl. Acad. Sci. U.S.A* **76**:4946–4950.

Fersht, A. R. 1985. *Enzyme structure and mechanism.* W. H. Freeman, New York.

Fersht, A. R., J. W. Knill-Jones, and W. C. Tsui. 1982. Kinetic basis of spontaneous mutation. Misinsertion frequencies, proofreading specificities and cost of proofreading by DNA polymerases of *Escherichia coli. J. Mol. Biol.* **156**:37–51.

Fersht, A. R., J. Shi, and W. Tsui. 1983. Kinetics of base misinsertion by DNA polymerase I of *Escherichia coli. J. Mol. Biol.* **165**:655–667.

Goodman, M. F., S. Keener, S. Guidotti, and E. W. Branscomb. 1982. On the enzymatic basis for mutagenesis by manganese. *J. Biol. Chem.* **258**:3469–3475.

Goodman, M. F., S. Creighton, L. B. Bloom, and J. Petruska. 1993. Biochemical basis of DNA replication fidelity. *Crit. Rev. Biochem. Molec. Biol.* **28**:83–126.

Huang, M.-M., N. Arnheim, and M. F. Goodman. 1992. Extension of base mispairs by *Taq* DNA polymerase: implications for single nucleotide discrimination in PCR. *Nucleic Acids Res.* **20**:4567–4573.

Joyce, C. M. 1989. How DNA travels between the separate polymerase and 3'→5'-exonuclease sites of DNA polymerase I (Klenow fragment). *J. Biol. Chem.* **264**:10858–10866.

Kornberg, A., and T. A. Baker. 1992. *DNA replication.* W. H. Freeman, New York.

Kunkel, T. A. 1986. Frameshift mutagenesis by eucaryotic DNA polymerases *in vitro. J. Biol. Chem.* **261**:13581–13587.

Kunkel, T. A., and K. Bebenek. 1988. Recent studies of the fidelity of DNA synthesis. *Biochim. Biophys. Acta* **951**:1–15.

Kunkel, T. A., and A. Soni. 1988. Mutagenesis by transient misalignment. *J. Biol. Chem.* **263**:14784–14789.

Kwok, S., D. E. Kellogg, N. McKinney, D. Spasic, L. Goda, C. Levenson, and J. J. Sninsky. 1990. Effects of primer-template mismatches on the polymerase chain reaction: human immunodeficiency virus type 1 model studies. *Nucleic Acids Res.* **18**:999–1005.

Lawley, P. D., and P. Brooks. 1962. Ionization of DNA bases or base analogues as a possible explanation of mutagenesis. *J. Mol. Biol.* **4**:216–219.

Mendelman, L. V., M. S. Boosalis, J. Petruska, and M. F. Goodman. 1989. Nearest neighbor influences on DNA polymerases insertion fidelity. *J. Biol. Chem.* **264**:14415–14423.

Mendelman, L. V., J. Petruska, and M. F. Goodman. 1990. Base mispair extension kinetics: Comparison of DNA polymerase α and reverse transcriptase. *J. Biol. Chem.* **265**:2338–2346.

Muzyczka, N., R. L. Poland, and M. J. Bessman. 1972. Studies on the biochemical basis of spontaneous mutation, I. A comparison of the deoxyribonucleic acid polymerase of mutator, antimutator, and wild type strains of bacteriophage T4. *J. Biol. Chem.* **247**:7116–7122.

Newton, C. R., A. Graham, L. E. Heptinstall, S. J. Powell, C. Summers, N. Kalsheker, J. C. Smith, and A. F. Markham. 1989. Analysis of any point mutation in DNA.

The amplification refractory mutation system (ARMS). *Nucleic Acids Res.* **17**:2503–2516.

Perrino, F. W., and L. A. Loeb. 1989. Differential extension of 3' mispairs is a major contribution to the high fidelity of calf thymus DNA polymerase-alpha. *J. Biol. Chem.* **264**:2898–2905.

Petruska, J., M. F. Goodman, M. S. Boosalis, L. C. Sowers, C. Cheong, and I. Tinoco Jr. 1988. Comparison between DNA melting thermodynamics and DNA polymerase fidelity. *Proc. Natl. Acad. Sci. U.S.A* **85**:6252–6256.

Randall, S. K., R. Eritja, B. E. Kaplan, J. Petruska, and M. F. Goodman. 1987. Nucleotide insertion kinetics opposite abasic lesions in DNA. *J. Biol. Chem.* **262**:6864–6870.

Reddy, M. K., S. E. Weitzel, and P. H. von Hippel. 1992. Processive proofreading is intrinsic to T4 DNA polymerase. *J. Biol. Chem.* **267**:14157–14166.

Schaaper, R. M. 1988. Mechanisms of mutagenesis in the *Escherichia coli* mutator mutD5: role of DNA mismatch repair. *Proc. Natl. Acad. Sci. U.S.A* **85**:8126–8130.

Sinha, N. K. 1987. Specificity and efficiency of editing of mismatches involved in the formation of base substitution mutations by the 3' → 5' exonuclease activity of phage T4 DNA polymerase. *Proc. Natl. Acad. Sci. U.S.A* **84**:915–919.

Sowers, L. C., G. V. Fazakerley, R. Eritja, B. E. Kaplan, and M. F. Goodman. 1986a. Base pairing and mutagenesis: observation of a protonated base pair between 2-aminopurine and cytosine in an oligonucleotide by proton NMR. *Proc. Natl. Acad. Sci. U.S.A* **83**:5434–5438.

Sowers, L. C., G. V. Fazakerley, H. Kim, L. Dalton, and M. F. Goodman. 1986b. Variation of nonexchangeable proton resonance chemical shifts as a probe of aberrant base pair formation in DNA. *Biochemistry* **25**:3983–3988.

Sowers, L. C., B. Ramsay Shaw, M. L. Veigl, and W. D. Sedwick. 1987. DNA base modification: ionized base pairs and mutagenesis. *Mutat. Res.* **177**:201–218.

Sowers, L. C., M. F. Goodman, R. Eritja, B. Kaplan, and G. V. Fazakerley. 1989. Ionized and wobble base-pairing for bromouracil-guanine in equilibrium under physiological conditions. A nuclear magnetic resonance study on an oligonucleotide containing a bromouracil-guanine base-pair as a function of pH. *J. Mol. Biol.* **205**:437–447.

Ugozzoli, L., and R. B. Wallace. 1991. Allele-specific polymerase chain reaction. *METHODS: A companion to methods in enzymology* **2**:42–48.

Wong, I., S. S. Patel, and K. A. Johnson. 1991. An induced-fit kinetic mechanism for DNA replication fidelity: direct measurement by single-turnover kinetics. *Biochemistry* **30**:526–537.

Wu, D. Y., L. Ugozzoli, B. K. Pal, and R. B. Wallace. 1989. Allele-specific enzymatic amplification of beta-globin genomic DNA for diagnosis of sickle cell anemia. *Proc. Natl. Acad. Sci. U.S.A* **86**:2757–2760.

Yu, H., and M. F. Goodman. 1992. Comparison of HIV-1 and avian myeloblastosis virus reverse transcriptase fidelity on RNA and DNA templates. *J. Biol. Chem.* **267**:10888–10896.

Yu, H., R. Eritja, L. B. Bloom, and M. F. Goodman. 1993. Ionization of bromouracil and fluorouracil stimulates base mispairing frequencies with guanine. *J. Biol. Chem.* **268**:15935–15943.

3

EXTRACTION OF NUCLEIC ACIDS: SAMPLE PREPARATION FROM PARAFFIN-EMBEDDED TISSUES

Hiroko Shimizu and Jane C. Burns

The use of PCR to analyze nucleic acid extracted from fixed, paraffin-embedded tissues has become a well-established research technique (Impraim et al., 1987; Shibata et al., 1988a). The principal advantage of using gene amplification to detect specific DNA and RNA sequences in archival tissue is the ability to analyze small amounts (≤ 100 μg) of routinely handled and preserved tissues which may contain partially degraded nucleic acid.

Extensive information is now available on the optimal methods of tissue fixation and design of primers for PCR detection of nucleic acid sequences in preserved tissue (Greer et al., 1991a,b; Ben-Ezra et al., 1991; Rogers et al., 1990; Jackson et al., 1990; Weizsacker et al., 1991; Coates et al., 1991; Woodall et al., 1993; Crisan et al., 1990). In general, acetone, 95% ethanol, and 10% buffered formalin are the preferred fixatives (Greer et al., 1991a,b; Ben-Ezra et al., 1991; Rogers et al., 1990; Jackson et al., 1990). Fixation times of $\leq$24 hr before embedding in paraffin are less likely to result in extensive loss of the RNA or DNA template into the fixative (Greer et al., 1991a,b; Ben-Ezra et al., 1991; Rogers et al., 1990; Jackson et al., 1990). Choosing primer sequences that yield an amplicon $\leq$ 300 bp

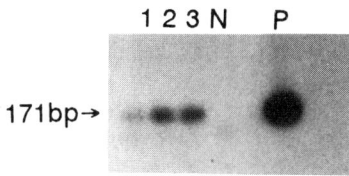

Figure 1 Detection of Coxsackie B3 genome in formalin-fixed, paraffin-embedded mouse myocardium. Mice were experimentally infected with CBV 3 and the hearts harvested for determination of virus titer and for histologic and PCR analysis. Half of the heart was fixed in 10% buffered formalin for 24 hr and RNA was extracted from a 10-μm section (100 μg of tissue) according to the protocol presented here [reverse transcription (RT)-PCR conditions as in Shimizu, 1994]. Lanes 1–3, amplification product from RT-PCR analysis of infected myocardium containing 5, 25, and 50 infectious units of CBV 3 genome; N, no template; P, positive control (RT)-PCR amplification of purified CBV 3 virus.

allows amplification of templates that are partially degraded during the fixation process (Greer et al., 1991a,b).

The sensitivity of PCR for the detection of low copy-number viral DNA and RNA sequences in formalin-fixed, paraffin-embedded tissues has been documented (Shibata et al. 1988a,b,1989; Shimizu et al. 1988,1993; Lozinski et al., 1994; Lo et al. 1989; Sallie et al. 1992). Modifications that have simplified the extraction of nucleic acid and increased the efficiency of detecting target sequence include (1) the direct digestion of tissue samples without pretreatment to remove the paraffin and (2) addition of 25 mM EDTA to the proteinase K–sodium dodecyl sulfate (SDS) digestion buffer. The protocol which follows permits the detection of five infectious units (50–5,000 templates based on expected ratio of infectious to noninfectious particles) of enteroviral genome in RNA extracted from a 10-μm section of formalin-fixed, paraffin-embedded myocardium (Shimizu et al., 1994 and Fig. 1).

Protocols

Reagents

Tissue sections
Digestion buffer: 10 mM Tris (pH 7.4) 100 mM NaCl, 25 mM disodium EDTA, 0.5% sodium dodecyl sulfate

Proteinase K (10 mg/ml stock solution)
Tris-equilibrated phenol (pH 7.4) for DNA extraction
Water-equilibrated phenol (pH 4.4) for RNA extraction
8 M ammonium acetate
Isopropanol
Yeast tRNA (10 mg/ml) or glycogen (10 mg/ml)

Preparation of Tissue Sections

For ease of handling samples, tissue can be digested in 10-μm sections. Use of thinner sections is awkward because the tissue is very fragile. Thicker sections impede efficient digestion. If more tissue must be sampled to detect rare target sequences, multiple 10-μm sections can be digested in a single tube. Care must be taken to avoid cross-contamination of tissue sections by the microtome blade or the forceps used to handle the paraffin sections. Instruments and blades can be cleaned with xylene to remove residual specimens. If possible, trim excess paraffin from the 10-μm sections before placing them in a 1.5-ml microcentrifuge tube.

Digestion of Tissue

1. Digest up to five 10-μm sections in 500 μl of digestion buffer with 10 μl of proteinase K stock solution added (200 μg/ml final concentration). The volume of buffer required will depend to a certain extent on the cross-sectional dimensions of the tissue sample.
2. Incubate at 37°C overnight.
3. Extract nucleic acids in an equal volume of phenol: chloroform: isoamyl alcohol (25 : 24 : 1). Use acid phenol for RNA extraction and Tris-buffered phenol (pH 7.4) for DNA extractions.
4. Repeat extraction of the aqueous phase with an equal volume of chloroform:isoamyl alcohol (24 : 1).
5. Precipitate the nucleic acid from the aqueous phase by addition of 1/4 volume 8 M ammonium acetate and an equal volume of isopropanol. Addition of carrier to increase the yield of precipitated nucleic acid is recommended for most applications. Use 2 μl of either tRNA or glycogen (10 mg/ml) per tube. After incubation at -20°C for 1 h, concentrate the precipitate by centrifugation at 13,000 rpm for 10 min.
6. Wash the pellet in 70% ethanol to remove salts and resuspend in 10 μl sterile water.

PCR

The amount of extracted nucleic acid that must be used for the PCR reaction will depend on a number of factors, including the anticipated concentration of the intact target sequence and the presence of inhibitors copurified in the extraction procedure. To screen for the quantity of amplifiable nucleic acid in the extract, a control amplification for a single-copy housekeeping gene or other suitable target can be performed. Amplification with β-actin primers for DNA extraction (Ben-Ezra et al., 1991) and porphobilinogen deaminase primers for RNA extraction (Shimizu et al., 1994) represent suitable controls when working with mammalian samples.

To screen for the presence of inhibitors to the PCR reaction, target sequences can be spiked into a reaction mixture containing the sample to be amplified. In the event of failure to amplify this sequence, which suggests the presence of an inhibitor in the extracted sample, further purification of the nucleic acid can be attempted with commercially available glass or silica powder suspensions. In an extensive survey of many different formalin-fixed, paraffin-embedded tissues, however, no inhibitors to PCR were detected when nucleic acid was extracted according to the above protocol (Lozinski et al., 1994). RNA extracted by the above protocol is suitable for use in a one-step RT-PCR reaction in which the cDNA synthesis step and the target sequence amplification step are combined (Shimizu et al., 1994).

Technical Comments

Dewaxing of the sample with solvents such as mixed xylenes has been recommended in some protocols (Greer et al., 1991a,b; Ben-Ezra et al. 1991). No advantage to deparaffinization was detected in a PCR analysis of RNA viral genome in human tissues (Fig. 3, lane 4). It is therefore recommended that this step, which is time-consuming and involves the use of toxic and volatile reagents, be deleted.

Digestion of 10-μm tissue sections for periods up to 5 days did not appreciably increase the yield of amplifiable β-actin DNA extracted from mouse myocardium (Fig. 2). The optimal digestion time for each target sequence must be determined empirically, but overnight digestion has proven adequate for most applications.

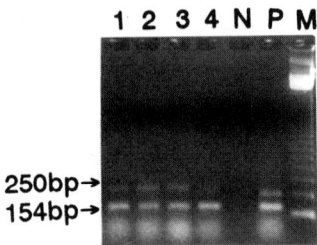

Figure 2 Detection of β-actin DNA in mouse myocardium digested for variable times. Nucleic acid was extracted according to the protocol described, with incubation times in the digestion buffer ranging from 3 hr to 5 days at 37°C. The primers amplify both the single-copy genomic DNA template (250 bp) and the multiple processed pseudogenes (154 bp) (Ben-Ezra et al., 1991). Digestion times were as follows: Lane 1, 5 days; lane 2, 3 days; lane 3, 16 hr; lane 4, 3 hr; N, no template; P, positive control amplification of 1 μg of DNA extracted from human peripheral blood cells.

Many different digestion buffers have been reported for this application (Greer et al., 1991a,b; Ben-Ezra et al., 1991; Rogers et al., 1990; Jackson et al., 1990). A systematic comparison of a chaotropic agent (guanidinium isothiocyante) and different detergents combined with proteinase K digestion showed that the above protocol was the most sensitive for the extraction of viral RNA from mammalian tissues (Shimizu et al., 1994). The use of a high concentration of EDTA (25 mM) was critical to the success of this method (Fig. 3, lane 3). The undesirable coprecipitation of the EDTA with the nucleic acid that may occur with sodium acetate and ethanol can be avoided with the use of ammonium acetate and isopropanol.

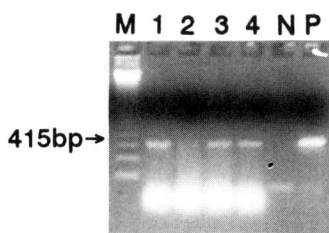

Figure 3 Detection of measles virus genome following digestion of tissues under different conditions. Lanes 1–4 show (RT)-PCR amplification of RNA extracted from formalin-fixed, paraffin-embedded sections of human lung from a patient with a fatal measles infection. M, molecular-size standard; lane 1, 100 mM NaCl and 25 mM EDTA, no deparaffinization step; lane 2, no salt and 5 mM EDTA; lane 3, no salt and 25 mM EDTA; lane 4, conditions as for lane 1, tissue section treated with mixed xylenes to remove paraffin; N, no template; P, positive control amplification of purified measles virus ((RT)-PCR conditions as in Shimizu et al., 1993).

To increase the sensitivity of the PCR to compensate for the loss or degradation of template during the fixation or extraction procedures, both the total number of PCR cycles and the time at each temperature in the thermal cycling profile can be increased. The need for these modifications must be determined empirically for each PCR application.

Discussion

The protocol for extracting nucleic acid for PCR analysis presented here involves multiple manipulations of the tissue sample, and introduces many opportunities for contaminating the reaction with exogenous target sequences. Great care must be exercised to avoid such contamination. Ultimately, a compromise must be reached which permits the most complete extraction of intact nucleic acid with the least manipulation of the sample. For applications in which the target sequence for PCR is abundant, simple boiling of the sample may suffice (Coates *et al.*, 1991). However, for analysis of RNA or low copy-number DNA targets, the protocol described above is preferred.

Acknowledgments

We thank David P. Schnurr and Henry F. Krous for providing infected tissues, and Grace Lozinski for technical assistance. We also thank John J. Sninsky for valuable discussion.

Literature Cited

Ben-Ezra, J., D. A. Johnson, J. Rossi, N. Cook, and A. Wu. 1991. Effect of fixation on the amplification of nucleic acids from paraffin-embedded material by the polymerase chain reaction. *J. Histochem. Cytochem.* **39**:351–354.

Coates, P. J., A. J. d'Ardenne, G. Khan, H. O. Kangro, and G. Slavin. 1991. Simplified procedures for applying the polymerase chain reaction to routinely fixed paraffin wax sections. *J. Clin. Pathol.* **44**:115–118.

Crisan, D., E. M. Cadoff, J. C. Mattson, and K. A. Hartle. 1990. Polymerase chain reaction: amplification of DNA from fixed tissue. *Clin. Biochem.* **23**:489–495.

Greer, C. E., S. L. Peterson, N. B. Kiviat, and M. M. Manos. 1991a. PCR amplification from paraffin-embedded tissues. *Am. J. Clin. Pathol.* **95**:117–124.

Greer, C. E., J. K. Lund, and M. M. Manos. 1991b. PCR amplification from paraffin-embedded tissues: recommendations on fixatives for long-term storage and prospective studies. *PCR Meth. Appl.* **1**:46–50.

Impraim, C. C., R. K. Saiki, H. A. Erlich, and R. L. Teplitz. 1987. Analysis of DNA extracted from formalin-fixed, paraffin-embedded tissues by enzymatic amplification and hybridization with sequence-specific oligonucleotides. *Biochem. Biophys. Res. Commun.* **142**:710–716.

Jackson, D. P., F. A. Lewis, G. R. Taylor, A. W. Boylston, and P. Quirke. 1990. Tissue extraction of DNA and RNA and analysis by the polymerase chain reaction. *J. Clin. Pathol.* **43**:499–504.

Lo, Y-M., W. Z. Mehal, and K. A. Fleming. 1989. *In vitro* amplification of hepatitis B virus sequences from liver tumour DNA and from paraffin wax embedded tissues using the polymerase chain reaction. *J. Clin. Pathol.* **42**:840–846.

Lozinski, G. M., G. G. Davis, H. F. Krous, G. F. Billman, H. Shimizu, and J. C. Burns. 1994. Adenovirus myocarditis: retrospective diagnosis by gene amplification from formalin-fixed, paraffin-embedded tissues. *Hum. Pathol.* **25**:831–834.

Rogers, B. B., L. C. Alpert, E. A. S. Hine, and G. J. Buffone. 1990. Analysis of DNA in fresh and fixed tissue by the polymerase chain reaction. *Am. J. Pathol.* **136**:541–548.

Sallie, R., A. Rayner, B. Portmann, A. L. W. F. Eddleston, and R. Williams. 1992. Detection of hepatitis C virus in formalin-fixed liver tissue by nested polymerase chain reaction. *J. Med. Virol.* **37**:310–314.

Shibata, D. K., N. Arnheim, and W. J. Martin. 1988a. Detection of human papilloma virus in paraffin-embedded tissue using the polymerase chain reaction. *J. Exptl. Med.* **167**:224–230.

Shibata, D., W. J. Martin, and N. Arnheim. 1988b. Analysis of forty-year-old paraffin-embedded thin-tissue sections: a bridge between molecular biology and classical histology. *Cancer Res.* **48**:4564–4566.

Shibata, D., R. K. Brynes, B. Nathwani, S. Kwok, J. Sninsky, and N. Arnheim. 1989. Human immunodeficiency viral DNA is readily found in lymph node biopsies from seropositive individuals. *Am. J. Pathol.* **135**:697–702.

Shimizu, H., D. P. Schnurr, and J. C. Burns. 1994. Comparison of methods to detect enteroviral genome in frozen and fixed myocardium by polymerase chain reaction. *Lab. Invest.* **71**:612–615.

Shimizu H., C. A. McCarthy, M. F. Smaron, and J. C. Burns. 1993. Polymerase chain reaction for detection of measles virus in clinical samples. *J. Clin. Micro.* **31**:1034–1039.

Weizsacker, F. v., S. Labeit, H. K. Koch, W. Oehlert, W. Gerok, and H. E. Blum. 1991. A simple and rapid method for the detection of RNA in formalin-fixed, paraffin-embedded tissues by PCR amplification. *Biochem. Biophys. Res. Commun.* **174**:176–180.

Woodall, C. J., N. J. Watt, and G. B. Clements. 1993. Simple technique for detecting RNA viruses by PCR in single sections of wax embedded tissue. *J. Clin. Pathol.* **46**:276–277.

4

THERMOSTABLE DNA POLYMERASES

Richard D. Abramson

The isolation and characterization of DNA polymerases from mesophilic microorganisms has been extensively studied (Kornberg and Baker, 1992). Much less is known about the properties of thermostable DNA polymerases. With the introduction of the polymerase chain reaction (PCR) method of DNA amplification (Mullis and Faloona, 1988; Saiki et al., 1988), considerable interest has been focused on the DNA polymerases of thermophilic organisms. Many of these DNA polymerases are both thermoactive and thermostable. They are capable of catalyzing polymerization at the high temperatures required for stringent and specific DNA amplification as well as withstanding the even higher temperatures necessary to separate DNA strands (see Table 1). The most extensively studied thermostable DNA polymerase used in PCR amplification is that isolated from the thermophilic eubacterium *Thermus aquaticus* (*Taq*) strain YT1 (Lawyer et al., 1989, 1993). More recently, other thermostable DNA polymerases have become commercially available, including those derived from *Thermus thermophilus* (*Tth*), *Thermotoga maritima* (*Tma*) (Lawyer and Gelfand, 1992), *Thermococcus litoralis* (*Tli*) (Kong et al., 1993), and *Pyrococcus furiosus* (*Pfu*) (Lundberg et al., 1991) (see Table 1).

Table 1
Properties of Commercially Available Thermophilic DNA Polymerases

Enzyme	Optima				Molecular Weight ($\times 10^{-3}$)	Specific Activity[a]	Exonuclease Activity		Processivity (nucleotides)	Extension Rate[b] (nucleotides/sec)	Thermostability (min)
	$MgCl_2$ (mM)	pH	KCl (mM)	Temp (°C)			5'–3'	3'–5'			
Taq Pol	2–3	9.4	10–55	75–80	94	292,000	+	–	50–60	75	9[c]
Stoffel fragment	3.5–4	8.3	0–10	75–80	61.3	369,000	–	–	5–10	50	21[c]
Tth Pol	1.5–2.5	9.4	35–100	75–80	94	148,000	+	–	20–30	60	2[c]
Met$_{284}$-Tma Pol	1.5–2	8.3	10–35	75–80	70	115,000	–	+	ND	ND	40–50[c]
Tli Pol	ND[f]	ND	ND	75–80	90–93	23,000	–	+	7	67	120[d]
Pfu Pol	ND	ND	ND	ND	90	ND	–	+	ND	ND	60[e]

[a] Specific activities have been normalized such that one unit equals the amount of activity which will incorporate 10 nmol of dNTPs into product in 30 min.
[b] Determined at 70°C.
[c] Half-life of activity at 97.5°C.
[d] Half-life of activity at 100°C.
[e] 95% activity at 95°C
[f] ND, Not determined.

Taq DNA Polymerase

Taq DNA polymerase is an 832-amino acid protein with an inferred molecular weight of 93,920 and a specific activity of 292,000 units/ mg; optimal polymerization activity is achieved at 75–80° C, with half-maximal activity at 60–70° C (Lawyer et al., 1993; see also Table 1). It's thermostability as measured as a half-life of activity is between 45 and 96 min at 95° C and 9 min at 97.5° C (Lawyer et al., 1993; Kong et al., 1993). Maximal enzymatic activity is achieved in a N-tris[hydroxymethyl]methyl-3-aminopropanesulfonic acid (Taps)–KOH (25 mM), pH 9.4 buffer containing 10–55 mM KCl and 2–3 mM MgCl$_2$ (Lawyer et al., 1993). Under conditions of enzyme excess, the polymerase extends a primer at a maximum rate of 75 nucleotides per second at 70° C (Innis et al., 1988; Abramson et al., 1990). The enzyme is moderately processive, extending a primer an average of 50–60 nucleotides before it dissociates (King et al., 1993; Abramson et al., 1990). The fidelity of polymerase base substitution has been estimated to range between 3×10^{-4} and 3×10^{-6} errors per nucleotide polymerized, depending on reaction conditions (Eckert and Kunkel, 1992). In a standard PCR, amplification with *Taq* DNA polymerase is optimal in a Tris–HCl (10 mM), pH 8.3 buffer containing 50 mM KCl and 1.5 mM MgCl$_2$, although MgCl$_2$ concentration may need to be adjusted for any specific primer pair–template combination (Innis and Gelfand, 1990). Based on sequence homology to *Escherichia coli* DNA polymerase I, *Taq* DNA polymerase has been included in a family of DNA polymerases known as Family A (Braithwaite and Ito, 1993). Like *E. coli* DNA polymerase I, *Taq* DNA polymerase possesses an inherent 5' to 3' exonuclease or "nick-translation" activity (Longley et al., 1990; Holland et al., 1991; Lawyer et al., 1993). It does not, however, possess a 3' to 5' exonuclease or "proofreading" activity (Tindall and Kunkel, 1988; Lawyer et al., 1989).

Similar to the Klenow or large fragment of *E. coli* DNA polymerase I (Brutlag et al., 1969; Klenow and Henningsen, 1970), amino-terminal truncated forms of the enzyme lacking 5' to 3' exonuclease activity have also been described (Lawyer et al., 1989, 1993; Barnes, 1992). One such form, the Stoffel fragment, was constructed by deleting the first 867 bp of the *Taq* DNA polymerase gene, to yield a 544-amino acid protein with an inferred molecular weight of 61,300 and a specific activity of 369,000 U/mg (Lawyer et al., 1993; see also Table 1). The enzyme has a temperature optimum for activity of

75–80° C, with half-maximal activity at 55–65° C, and a thermostability half-life of 80 min at 95° C and 21 min at 97.5° C (Lawyer et al., 1993). Maximal enzymatic activity is achieved in a Tris–HCl (25 mM), pH 8.3 buffer containing 0–10 mM KCl and 3.5–4 mM MgCl$_2$ (Lawyer et al., 1993). Under conditions of moderate enzyme excess, the polymerase extends a primer at a rate of approximately 50 nucleotides per second at 70° C; however, rates as high as 180 nucleotides per second are possible with a very large excess of enzyme compared with template molecules (R. D. Abramson, unpublished data). The enzyme possesses low processivity, extending a primer an average of only 5–10 nucleotides before it dissociates (R. D. Abramson, unpublished data). In a standard PCR, amplification with the Stoffel fragment is optimal in a Tris–HCl (10 mM), pH 8.3 buffer containing 10 mM KCl and 3.0 mM MgCl$_2$ (D. H. Gelfand, personal communication). A similarly truncated 67-kDa enzyme has been shown to have a two-fold greater fidelity than the full-length protein (Barnes, 1992), suggesting that the Stoffel fragment may also have improved fidelity, possibly as a result of its lower processivity (see fidelity discussion).

Subtle differences exist between the polymerase activity of the Stoffel fragment and full-length *Taq* DNA polymerase (Lawyer et al., 1993). Whereas the Stoffel fragment is more thermostable than *Taq* DNA polymerase, it is not as active at temperatures greater than 80° C. This may be due to a difference in the ability of the enzyme to bind to DNA at high temperatures. That a difference in DNA binding exists between the two enzymes is reflected in the lower processivity of the Stoffel fragment. This difference in processivity may also explain the observed difference in sensitivity to ionic strength between the enzymes, as well as possible differences in polymerase fidelity. Differences are also apparent in the response of the two enzymes to divalent cations. The Stoffel fragment is optimally active over a broader range of Mg^{2+} concentrations, but is less active when either Mn^{2+} or Co^{2+} is substituted for Mg^{2+}.

Tth DNA Polymerase

Similar to *Taq* DNA polymerase, *Tth* DNA polymerase is an 834-amino acid protein with an inferred molecular weight of 94,000 and a specific activity of 148,000 units/mg (S. Stoffel and D. H. Gelfand, personal communication; see also Table 1). The two proteins are 93% similar and 88% identical, based on amino acid sequence. The enzyme is as thermoactive as *Taq* DNA polymerase; optimal polymerization activity is achieved at 75–80° C, with half-maximal activity at 60–65° C (S. Stoffel and D. H. Gelfand, personal communication). Its thermostability, however, is lower, with a half-life of activity of 20 min at 95° C and 2 min at 97.5° C (P. Landre, personal communication). Maximal DNA polymerase activity is achieved in a Taps/KOH (25 mM), pH 9.4 buffer containing 35–100 mM KCl and 1.5–2.5 mM MgCl$_2$ (S. Stoffel and D. H. Gelfand, personal communication). Under conditions of enzyme excess, the polymerase extends a primer at a maximum rate of 60 nucleotides per second at 70° C (R. D. Abramson, unpublished data). The enzyme is moderately processive, extending a primer an average of 20–30 nucleotides (in 100 mM KCl) before it dissociates (R. D. Abramson, unpublished data). In a standard PCR, amplification with *Tth* DNA polymerase is optimal in a Tris–HCl (10 mM), pH 8.3 buffer containing 100 mM KCl and 2.0 mM MgCl$_2$ (P. Landre, personal communication). Like *Taq* DNA polymerase, *Tth* DNA polymerase possesses an inherent 5′ to 3′ exonuclease activity (Lyamichev *et al.*, 1993; Aver *et al.*, 1995) and is devoid of 3′ to 5′ exonuclease activity. The most striking difference between the two enzymes is the ability of *Tth* DNA polymerase to efficiently catalyze high-temperature reverse transcription (RT) in the presence of MnCl$_2$, with *Tth* DNA polymerase proving to be greater than 100-fold more efficient in a coupled RT-PCR than *Taq* DNA polymerase (Myers and Gelfand, 1991; see also Chapter 5).

Tma DNA Polymerase

Thermotoga maritima is an anaerobic hyperthermophilic eubacterium isolated from geothermally heated marine surfaces, growing at temperatures up to 90° C (Huber *et al.*, 1986). *Tma* DNA polymerase is an 893-amino acid protein with an inferred molecular weight of 103,000 (Lawyer and Gelfand, 1992). *Tma* DNA polymerase is only 61% similar and 44% identical to *Taq* DNA polymerase, based on amino acid sequence. Like *E. coli* DNA polymerase I, *Tma* DNA polymerase contains both a 5′ to 3′ exonuclease (R. D. Abramson, unpublished data) and a 3′ to 5′ exonuclease (Lawyer and Gelfand,

1992). Similar to the Stoffel and Klenow fragments, an amino terminal truncated form of the enzyme, lacking 5' to 3' exonuclease activity (commercially available as Ultma DNA polymerase from Perkin–Elmer Corp., Norwalk, CT) has also been described (Lawyer and Gelfand, 1992; see also Table 1). This enzyme was produced by engineering translation to initiate at the Met_{284} codon, resulting in a 610-amino acid, 70-kDa protein which retains polymerase activity, with a specific activity of 115,000 U/mg, as well as 3' to 5' exonuclease activity. The enzyme is as thermoactive as *Taq* DNA polymerase; optimal polymerization activity is achieved at 75–80° C, with half-maximal activity at 55–60° C (D. Bost, S. Stoffel, and D. H. Gelfand, personal communication). Its thermostability, however, is considerably higher, with a half-life of activity of 40–50 min at 97.5° C (P. Landre, personal communication). Maximal DNA polymerase activity is achieved in a Tris–HCl (25 mM), pH 8.3 buffer containing 10–35 mM KCl and 1.5–2.0 mM MgCl_2 (D. Bost, S. Stoffel, and D. H. Gelfand, personal communication). Preliminary studies suggest that the inherent fidelity of the enzyme is such that nearly 100% of the PCR product is corrected by the proofreading activity in a 3' mismatched primer correction assay (F. C. Lawyer, personal communication). In a standard PCR, amplification with Met_{284}-*Tma* DNA polymerase is optimal in a Tris–HCl (10 mM), pH 8.8 buffer containing 10 mM KCl and 2.0 mM MgCl_2 (F. C. Lawyer and D. H. Gelfand, personal communication).

Tli DNA Polymerase

Thermococcus litoralis is an extremely thermophilic marine archaebacterium obtained from a shallow submarine thermal spring in Italy and grows at temperatures up to 98° C (Neuner *et al.*, 1990). The Archaea constitute a group of prokaryotes with an intermediate phylogenetic position between eukaryotes and eubacteria. They comprise the most extreme thermophilic organisms known. In contrast to the eubacterial DNA polymerases discussed earlier, *Tli* DNA polymerase (commercially available as Vent DNA polymerase from New England Biolabs, Inc., Beverly, MA) is a Family B DNA polymerase, with sequence similarity to human DNA polymerase α and *E. coli* DNA polymerase II, rather than *E. coli* DNA polymerase I (Braithwaite and Ito, 1993). The enzyme is a 90–93-kDa protein with a specific activity of 23,000 U/mg; optimal polymerization activity is achieved at 75–80° C, with half-maximal activity at 60–65° C (Kong et al., 1993; see also Table I). The polymerase possesses exceptional thermostability, with an activity half-life of 8 hr at 95° C and approximately 2 hr at 100° C (Kong et al., 1993). The polymerase extends a primer at a rate of 67 nucleotides per second with only limited processivity, approximately 7 nucleotides per binding event (Kong et al., 1993). The fidelity of polymerase base substitution has been estimated to be $3–5 \times 10^{-5}$ (Cariello et al., 1991; Mattila et al., 1991; Keohavong et al., 1993). Tli DNA polymerase possesses an inherent 3' to 5' exonuclease activity (Mattila et al., 1991; Kong et al., 1993). It does not, however, possess a 5' to 3' exonuclease activity (Kong et al., 1993; Lyamichev et al., 1993).

Pfu DNA Polymerase

Pyrococcus furiosus is a hyperthermophilic archaea obtained from geothermally heated marine sediments in Italy and grows optimally at 100° C (Fiala and Stetter, 1986). Like *Tli* DNA polymerase, *Pfu* DNA polymerase belongs to Family B (Mathur et al., 1991; Uemori et al., 1993a; Braithwaite and Ito, 1993). The enzyme is a 775-amino acid protein with a predicted molecular weight of 90,109 (Uemori et al., 1993a; see also Table I). The polymerase is extremely thermostable, retaining greater than 95% of its initial activity following 1 hour at 95° C (Mathur et al., 1992). Polymerase base substitution fidelity has been estimated to be 2×10^{-6} (Lundberg et al., 1991). *Pfu* DNA polymerase possesses an inherent 3' to 5' exonuclease activity (Lundberg et al., 1991) but does not possess a 5' to 3' exonuclease activity (Lyamichev et al., 1993).

Other Thermostable DNA Polymerases

Another DNA polymerase isolated from *Thermus aquaticus* has been described (Chien et al., 1976; Kaledin et al., 1980). This enzyme has an approximate molecular weight of 62,000–68,000, a specific activity between 500 and 5200 U/mg, a temperature optimum of

70–80° C, and a pH optimum in the range of 7.8 to 8.3 (see Table 2). Optimal activity is obtained with 60–200 mM KCl and 10 mM Mg^{2+}. Based on the physical and biochemical characterization of the polymerases purified by Chien et al. (1976) and Kaledin et al. (1980), it is unclear whether these proteins are the products of a gene (or genes) distinct from that which encodes the DNA polymerase described by Lawyer et al. (1989, 1993), or are partially purified proteolytic degradation fragments of the same translation product.

Similarly, additional DNA polymerases have also been isolated from *Thermus thermophilus* (*Tth*) HB-8 (Rüttimann et al., 1985; Carballeira et al., 1990; see also Table 2). Rüttimann et al. (1985) characterized three DNA polymerase isoenzymes with molecular masses in the range of 110,000–120,000. These enzymes all lacked exonuclease activity. The three enzymes differ in their heat stability, thermal activity profile, and their ability to use manganese as a cofactor. A fifth *Tth* DNA polymerase reported by Carballeira et al. (1990) differs in molecular weight from the four described above. This 67,000-dalton protein lacks a 5'-to 3'-exonuclease and has a specific activity of 4000 U/mg.

The DNA polymerases from a number of other thermophilic eubacteria have also been isolated and partially characterized (Table 2). These include *Bacillus stearothermophilus* (*Bst*) (Stenesh and Roe, 1972; Kaboev et al., 1981), *Bacillus caldotenax* (*Bca*) (Uemori et al., 1993b), *Thermus ruber* (*Tru*) (Kaledin et al., 1982), *Thermus flavus* (*Tfl*) (Kaledin et al., 1981), and *Thermotoga* sp. (*Tsp*) strain FjSS3-B.1 (Simpson et al., 1990). In addition, other thermophilic archaeal DNA polymerases have been identified. A DNA polymerase has been purified and characterized from the thermoacidophilic archaeon *Sulfolobus acidocaldarius* (*Sac*) (Klimczak et al., 1985; Elie et al., 1988; Salhi et al., 1989). Similarly, DNA polymerases from *Sulfolobus solfataricus* (*Sso*) (Rella et al., 1990) and *Thermoplasma acidophilum* (*Tac*) (Hamal et al., 1990) have been isolated. The purification and characterization of a DNA polymerase from a methanogenic archaeon *Methanobacterium thermoautotrophicum* (*Mth*) (Klimczak et al., 1986) has also been reported.

Tables 1 and 2 show that considerable variation exists among those thermostable DNA polymerases that have been characterized in the literature. A distinct feature of these DNA polymerases is that they all appear to be monomeric. It is not known whether a multimeric DNA polymerase similar to *E. coi* DNA polymerase III exists in thermophilic bacteria. It is also not known whether these polymerases are involved in replication or repair. Two distinct DNA

Table 2
Properties of Other Known Thermophilic DNA Polymerases

Enzyme	Optima MgCl$_2$ (mM)	pH	KCl (mM)	Temp (°C)	Molecular Weight ($\times 10^{-3}$)	Specific Activity[a]	Exonuclease Activity 5'–3'	3'–5'
Taq Pol (Chien et al., 1976)	10	7.8	60	80	63–68	>475	–	–
Taq Pol (Kaledin et al., 1980)	10	8.3	100–200	70	60–62	1,600–5,200	–	–
Tth Pol (Rüttimann et al., 1985)								
A	ND[c]	ND	ND	50	110–150	58	–	–
B	ND	ND	ND	63	~110	922	–	–
C	ND	ND	ND	63	~110	1,380	–	–
Tth Pol (Carballeira et al., 1990)	ND	ND	ND	ND	67	4,000	–	ND
Bst Pol (Stenesh and Roe, 1972)	30	9.0	ND	65	ND	424	–	–
Bst Pol (Kaboev et al., 1982)	20	8–9	270	60	76	16,000	±	–
Bca Pol (Uemori et al., 1993b)	ND	7.5	ND	65–70	99.5	ND	+	+
Tru Pol (Kaledin et al., 1982)	2.5	9.0	15	70	70	8,000	–	–
Tfl Pol (Kaledin et al., 1981)	10–40	10.0	50	70	66	12,000	–	–
Tsp Pol (Simpson et al., 1990)	10	7.5–8.0	0	>80	85	16	ND	ND
Sac Pol (Klimczak et al., 1985)	0.1–1	6.0–8.0	0	65	100	71,500	–	+
Sac Pol (Elie et al., 1988)	ND	ND	ND	70	100	60,000	–	–
Sso Pol (Rella et al., 1990)	3	6.8	0	75	110	42,360	–	–
Tac Pol (Hamal et al., 1990)	4	8.0	5[b]	65	88	17,500	–	+
Mth Pol (Klimczak et al., 1986)	10–20	8.0	100	65	72	4,720	+	+

[a] Specific activities have been normalized such that one unit equals the amount of activity which will incorporate 10 nmol of dNTPs into product in 30 min.
[b] NH$_4$Cl.
[c] ND, Not determined.

polymerase genes have been identified from *Sso* (Pisani et al., 1992; Prangishvili and Klenk, 1993), indicating that, like mesophilic organisms, multiple DNA polymerases are inherent to at least some thermophiles. A substantial difference among polymerases is the presence or absence of associated exonuclease activities. Only *Taq* DNA polymerase (Longley et al., 1990; Holland et al., 1991; Lawyer et al., 1993), *Tth* DNA polymerase (as isolated by Stoffel and Gelfand; Lyamichev et al., 1993; Aver et al., 1995), *Tma* DNA polymerase (Lawyer and Gelfand, 1992), *Bca* DNA polymerase (Uemori et al., 1993b), and *Mth* DNA polymerase (Klimczak et al., 1986) have been shown to contain an inherent 5' to 3'-exonuclease activity, whereas only *Tac* DNA polymerase (Hamal et al., 1990), *Tma* DNA polymerase (Lawyer and Gelfand, 1992), *Bca* DNA polymerase (Uemori et al., 1993b), *Mth* DNA polymerase (Klimczak et al., 1986), *Tli* DNA polymerase (Kong et al., 1993), and *Pfu* DNA polymerase (Lundgren et al., 1991) have been shown to contain an inherent 3'-to 5'-exonuclease activity. Of the characterized thermostable DNA polymerases, only *Mth* DNA polymerase, *Tma* DNA polymerase, and *Bca* DNA polymerase resemble *E. coli* DNA polymerase I, in that they contain both a 5'- to 3'-exonuclease and a 3'- to 5'-exonuclease. The lack of any 5'- to 3'-exonucleolytic activity associated with many of the thermostable DNA polymerases may reflect a real difference among the enzymes (e.g., Family A- vs Family B-type polymerases), differences in the sensitivity of the assay procedures (see structure-dependent enhancement of activity, described later), or alternatively, may be the consequence of proteolytic degradation resulting in an amino terminal truncation of the protein. The absence of an inherent 3'- to 5'-exonuclease activity in a number of the polymerases may again reflect a real difference between species or type of polymerase characterized; alternatively a 3'- to 5'-exonuclease may exist as a separate subunit or auxiliary protein.

5'- to 3'-Exonuclease Activity

The 5'- to 3'-exonuclease activity of certain DNA polymerases, for example, *E. coli* DNA polymerase I, provides important functions in DNA biosynthesis, participating in the removal of RNA primers in lagging strand synthesis during replication and in the removal of damaged nucleotides during repair (Kornberg and Baker, 1992). Similar to *E. coli* DNA polymerase I (Cozzarelli *et al.*, 1969; Deutscher and Kornberg, 1969; Kelly *et al.*, 1969, 1970; Lundquist and Olivera, 1982), *Taq* DNA polymerase cleaves 5' terminal nucleotides of double-stranded DNA, releasing mono- and oligonucleotides (Longley *et al.*, 1990; Holland *et al.*, 1991; Lawyer *et al.*, 1993; Lyamichev *et al.*, 1993). The preferred substrate for cleavage is displaced single-stranded DNA, a fork-like structure, with hydrolysis occurring at the phosphodiester bond joining the displaced single-standed region with the base-paired portion of the strand (i.e., a structure-dependent single-stranded endonuclease or SDSSE activity) (Holland *et al.*, 1991; Lyamichev *et al.*, 1993). Such a substrate could be generated during strand displacement synthesis, or at a replication fork. Forked structures are also presumed to occur as PCR intermediates. Thus, the 5'- to 3'-exonuclease activity of *Taq* DNA polymerase can degrade single-stranded linear M13 DNA into distinct fragments at 70° C as a consequence of its SDSSE activity (R. D. Abramson, unpublished data), as well as cleave the template strand in a PCR, thereby shutting down the amplification reaction (F. C. Lawyer, unpublished data).

The mechanism of action of the 5'- to 3'-exonuclease of *Taq* DNA polymerase has been extensively studied (R. D. Abramson, unpublished data). In the absence of concurrent polymerization [i.e., in the absence of deoxyribonucleoside triphosphate (dNTP)], cleavage activity is 15–20-fold greater at a nick than at a one-base gap, and 45–60-fold greater at a nick than at a five-base gap. This activity is 15–20-fold greater on a noncomplementary single-stranded, forked structure than on duplex DNA. Activity at a nick in the presence of concurrent polymerization (where duplex DNA is first displaced by the action of the polymerase, then cleaved by the exonuclease) is stimulated an additional two to threefold. The extension rate of *Taq* DNA polymerase under conditions of nick translation synthesis (two to six nucleotides per second) is approximately 10-fold lower than the extension rate in the absence of a blocking strand. The 5'- to 3'-exonuclease activity of *Taq* DNA polymerase is 30–35-fold greater than that of *E. coli* DNA polymerase I on a molar basis. On a DNA polymerase unit basis, however, the enzymes have approximately equal activity. The mechanism of exonuclease action of *Taq* DNA polymerase is similar to that of *E. coli* DNA polymerase I; however, in contrast, the polymerization-independent 5' to 3' exonuclease activity of *E. coli* DNA polymerase I is only slightly more active at a nick than at a short gap (R. D. Abramson, unpublished data).

3'- to 5'-Exonuclease Activity and DNA Polymerase Fidelity

Proofreading—the 3'- to 5'-exonucleolytic removal of nucleotides that have been misinserted during a polymerization reaction—is one of several mechanisms that ensure the accurate replication and repair of DNA (Kornberg and Baker, 1992; see also Echols and Goodman, 1991; Kunkel, 1992; Goodman et al., 1993, for recent reviews on DNA replication fidelity). Proofreading exonucleases associated with DNA polymerases (either as part of the same polypeptide or as an associated subunit) typically prefer single-stranded to double-stranded DNA substrates. They preferentially remove mispaired rather than correctly paired nucleotides from the 3' terminus of primers, and act coordinately with the polymerase function to enhance the fidelity of DNA replication. The extent to which proofreading enhances fidelity has been estimated to be about 10-fold for *E. coli* DNA polymerase I (Kunkel, 1988). Thus, average base substitution error rates for proofreading-proficient polymerases are about 10^{-6} to 10^{-7}, whereas those for nonproofreading polymerases range from 10^{-2} to $\geq 10^{-6}$ (Kunkel, 1992).

A more complete understanding of DNA polymerase fidelity can be gained from examining the individual steps involved in the incorporation of a single nucleotide. During the polymerization reaction, there are several points at which discrimination against errors occurs, the effectiveness of which is highly dependent on reaction conditions. The first is base selection, which determines whether the correct or incorrect nucleotide is inserted. This begins with the binding of a dNTP to the polymerase–primer–template complex, followed by a conformational change in the ternary complex in order to position the nucleotide for phosphodiester bond formation. Depending on the particular DNA polymerase, either or both of these steps can be disfavored for the incorporation of the incorrect nucleotide. A second conformational change occurs in the ternary complex prior to pyrophosphate release. The disfavoring of this step allows for exonucleolytic removal of the incorrect nucleotide by proofreading polymerases, and may also be influenced by the concentration of pyrophosphate in the reaction. The final step in error discrimination results from the slower rate of incorporation of the next correct nucleotide onto a mispaired rather than a paired 3'-OH terminus, that is, misextension—a rate highly dependent on sequence context. This also affords an additional opportunity for proofreading, if 3'- to 5'-exonuclease activity is present.

The fidelity of any given PCR amplification depends on a number of elements, the enzyme being only one of these factors (see Eckert and Kunkel, 1991, 1992, for excellent reviews on PCR fidelity). Other factors include the buffer conditions, thermal cycling parameters, number of cycles, efficiency of amplification, and DNA template being amplified. Overall fidelity in a PCR (i.e., the cumulative error frequency) depends not only on the misincorporation rate (the errors per nucleotide polymerized) but also on the number of doublings. Thus a misincorporation rate of 1×10^{-5} results in a cumulative error frequency of 1/10,000 after 20 cycles at 100% efficiency, but falls to 1/5000 after 40 cycles. This translates into the difference between 95% of all PCR products being correct (assuming a 500 bp product) and only 90% being correct.

Individual components of the amplification reaction which can be manipulated in order to effect overall fidelity include: (1) dNTP concentration; a high dNTP concentration increases erorr rate by driving the reaction in the direction of DNA synthesis, thereby decreasing the amount of error discrimination at the extension step, a low dNTP concentration will increase fidelity by influencing the rate of mispair extension. (2) dNTP pool balance, which influences nucleotide selectivity at the insertion step; pool imbalance increases error rate. (3) Free Mg^{2+} concentration; low free magnesium increases fidelity (Eckert and Kunkel, 1992). (4) Reaction pH; pH can influence the rate of misinsertion, misextension, and proofreading, depending on the enzyme; for example, lower pH has been shown to increase the fidelity of *Taq* DNA polymerase (Eckert and Kunkel, 1990), whereas a higher pH was shown to increase the fidelity of Met_{284} − *Tma* DNA polymerase (F. C. Lawyer, personal communication. (5) Temperature; error rate increases with extension temperature (Eckert and Kunkel, 1992); errors due to DNA damage increase with high denaturation temperatures. (6) Extension time; minimum extension time increases fidelity by minimizing the time available for the polymerase to extend mispaired termini. (7) Denaturation time; minimuim denaturation time increases fidelity by minimizing DNA damage and the decreased fidelity observed for extension at high temperatures (during ramping). (8) Cycle number; performing the minimum number of cycles required to produce an adequate amount of product DNA will maximize fidelity. (9) Enzyme processivity; lower polymerase processivity may increase fidelity by decreasing mismatch extension efficiencies (Barnes, 1992; Huang et al., 1992);

however, other data suggest that increased processivity may increase fidelity (Eckert and Kunkel, 1992). (10) 3'- 5'-Exonuclease activity; proofreading polymerases generally have higher fidelity than exonuclease-deficient enzymes.

Thermostable DNA Polymerases and PCR

The differences among the various thermostable enzymes can be exploited for a variety of applications. The Stoffel fragment has proved superior in performing the polymerase chain reaction with certain templates that contain stable secondary structures. Extension of a primer on a template strand possessing a hairpin structure creates an ideal substrate for the 5'- to 3'-exonuclease activity of *Taq* DNA polymerase at the site of the hairpin. Extension with *Taq* DNA polymerase can result in exonucleolytic cleavage of the template strand at the hairpin structure, rather than strand displacement, creating DNA fragments that are incapable of serving as templates in later cycles of the PCR. The Stoffel fragment is also useful in polymerase chain reactions where a large amount of product is desired. Standard polymerase chain reactions with *Taq* DNA polymerase reach a plateau stage in late cycle, where product accumulation is no longer exponential. An effect contributing to this plateau is the renaturation of product strands during extension. This can result in a substrate for 5'- to 3'-exonuclease activity, in which the renatured strand is cleaved, and synthesis proceeds under nick translation conditions, resulting in no net gain in product. The Stoffel fragment, on the other hand, can proceed under strand displacement conditions, with no destruction of product, albeit at a slower rate of synthesis.

The increased thermostability of the Stoffel fragment, Met_{284} – *Tma* DNA polymerase, *Tli* DNA polymerase, or *Pfu* DNA polymerase, allows for the amplification of exceptionally G + C-rich targets, where high denaturation temperatures are required. The broader Mg^{2+} optimum for activity of the Stoffel fragment can prove useful in multiplex PCR, where multiple templates are being amplified simultaneously. Alternatively, in applications such as random mutagenesis PCR (Leung *et al.*, 1989), where Mn^{2+} is desired as the divalent cation cofactor, full-length *Taq* DNA polymerase is superior. The 5'- to 3'-exonuclease activity of *Taq* or *Tth* DNA polymerase can be used in a PCR-based detection system to generate a specific detectable signal concomitantly with amplification (Holland et al., 1991). *Taq* or *Tth* DNA polymerase also may be preferable in the amplification of long templates, where increased processivity may be an advantage, whereas the Stoffel fragment, if shown to have higher fidelity than *Taq* DNA polymerase, may be preferable for high-fidelity amplifications. Recently, *Tth* DNA polymerase has been shown to be efficient in amplifying large fragments of DNA (Ohler and Rose, 1992; Cheng et al., 1994). The lower processivity of the Stoffel fragment also makes the enzyme useful for amplifying rare mutant alleles in a background of normal DNA, using allele-specific primers where 3' mismatch extension is suppressed relative to that of the full-length polymerase (Huang et al., 1992; Tada et al., 1993). In addition, the Stoffel fragment has been shown to be superior for performing random amplified polymorphic DNA (RAPD) or arbitrarily primed (AP)-PCR (Sobral and Honeycutt, 1993). For high-temperature reverse transcription or combined RT-PCR, *Tth* DNA polymerase is optimal (Myers and Gelfand, 1991; Young et al., 1993).

A thermostable DNA polymerase possessing proofreading activity is desirable for high-fidelity amplifications—for example, (1) when amplifying large segments of DNA (>1 kb) that are found at low copy-number (less than 1000 copies) and each copy may be a sequence variant relative to the others, (2) when amplifying genes where the exact sequence is known only for a related gene, or (3) for expression studies. A polymerase with proofreading activity is also required for efficient blunt-end cloning of PCR amplicon without further treatment. Proofreading polymerases introduce additional limitations to the PCR, however. The ability of the exonuclease to degrade primers from the 3' end can impair the correct functioning of the primers, resulting in either a lower yield or nonspecific products (Skerra, 1992). This can be attenuated by a phosphorothioate bond at the 3' terminus of the primer (Skerra, 1992). Whether a phosphorothioate bond will also protect a mismatched base at the 3' end, allowing sequence-specific priming reactions, has yet to be demonstrated. An additional complication of the presence of a proofreading exonuclease activity arises in the amplification of genomic DNA, where a large number of 3' ends are present for the enzyme to bind to and subsequently degrade (Cariello et al., 1991).

Literature Cited

Abramson, R. D., S. Stoffel, and D. H. Gelfand. 1990. Extension rate and processivity of *Thermus aquaticus* DNA polymerase. *FASEB J.* **4**:A2293.

Auer, T., P. R. Landre, and T. W. Myers. 1995. Properties of the 5' to 3' exonuclease/ribonuclease H activity of *Thermus Thermophilus* DNA polymerase. *Biochemistry*, in press.

Barnes, W. M. 1992. The fidelity of *Taq* polymerase catalyzing PCR is improved by an N-terminal deletion. *Gene* **112**:29–35.

Braithwaite, D. K., and J. Ito. 1993. Compilation, alignment, phylogenetic relationships of DNA polymerases. *Nucleic Acids Res.* **21**:787–802.

Brutlag, D., M. R. Atkinson, P. Setlow, and A. Kornberg. 1969. An active fragment of DNA polymerase produced by proteolytic cleavage. *Biochem. Biophys. Res. Commun.* **37**:982–989.

Carballeira, N., M. Nazabal, J. Brito, and O. Garcia. 1990. Purification of a thermostable DNA polymerase from *Thermus thermophilus* HB8, useful in the polymerase chain reaction. *BioTechniques* **9**:276–281.

Cariello, N. F., J. A. Swenberg, and T. R. Skopek. 1991. Fidelity of *Thermococcus litoralis* DNA polymerase (Vent™) in PCR determined by denaturing gradient gel electrophoresis. *Nucleic Acids Res.* **19**:4193–4198.

Cheng, S., C. Fockler, W. B. Barnes, and R. Higuchi. 1994. Effective amplification of long targets from cloned inserts and human genomic DNA. *Proc. Natl. Acad. Sci. U.S.A.* **91**:5695–5699.

Chien, A., D. B. Edgar, and J. M. Trela. 1976. Deoxribonucleic acid polymerase from the extreme thermophile *Thermus aquaticus. J. Bacteriol.* **127**:1550–1557.

Cozzarelli, N. R., R. B. Kelly, and A. Kornberg. 1969. Enzymic synthesis of DNA: XXXIII. Hydrolysis of a 5'-triphosphate-terminated polynucleotide in the active center of DNA polymerase. *J. Mol. Biol.* **45**:513–531.

Deutscher, M. P. and A. Kornberg. 1969. Enzymatic synthesis of deoxyribonucleic acid: XXIX. Hydrolysis of deoxyribonucleic acid from the 5' terminus by an exonuclease function of deoxyribonucleic acid polymerase. *J. Biol. Chem.* **244**:3029–3037.

Echols, H., and M. F. Goodman. 1991. Fidelity mechanisms in DNA replication. *Annu. Rev. Biochem.* **60**:477–511.

Eckert, K. A., and T. A. Kunkel. 1990. High fidelity DNA synthesis by the *Thermus aquaticus* DNA polymerase. *Nucleic Acids Res.* **18**:3739–3744.

Eckert, K. A., and T. A. Kunkel. 1991. DNA polymerase fidelity and the polymerase chain reaction. *PCR Methods App.* **1**:17–24.

Eckert, K. A., and T. A. Kunkel. 1992. The fidelity of DNA polymerases used in the polymerase chain reactions. In *PCR: A practical approach* (ed. M. J. McPherson, P. Quirke, and G. R. Taylor), pp. 225–244. Oxford University Press, New York. (Reprinted with corrections edition.)

Elie, C., S. Salhi, J.-M. Rossignol, P. Forterre, and A.-M. de Recondo. 1988. A DNA polymerase from a thermoacidophilic archaebacterium: Evolutionary and technological interests. *Biochem. Biophys. Acta* **951**:261–267.

Fiala, G., and K. O. Stetter. 1986. *Pyrococcus furiosus* sp. nov. represents a novel genus of marine heterotrophic archbacteria growing optimally at 100°C. *Arch. Microbiol.* **145**:56–61.

Goodman, M. F., S. Creighton, L. B. Bloom, and J. Petruska. 1993. Biochemical basis of DNA replication fidelity. *Crit. Rev. Biochem. Mol. Biol.* **28**:83–126.

Hamal, A., P. Forterre, and C. Elie. 1990. Purification and characterization of a DNA polymerase from the archaebacterium *Thermoplasma acidophilum. Eur. J. Biochem.* **190**:517–521.

Holland, P. M., R. D. Abramson, R. Watson, and D. H. Gelfand. 1991. Detection of specific polymerase chain reaction product by utilizing the 5' to 3' exonuclease activity of *Thermus aquaticus* DNA polymerase. *Proc. Natl. Acad. Sci. U.S.A.* **88**:7276–7280.

Huang, M-M., N. Arnheim, and M. Goodman. 1992. Extension of base mispairs by *Taq* DNA polymerase: Implications for single nucleotide discrimination in PCR. *Nucleic Acids Res.* **20**:4567–4573.

Huber, R., T. A. Langworthy, H. Konig, M. Thomm, C. R. Woese, U. B. Sleytr, and K. O. Stetter. 1986. *Thermotoga maritima* sp. nov. represents a new genus of unique extremely thermophilic eubacteria growing up to 90°C. *Arch. Microbiol.* **144**:324–333.

Innis, M. A., K. B. Myambo, D. H. Gelfand, and M. A. D. Brow. 1988. DNA sequencing with *Thermus aquaticus* DNA polymerase and direct sequencing of polymerase chain reaction-amplified DNA. *Proc. Natl. Acad. Sci. U.S.A.* **85**:9436–9440.

Innis, M. A., and D. H. Gelfand. 1990. Optimization of PCRs. In *PCR protocols: A guide to methods and applications* (ed. M. A. Innis, D. H. Gelfand, J. J. Sninsky, and T. J. White), pp. 3–12. Academic Press, San Diego.

Kaboev, O. K., L. A. Luchkina, A. T. Akhmedov, and M. L. Bekker. 1981. Purification and properties of deoxyribonucleic acid polymerase from *Bacillus stearothermophilus. J. Bacteriol.* **145**:21–26.

Kaledin, A. S., A. G. Slyusarenko, and S. I. Gorodetskii. 1980. Isolation and properties of DNA polymerase from extremely thermophilic bacterium *Thermus aquaticus* YT1. *Biokhimiya* **45**:644–651.

Kaledin, A. S., A. G. Slyusarenko, and S. I. Gorodetskii. 1981. Isolation and properties of DNA polymerase from extremely thermophilic bacterium *Thermus flavus. Biokhimiya* **46**:1576–1584.

Kaledin, A. S., A. G. Slyusarenko, and S. I. Gorodetskii. 1982. Isolation and properties of DNA polymerase from extremely thermophilic bacterium *Thermus ruber. Biokhimiya* **47**:1785–1791.

Kelly, R. B., M. R. Atkinson, J. A. Huberman, and A. Kornberg. 1969. Excision of thymine dimers and other mismatched sequences by DNA polymerase of *Escherichia coli. Nature* **224**:495–501.

Kelly, R. B., N. R. Cozzarelli, M. P. Deutscher, I. R. Lehman, and A. Kornberg. 1970. Enzymatic synthesis of deoxyribonucleic acid: XXXII. Replication of duplex deoxyribonucleic acid by polymerase at a single strand break. *J. Biol. Chem.* **245**:39–45.

Keohavong, P., L. Ling, C. Dias, and W. G. Thilly. 1993. Predominant mutations induced by the Thermococcus litoralis, Vent DNA polymerase during DNA amplification *in vitro. PCR Methods App.* **2**:288–292.

Klenow, H. and I. Henningsen. 1970. Selective elimination of the exonuclease activity of the deoxyribonucleic acid polymerase from *Escherichia coli* B by limited proteolysis. *Proc. Natl. Acad. Sci. U.S.A.* **65**:168–175.

Klimczak, L. J., F. Grummt, and K. J. Burger. 1985. Purification and characterization of DNA polymerase from the archaebacterium *Sulfolobus acidocaldarius. Nucleic Acids Res.* **13**:5269–5282.

Klimczak, L. J., F. Grummt, and K. J. Burger. 1986. Purification and characterization of DNA polymerase from the archaebacterium *Methanobacterium thermoautotrophicum. Biochemistry* **25**:4850–4855.

Kong, H., R. B. Kucera, and W. E. Jack. 1993. Characterization of a DNA polymerase from the hyperthermophile Archaea *Thermococcus litoralis. J. Biol. Chem.* **268**: 1965–1975.

Kornberg, A., and T. A. Baker. 1992. *DNA Replication* (Second Ed.). W. H. Freeman and Co., New York.

Kunkel, T. A. 1988. Exonucleolytic proofreading. *Cell* **53**:837–840.

Kunkel, T. A. 1992. DNA replication fidelity. *J. Biol. Chem.* **267**:18251–18254.

Lawyer, F. C., S. Stoffel, R. K. Saiki, K. Myambo, R. Drummond, and D. H. Gelfand. 1989. Isolation, characterization, and expression in *Escherichia coli* of the DNA polymerase gene from *Thermus aquaticus. J. Biol. Chem.* **264**:6247–6437.

Lawyer, F. C., and D. H. Gelfand. 1992. The DNA polymerase I gene from the extreme thermophile *Thermotoga maritima:* Identification, cloning, and expression of full-length and truncated forms in *Escherichia coli. Abs. 92nd Gen Meet. Am. Soc. Microbiology*, p. 200.

Lawyer, F. C., R. K. Saiki, S.-Y. Chang, P. A. Landre, R. D. Abramson, and D. H. Gelfand. 1993. High-level expression, purification, and enzymatic characterization of full-length *Thermus aquaticus* DNA polymerase and a truncated form deficient in 5' to 3' exonuclease activity. *PCR Methods App.* **2**:275–287.

Leung, D. W., E. Chen, and D. V. Goeddel. 1989. A method for random mutagenesis of a defined DNA segment using a modified polymerase chain reaction. *Technique* **1**:11–15.

Longley, M. J., S. E. Bennett, and D. W. Mosbaugh. 1990. Characterization of the 5' to 3' exonuclease associated with *Thermus aquaticus* DNA polymerase. *Nucleic Acids Res.* **18**:7317–7322.

Lundberg, K. S., D. D. Shoemaker, M. W. W. Adams, J. M. Short, J. A. Sorge, and E. J. Mathur. 1991. High-fidelity amplification using a thermostable DNA polymerase isolated from *Pyrococcus furiosus. Gene* **108**:1–6.

Lundquist, R. C., and B. M. Olivera. 1982. Transient generation of displaced single-stranded DNA during nick translation. *Cell* **31**:53–60.

Lyamichev, V., M. A. D. Brow, and J. E. Dahlberg. 1993. Structure-specific endonucleolytic cleavage of nucleic acids by eubacterial DNA polymerases. *Science* **260**:778–783.

Mathur, E. J., M. W. W. Adams, W. N. Callen, and J. M. Cline. 1991. The DNA polymerase gene from the hyperthermophilic marine archaebacterium, *Pyrococcus furiosus,* shows sequence homology with α-like DNA polymerases. *Nucleic Acids Res.* **19**:6952.

Mathur, E., J. Cline, K. Nielson, W. Schoetllin, W. Callen, K. Kretz, and J. Sorge. 1992. Characterization and cloning of DNA polymerase I and DNA ligase I from *Pyrococcus furiosus. Abs. 1992 San Diego Conf. Nucl. Acids: Gene Recog.,* p. 10.

Mattila, P., J. Korpela, T. Tenkanen, and K. Pitkänen. 1991. Fidelity of DNA synthesis by the *Thermococcus litoralis* DNA polymerase—an extremely heat stable enzyme with proofreading activity. *Nucleic Acids Res.* **19**:4967–4973.

Mullis, K. B., and F. Faloona. 1987. Specific synthesis of DNA *in vitro* via a polymerase-catalyzed chain reaction. *Methods Enzymol.* **155**:335–350.

Myers, T. W., and D. H. Gelfand. 1991. Reverse transcription and DNA amplification by a *Thermus thermophilus* DNA polymerase. *Biochemistry* **30**:7661–7666.

Neuner, A., H. W. Jannasch, S. Belkin, and K. O. Stetter. 1990. *Thermococcus litoralis*

sp. nov.: A new species of extremely thermophilic marine archaebacteria. *Arch Microbiol.* **153**:205–207.
Ohler, L. D., and E. A. Rose. 1992. Optimization of long-distance PCR using a transposon-based model system. *PCR Methods App.* **2**:51–59.
Pisani, F. M., C. De Martino, and M. Rossi. 1992. A DNA polymerase from the archaeon *Sulfolobus solfataricus* shows sequence similarity to family B DNA polymerases. *Nucleic Acids Res.* **20**:2711–2716.
Prangishvili, D., and H.-P. Klenk. 1993. Nucleotide sequence of the gene for a 74 kDa DNA polymerase from the archaeon *Sulfolobus solfataricus*. *Nucleic Acids Res.* **21**:2768.
Rella, R., C. A. Raia, F. M. Pisani, S. D'Auria, R. Nucci, A. Gambacorta, M. De Rosa, and M. Rossi. 1990. Purification and properties of a thermophilic and thermostable DNA polymerase from the archaebacterium *Sulfolobus solfataricus*. *Ital. J. Biochem.* **39**:83–99.
Rüttimann, C., M. Cotorás, J. Zaldívar, and R. Vicuña. 1985. DNA polymerases from the extremely thermophilic bacterium *Thermus thermophilus* HB-8. *Eur. J. Biochem.* **149**:41–46.
Saiki, R. K., D. H. Gelfand, S. Stoffel, S. Scharf, R. Higuchi, G. T. Horn, K. B. Mullis, and H. A. Erlich. 1988. Primer-directed enzymatic amplification of DNA with a thermostable DNA polymerase. *Science* **239**:487–491.
Salhi, S., C. Elie, P. Forterre, A.-M. de Recondo, and J.-M. Rossignol. 1989. DNA polymerase from *Sulfolobus acidocaldarius:* Replication at high temperature of long stretches of single-stranded DNA. *J. Mol. Biol.* **209**:635–644.
Simpson, H. D., T. Coolbear, M. Vermue, and R. M. Daniel. 1990. Purification and some properties of a thermostable DNA polymerase from a *Thermotoga* species. *Biochem. Cell Biol.* **68**:1292–1296.
Skerra, A. 1992. Phosphorothioate primers improve the amplification of DNA sequences by DNA polymerases with proofreading activity. *Nucleic Acids Res.* **20**:3551–3554.
Sobral, B. W. S., and R. J. Honeycutt. 1993. High output genetic mapping of polyploids using PCR-generated markers. *Theor. Appl. Genet.* **86**:105–112.
Stenesh, J., and B. A. Roe. 1972. DNA polymerase from mesophilic and thermophilic bacteria: I. Purification and properties of DNA polymerase from *Bacillus licheniformis* and *Bacillus stearothermophilus*. *Biochim. Biophys. Acta* **272**:156–166.
Tada, M., M. Omata, S., Kawai, H. Saisho, M. Ohto, R. K. Saiki, and J. J. Sninsky. 1993. Detection of *ras* gene mutations in pancreatic juice and peripheral blood of patients with pancreatic adenocarcinoma. *Cancer Res.* **53**:2472–2474.
Tindall, K. R., and T. A. Kunkel. 1988. Fidelity of DNA synthesis by the *Thermus aquaticus* DNA polymerase. *Biochemistry* **27**:6008–6013.
Uemori, T., Y. Ishino, H. Toh, K. Asada, and I. Kato. 1993a. Organization and nucleotide sequence of the DNA polymerase gene from the archaeon *Pyrococcus furiosus*. *Nucleic Acids Res.* **21**:259–265.
Uemori, T., Y. Ishino, K. Fujita, K. Asada, and I. Kato. 1993b. Cloning of the DNA polymerase gene of *Bacillus caldotenax* and characterization of the gene product. *J. Biochem.* **113**:401–410.
Young, K. K., R. M. Resnick, and T. W. Myers. 1993. Detection of hepatitis C virus RNA by a combined reverse transcription-polymerase chain reaction assay. *J. Clin. Microbiol.* **31**:882–886.

5

AMPLIFICATION OF RNA: HIGH-TEMPERATURE REVERSE TRANSCRIPTION AND DNA AMPLIFICATION WITH *THERMUS THERMOPHILUS* DNA POLYMERASE

Thomas W. Myers and Christopher L. Sigua

Standard methods for the detection and analysis of RNA molecules are limited in their usefulness either because of the large quantity of RNA needed for detection or the degree of technical expertise required to perform the methodologies. The technique of DNA amplification by PCR (Saiki *et al.*, 1985; Mullis and Faloona, 1987) has been extended to include RNA as the starting template by first converting RNA to cDNA by a retroviral reverse transcriptase (Veres *et al.*, 1987; Powell *et al.*, 1987). Detailed methodologies for coupling reverse transcription and PCR (RT-PCR) can be found in several laboratory manuals (Kawasaki and Wang, 1989; Kawasaki, 1990). The rapid incorporation of this technology into the standard repertoire of molecular biology techniques has enabled researchers in diverse fields to use and extend the original method. The process of RT-PCR has proved invaluable for detecting gene expression, generating cDNA for cloning, and diagnosing infectious agents or genetic diseases.

Although the RT–PCR process has become a standard laboratory practice, the procedure is not without pitfalls. The mesophilic viral reverse transcriptases typically used for the RT step have optimum enzymatic activity between 37 and 42°C. These temperatures result in decreased specificity of primer binding and increased stability of duplex RNA. The use of a thermoactive reverse transcriptase allows for increased specificity of primer binding and alleviation of many secondary structures present in the RNA template, thus decreasing premature termination by the reverse transcriptase. In addition, as has been described for the removal of carryover contamination in a PCR (Longo et al., 1990), a thermal-labile uracil-N-glycosylase (UNG) can be used to prevent carryover contamination at lower temperatures prior to the high-temperature RT step, whereas the newly formed cDNA would be a substrate for UNG using a mesophilic reverse transcriptase at lower temperatures. A recombinant DNA polymerase from the thermophilic eubacterium *Thermus thermophilus* (rTth pol) was found to possess efficient reverse transcriptase activity in the presence of Mn^{2+} (Myers and Gelfand, 1991). The RT reaction was coupled to a PCR amplification and the resulting protocol provided for the amplification of RNA in a two-step coupled process that required only the addition of a single (thermostable) enzyme. The RT–PCR performed with rTth pol is both sensitive and specific, with the RT step being performed in the presence of Mn^{2+}, while the PCR amplification is activated by Mg^{2+} following preferential chelation of the Mn^{2+} with ethylenebis(oxyethylenenitrilo)tetraacetic acid (EGTA).

While increased specificity and ability to synthesize through regions of high G + C content are gained by using rTth pol, a concern in clinical diagnostic applications is that even though the RT reaction and PCR amplification are performed with the same enzyme, an alteration of buffer conditions is required between the two enzymatic steps. A buffer change is not critical for research laboratories performing a few reactions, but it is cumbersome and increases the likelihood of contamination when many reactions are being performed. Reaction conditions allowing rTth pol to perform both reverse transcription and DNA amplification in a single buffer containing Mn^{2+}, thus eliminating the requirement for a Mg^{2+}-dependent DNA amplification, have been determined and used to detect hepatitis C virus (Young et al., 1993). RT–PCR with rTth pol also has formed the basis of a quantitative HIV RNA assay (Mulder et al., 1994). However, the Mn^{2+} concentration optimum is different for RNA and DNA templates and the two reactions are performed under suboptimal conditions. The reaction is also very sensitive to the Mn^{2+} concentration and any reaction components that complex Mn^{2+} (deoxyribonucleoside triphosphates, template, primers, etc.), thus requiring strict control of their concentrations. In addition, the inability to readily modulate deoxynucleoside triphosphate (dNTP) concentrations makes it difficult to incorporate deoxyuridine triphosphate (dUTP) for UNG-mediated carryover prevention, since high concentrations of dUTP are typically used for recalcitrant A-rich targets, especially during cDNA synthesis by r*Tth* pol.

Key to the use of r*Tth* pol to copy readily both RNA and DNA templates under the same reaction conditions was the conversion of the buffer used in the reaction from the standard Tris–HCl buffer to a reagent capable of buffering the metal ion concentration. Bicine is known to possess metal buffering capacity (Good *et al.*, 1966) and its use has circumvented the narrow Mn^{2+} concentration tolerance (Myers *et al.*, 1994). The wide Mn^{2+} concentration range allowed in the reaction also enables the dNTP concentration to be varied substantially. The ability to use higher concentrations of dNTPs, in particular dUTP, has improved the incorporation of dUMP into cDNA by r*Tth* pol. The speed, sensitivity, robustness, and the ability to incorporate dUTP to prevent carryover contamination achieved by using r*Tth* pol for RT–PCR, further extends our ability to detect and quantitate cellular and viral RNA.

Protocols

Reagents and Supplies

Glycerol
Surfact-Amps 20: 10% solution of Tween 20 from Pierce
1 *M* Tris–HCl (pH 8.3)
1 *M* bicine–KOH (pH 8.3): adjust an approximately 1 *M* solution of N,N-bis(2-hydroxyethyl)glycine (bicine) to pH 8.3 with 45% potassium hydroxide (KOH) and dilute to 1 *M* with water. Filter sterilize, do not treat with diethyl pyrocarbonate.
1 *M* KCl
3 *M* potassium acetate (KOAc) (pH 7.5): adjust an approximately

3 M solution of KOAc to pH 7.5 with glacial acetic acid and dilute to 3 M with water. Note: very little acetic acid is required and pH decreases upon dilution, so adjust pH near final volume.

10 mM $MnCl_2$: Perkin-Elmer

100 mM $Mn(OAc)_2$: manganese(II)acetate tetrahydrate from Aldrich (#22,977-6)

25 mM $MgCl_2$: Perkin-Elmer

500 mM EGTA (pH 8.0)

Deoxynucleoside triphosphates: neutralized 10 or 20 mM (dUTP) solutions from Perkin-Elmer

Primers: 25 μM in 10 mM Tris–HCl (pH 8.3) from Perkin-Elmer

K562 RNA: total cellular RNA from the K562 cell line (Lozzio and Lozzio, 1975) was isolated as described (Sambrook et al., 1989).

E. coli rRNA: 16S and 23S rRNA from Boehringer Mannheim was phenol extracted, ethanol precipitated, and adjusted to 125 ng/μl in water.

pAW109 cRNA: positive control pAW109 cRNA from Perkin-Elmer at 5×10^3 copies/μl was diluted with 125 ng/μl Escherichia coli rRNA to provide a final concentration of 50 copies/μl pAW109 cRNA.

Uracil-N-glycosylase (UNG): 1 U/μl from Perkin-Elmer

rTth pol: 2.5 U/μl from Perkin-Elmer

Mineral oil: molecular biology grade from Sigma

Thin-walled GeneAmp tubes: Perkin-Elmer

DNA Thermal Cycler 480: Perkin-Elmer

NuSieve GTG agarose and SeaKem GTG agarose: FMC

1-kb DNA ladder: GIBCO BRL

5 μg/μl Ethidium bromide

Two-Buffer RT–PCR (Fig. 1)

1. Reverse transcription reaction (20 μl):
   10 mM Tris–HCl (pH 8.3)
   90 mM KCl
   200 μM each of deoxyadenosine triphosphate (dATP), deoxycytidine triphosphate (dCTP), deoxyguanosine triphosphate (dGTP), and deoxythymidine triphosphate (dTTP)
   0.75 μM RT "downstream" primer
   1 mM $MnCl_2$
   5 U rTth pol
   RNA ($\leq$250 ng)
   Overlay with 75 μl of mineral oil
   Incubate for 15–60 min at 55–70°C.

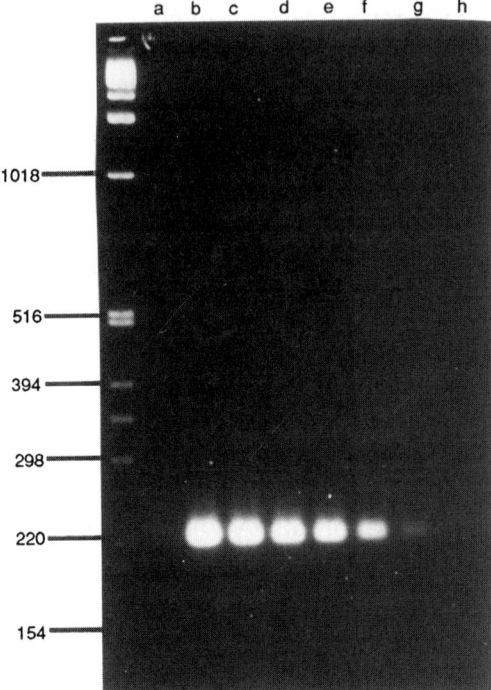

Figure 1 Two-buffer r*Tth* pol RT–PCR of PDGF-A mRNA from total cellular RNA. Reactions contained 250, 250, 50, 10, 2, 0.4, 0.08, and 0 ng of K562 total cellular RNA (lanes a–h, respectively). Reactions having less than 250 ng of K562 RNA also contained 60 ng of *E. coli* rRNA as carrier nucleic acid (lanes c–h). The sample in lane a was incubated at 4°C during the RT reaction while all other RT reactions were incubated 15 min at 70°C. All reactions contained AW117 as the RT primer and AW116 for the "upstream" PCR primer. This primer set is specific for the detection of PDGF-A mRNA and produces a 225-bp PCR product. Electrophoretic analysis of PCR products was performed with 5% of the PCR amplification on a 2% NuSieve GTG agarose/1% SeaKem GTG agarose gel stained with ethidium bromide.

2. Polymerase chain reaction (100 µl):
   Add 80 µl of the following solution to the RT reaction.
   10 mM Tris–HCl (pH 8.3)
   100 mM KCl
   0.05% Tween 20
   1.88 mM MgCl$_2$
   0.75 mM EGTA
   5% glycerol (v/v)
   0.19 µM PCR "upstream" primer.

The reactions are then amplified in a DNA Thermal Cycler 480 by using four linked files as follows:

File 1, STEP-CYCLE 2 min at 95°C for 1 cycle
File 2, STEP-CYCLE 1 min at 95°C and 1 min at 60°C for 35 cycles
File 3, STEP-CYCLE 7 min at 60°C for 1 cycle
File 4, SOAK 15°C.

Single-Buffer RT–PCR (Fig. 2)

1. Combined RT and PCR amplification (50 µl):
   50 mM bicine–KOH (pH 8.3)

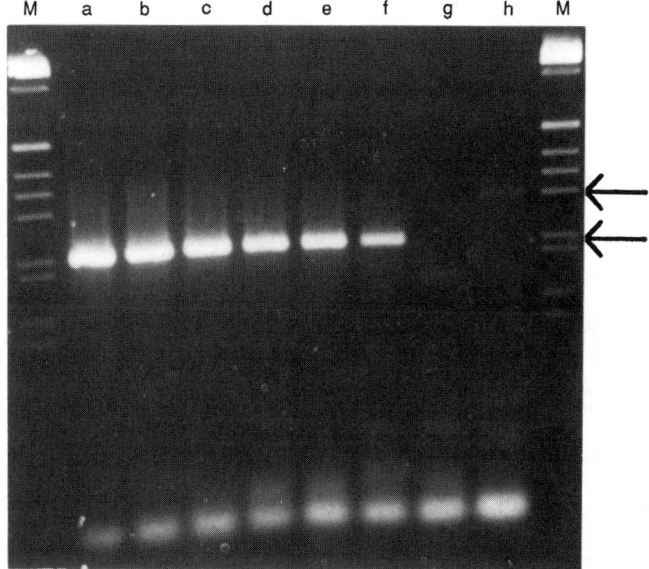

Figure 2 Single-buffer rTth pol RT–PCR of PDGF-A mRNA from total cellular RNA. Reactions contained 250, 50, 10, 2, 0.4, 0.08, and 0 ng of K562 total cellular RNA (lanes a–g, respectively). Lane h contains 100 copies of pAW109 cRNA. Reactions having less than 250 ng of K562 RNA also contained 250 ng E. coli rRNA as carrier nucleic acid (lanes b–h). All reactions incorporated dUTP and UNG to prevent carryover contamination. The RT reaction was incubated 25 min at 65°C. All reactions contained AW117 as the RT primer and AW116 for the "upstream" PCR primer. This primer set is specific for the detection of PDGF-A mRNA and produces a 225-bp PCR product with the K562 RNA or a 301-bp product with the pAW109 cRNA. Electrophoretic analysis of PCR products was performed with 10% of the PCR amplification on a 3% NuSieve GTG agarose/1% SeaKem GTG agarose gel stained with ethidium bromide. Reproduced from Myers et al., (1994). High temperature reverse transcription and PCR with *Thermus thermophilus* DNA polymerase. *Miami Short Reports* **4**:87. Permission granted by Oxford University Press.

100 mM KOAc (pH 7.5)

3% glycerol (v/v) [this 3% is in addition to the 2% contributed by the r*Tth* pol]

200 μM each of dATP, dCTP, dGTP, and 500 μM dUTP [use 200–300 μM each dNTP if using dTTP rather than dUTP]

0.30 μM RT "downstream" primer

0.30 μM PCR "upstream" primer

1 U UNG [omit if using dTTP rather than dUTP]

5 U r*Tth* pol

2.5 mM Mn(OAc)$_2$

RNA (≤1 μg)

Overlay with 75 μl of mineral oil.

The RT reaction and PCR were then performed consecutively in a DNA Thermal Cycler 480 by using five linked files as follows:

File 1, STEP-CYCLE 15–60 min at 55–70°C
File 2, STEP-CYCLE 2 min at 95°C for 1 cycle
File 3, STEP-CYCLE 1 min at 95°C and 1 min at 60°C for 40 cycles
File 4, STEP-CYCLE 7 min at 60°C for 1 cycle
File 5, SOAK 15°C.

Discussion and Helpful Hints

The use of a thermoactive DNA polymerase to perform a high-temperature reverse transcription reaction helps to alleviate several of the problems typically encountered in trying to amplify RNA targets. However, one must take into account several factors when using r*Tth* pol as a reverse transcriptase. Between 55 and 70°C (Myers et al., 1994), the enzyme has greater than 75% activity on RNA templates under both of the RT conditions described, thus the use of oligonucleotide primers with relatively high melting temperatures is advantageous for the RT step of the reaction. The use of oligo dT for poly(A+) RNA can be achieved by performing the RT step at a lower initial temperature to extend the primer, followed by elevated temperatures for the remainder of the reaction (Myers and Gelfand, 1991). Similar temperature ramping profiles would most likely allow the successful use of oligonucleotides with a lower melting temperature when using rTth pol, and would be preferred to simply performing the RT step at a single low temperature. Since rTth pol has less than 10% RT activity at 40°C, the use of random hexamers is not recommended.

The elevated temperatures used with rTth pol also result in increased hydrolysis of the RNA template in the presence of metal ions (Brown, 1974). The increase in specificity, decrease in RNA secondary structure, and increase in enzymatic activity must be weighed against destruction of the target. The generation of cDNA greater than 1.5 kb necessitates using lower reaction temperatures to minimize hydrolysis of the template, and modifying reaction parameters to optimize for longer products. Using the two-buffer RT–PCR protocol, the generation of cDNA greater than 3.5 kb has been achieved by performing the RT and PCR steps in a final glycerol concentration of 10% and incubating the RT reaction for 60 min at 60°C. Clearly, the extension time for the PCR amplification also will need to be increased for longer products. Increased concentrations of glycerol have improved both the RT and PCR steps for many targets. The inclusion of carrier RNA, such as poly(A) or *E. coli* rRNA, in RT reactions containing low concentrations of RNA helps to prevent nonspecific binding of target RNA to tubes and provides an alternative substrate for any RNase activity present in the sample.

The decision as to which protocol [two-buffer (Fig. 1) vs single-buffer RT–PCR (Fig. 2)] to use depends primarily on the length of the cDNA to be synthesized and whether the amplified cDNA is going to be cloned. The sensitivity of the two methods is quite similar (Fig. 1, lane g vs Fig. 2, lane f). However, the single-buffer format typically requires more cycles during the PCR to achieve the same sensitivity. For long cDNA synthesis and/or cloning purposes, the two-buffer protocol is preferred since the RT and PCR steps can be performed under their own optimal conditions. In addition to achieving increased enzymatic activity for the two independent reactions, conditions known to influence the fidelity of DNA polymerases can be altered much more easily than with the single-buffer method. The negative effect of Mn^{2+} on the fidelity of DNA synthesis has been documented for *E. coli* pol I (Beckman et al., 1985) and more recently for decreasing PCR fidelity with *Thermus aquaticus* DNA polymerase (*Taq* pol) (Leung et al., 1989). The addition of EGTA to preferentially chelate the Mn^{2+} in the two-buffer procedure is expected to decrease any effects that the Mn^{2+} may have on the fidelity of rTth pol during the PCR amplification. The low dNTP (160 μM total) and Mg^{2+} (1.5 mM) concentrations suggested for the two-buffer method also may be conducive to higher fidelity synthesis during the PCR, since similar conditions were shown to increase the fidelity of Taq pol (Eckert and Kunkel, 1990). First-strand cDNA synthesis, in addition to PCR, must be accurate for cloning purposes. Therefore, it is advisable that several independent RT–PCR amplifications be sequenced.

The utilization of dUTP-UNG to prevent carryover contamination works effectively in either the two-buffer or single-buffer format. RT reactions using the two-buffer approach with dUTP are often improved by adjusting the Mn^{2+} concentration to 1.2 mM. Increasing the concentration of dUTP higher than 200 μM for the RT step is not recommended, unless the Mn^{2+} concentration is also adjusted accordingly. In experiments not utilizing dUTP/UNG, procedures must be followed to minimize carryover of amplified DNA (Higuchi and Kwok, 1989). The simplicity, speed, and sensitivity of the single-buffer format makes this form of RT–PCR very amenable for analyzing large numbers of samples, and the incorporation of a carryover prevention strategy such as dUTP-UNG is highly recommended. The dUTP concentration used in the single-buffer RT–PCR can be adjusted depending on the percentage of riboadenylate (rA) in the template. High G + C content templates work fine with 200–300 μM dUTP, while templates having a large percentage of rA residues are reverse transcribed more efficiently with higher dUTP concentrations. The bicine buffer allows for these moderate fluctuations in dNTP concentrations without having to readjust the Mn^{2+} concentration.

While there are certain applications that may be more suitable for a mesophilic reverse transcriptase, especially when sequence information is lacking, a wide range of research applications would benefit from a high-temperature, reverse transcription and PCR amplification with rTth pol. Given the extreme sensitivity and immense amplification achieved by the RT–PCR process, one needs to remember that the generation of hundreds of nanograms of "full-length" cDNA is not required to obtain meaningful PCR results. Both rTth pol formats offer increased specificity, decreased premature termination as a result of enhanced denaturation of RNA secondary structure during cDNA synthesis, and the ability to incorporate dUTP-UNG to prevent carryover contamination. The single-buffer system combines all the attributes of the two-buffer method, but in a more robust, simplified manner. The coupling of the single-enzyme, single-buffer rTth pol RT–PCR (Myers et al., 1994) with a quantitative detection system (for example, Wang et al., 1989; Becker-André and Hahlbrock, 1989; Holland *et al.*, 1991; Mulder *et al.*, 1994) makes the monitoring of disease progression and therapeutic response feasible.

Literature Cited

Becker-André, M., and K. Hahlbrock. 1989. Absolute mRNA quantification using the polymerase chain reaction (PCR). A novel approach by a PCR aided transcript titration assay (PATTY). *Nucleic Acids Res.* **17**:9437–9446.

Beckman, R. A., A. S. Mildvan, and L. A. Loeb. 1985. On the fidelity of DNA replication: manganese mutagenesis *in vitro*. *Biochemistry* **24**:5810–5817.

Brown, D. M. 1974. Chemical reactions of polynucleotides and nucleic acids. In *Basic principles in nucleic acid chemistry* (ed. P. O. P. Ts'o), pp. 43–44. Academic Press, New York.

Eckert, K. A., and T. A. Kunkel. 1990. High fidelity DNA synthesis by the *Thermus aquaticus* DNA polymerase. *Nucleic Acids Res.* **18**:3739–3744.

Good, N. E., G. D. Winget, W. Winter, T. N. Connolly, S. Izawa, and R. M. M. Singh. 1966. Hydrogen ion buffers for biological research. *Biochemistry* **5**:467–477.

Higuchi, R., and S. Kwok. 1989. Avoiding false positives with PCR. *Nature* **339**:237–238.

Holland, P. M., R. D. Abramson, R. Watson, and D. H. Gelfand. 1991. Detection of specific polymerase chain reaction product by utilizing the 5′→3′ exonuclease activity of *Thermus aquaticus* DNA polymerase. *Proc. Natl. Acad. Sci. U.S.A.* **88**:7276–7280.

Kawasaki, E. S. 1990. Amplification of RNA. In *PCR protocols: a guide to methods and applications* (ed. M. A. Innis, D. H. Gelfand, J. J. Sninsky, and T. J. White), pp. 21–27. Academic Press, San Diego, CA.

Kawasaki, E. S., and A. M. Wang. 1989. Detection of gene expression. In *PCR technology: principles and applications for DNA amplification* (ed. H. A. Erlich), pp. 89–97. Stockton Press, New York.

Leung, D. W., E. Chen, and D. V. Goeddel. 1989. A method for random mutagenesis of a defined DNA segment using a modified polymerase chain reaction. *Technique* **1**:11–15.

Longo, M. C., M. S. Berninger, and J. L. Hartley. 1990. Use of uracil DNA glycosylase to control carry-over contamination in polymerase chain reactions. *Gene* **93**:125–128.

Lozzio, C. B., and B. B. Lozzio. 1975. Human chronic myelogenous leukemia cell-line with positive Philadelphia chromosome. *Blood* **45**:321–334.

Mulder, J., N. McKinney, C. Christopherson, J. Sninsky, L. Greenfield, and S. Kwok. 1994. A rapid and simple PCR assay for quantitation of HIV RNA: application to acute retroviral infection. *J. Clin. Microbiol.* **32**:292–300.

Mullis, K. B., and F. A. Faloona. 1987. Specific synthesis of DNA *in vitro* via a polymerase-catalyzed chain reaction. *Methods Enzymol.* **155**:335–350.

Myers, T. W., and D. H. Gelfand. 1991. Reverse transcription and DNA amplification by a *Thermus thermophilus* DNA polymerase. *Biochemistry* **30**:7661–7666.

Myers, T. W., C. L. Sigua, and D. H. Gelfand. 1994. High temperature reverse transcription and PCR with *Thermus thermophilus* DNA polymerase. *Miami Short Reports* **4**:87.

Powell, L. M., S. C. Wallis, R. J. Pease, Y. H. Edwards, T. H. Knott, and J. Scott. 1987. A novel form of tissue-specific RNA processing produces apolipoprotein-b48 in intestine. *Cell* **50**:831–840.

Saiki, R. K., S. Scharf, F. Faloona, K. B. Mullis, G. T. Horn, H. A. Erlich, and N. Arnheim. 1985. Enzymatic amplification of β-globin genomic sequences and restriction site analysis for diagnosis of sickle cell anemia. *Science* **230**:1350–1354.

Sambrook, J., E. F. Fritsch, and T. Maniatis. 1989. In *Molecular cloning: A laboratory manual* 2nd ed. (ed. C. Nolan), pp. 7.19–7.22. Cold Spring Harbor Laboratory, Cold Spring Harbor, New York.

Veres, G., R. A. Gibbs, S. E. Scherer, and C. T. Caskey. 1987. The molecular basis of the sparse fur mouse mutation. *Science* **237**:415–417.

Wang, A. M., M. V. Doyle, and D. F. Mark. 1989. Quantitation of mRNA by the polymerase chain reaction. *Proc. Natl. Acad. Sci. U.S.A.* **86**:9717–9721.

Young, K. K. Y., R. M. Resnick, and T. W. Myers. 1993. Detection of hepatitis C virus RNA by a combined reverse transcription-polymerase chain reaction assay. *J. Clin. Microbiol.* **31**:882–886.

6

NUCLEIC ACID HYBRIDIZATION AND UNCONVENTIONAL BASES

James G. Wetmur and John J. Sninsky

Modified deoxynucleotides have been used with PCR for three major purposes: (1) to label the PCR product for detection or affinity purification, (2) to change the thermodynamic properties of the PCR product, and (3) to alter the ability of the PCR product to serve as a substrate for various enzymes. Modified deoxynucleotides may be introduced into PCR products either by using chemically synthesized PCR primers containing the modifications or by polymerization using modified deoxynucleoside triphosphates. The aim of this chapter is to introduce the general considerations necessary for successful production of modified PCR products, including not only those synthesized to date, but also new modifications which may occur to the reader.

The PCR process involves hybridization, polymerization, and denaturation steps. Use of a modified PCR primer may affect hybridization, with the melting temperature of the modified primer–template hybrid increased or decreased relative to an unmodified primer–template hybrid. A modified PCR primer may also affect polymerization because the modified deoxynucleotides in the primer must act as template deoxynucleotides unless the modification is at the 5' end. Enzymatic incorporation of a modified deoxynucleotide may affect all three steps. Both the melting temperature of the primer–template hybrid and the denaturation temperature of the PCR product may change. Polymerization may be indirectly affected by the thermodynamics if there is an increase or decrease in secondary structure in the template. Polymerization may be directly affected both by the ability of the modified deoxynucleoside triphosphate to act as a substrate and the modified PCR product to act as a template.

Thermodynamics of Modified Nucleic Acids

For convenience, we divide the effects of modified bases on PCR into thermodynamic and enzymatic. First, consider the thermodynamics (reviewed by Wetmur, 1991).

The term "melting temperature" has been used for three different temperatures. These temperatures (in °C) and their characteristics are:

t_m^∞, the melting temperature of a polynucleotide duplex such as a PCR product.

t_m, the melting temperature where at least one duplex strand is an oligonucleotide, such as a PCR primer–template hybrid, and

t_d, the dissociation temperature for an oligonucleotide bound to a polynucleotide on a solid support, such as a blot.

The latter temperature, t_d, is not important for PCR. We will consider first t_m^∞, for the denaturation step, and then t_m, for the hybridization step of PCR.

Denaturation

An empirical formula for calculating the melting temperature of a PCR product, t_m^∞, is given in Appendix 1. This formula is valid for an extended range of monovalent (Na^+ or K^+) and divalent (Mg^{2+}) cation concentrations, including PCR buffers. For example, for a 1-kb 50% G + C PCR product in 0.05 M K^+ and 0.0015 M Mg^{2+}, $t_m^\infty = 89°C$. Strand separation may be ensured by achieving a denaturation temperature 3–4°C higher than t_m^∞, or 92–93°C.

Polymerization of modified bases into a PCR product may result in substantial increases or decreases in the melting temperature, Δt_m^∞, which may be directly added to the previously calculated t_m^∞. For each different modification, one may define a characteristic parameter, S, for, S, which may be used to calculate Δt_m^∞ for PCR products containing that modification. For a PCR product with a composition defined by (% G + C), modification of M_{GC} percent of the G + C base pairs or M_{AT} of the A + T base pairs will lead to:

$$\Delta t_m^\infty = S(\% \text{ G + C})(M_{GC}/100)$$

or

$$\Delta t_m^\infty = S(\% \text{ A + T})(M_{AT}/100)$$

An example of a deoxynucleotide substitution which lowers t_m^∞ (negative Δt_m^∞) is 7-deaza-2'-deoxyguanosine (c^{7}dG; McConlogue *et al.*, 1988) replacing deoxyguanosine (dG). Decreases in t_m^∞ permit use of lower denaturation temperatures after the first few cycles, when the original template no longer contributes significantly to product accumulation. An example of a deoxynucleotide substitution which raises t_m^∞ (positive Δt_m^∞) is 5-methyldeoxycytidine (5-MedC) replacing deoxycytidine (dC). The most extensive thermodynamic data are available for this modification (D. M. Wong *et al.*, 1991). Spectrophotometric determination of Δt_m^∞ was carried out with *Xanthamonas* bacteriophage XP-12 (Ehrlich *et al.*, 1975), a 51% G + C DNA containing 100% 5-MedC replacing dC. The Δt_m^∞ was found to be 6.1°C, corresponding to a value of 0.12 for the parameter S. Similar results were obtained with a fully substituted PCR product of known base composition. If long, modified duplex DNA is unavailable, S may also be extrapolated from t_m data obtained with synthetic oligonucleotides with different percentages of modified deoxynucleotides (see Appendix 3). Such an extrapolation has also been carried out for 5-MedC.

Just as with templates of high G + C percent, increases in t_m^∞ resulting from incorporation of modified bases require the use of a higher denaturation temperature and/or the use of an organic cosolvent which decreases t_m^∞. For example, t_m^∞ is decreased 0.25°C per percent glycerol, 0.4°C per percent ethylene glycol, and 0.63°C per percent formamide. Note that the choice of organic solvent is dictated both by melting temperature and the effect of the organic solvent on the enzymatic activity of the particular thermostable DNA polymerase to be employed.

Hybridization

An empirical calculation of the PCR primer–template hybrid melting temperature, t_m, based on nearest-neighbor thermodynamic data

(Breslauer et al., 1986; Quartin and Wetmur, 1989), is given in Appendix 2. It is important to note that there is no single t_m for a given primer, because t_m depends on primer concentration, C. The thermodynamic data also include an estimate for the contribution of "dangling ends" where the template strand extends beyond the primer strand. Initially, PCR priming involves dangling ends at both ends of the primer. Later in the amplification process, only one dangling end stabilizes the primer–template hybrid. These dangling end contributions become less important as the primer length is increased.

Oligodeoxynucleotide melting data sets for a modified deoxynucleotide are rarely complete enough to determine all of the different nearest-neighbor interactions for the modified base. A simplification is made. The entropy is assumed to be constant, and the same $\Delta\Delta G°$ term is added to the enthalpy, $\Delta H°$, and free energy, $\Delta G°$, terms in Appendix 2 for each of N modified base pairs. The best value for $\Delta\Delta G°$ may be determined by analyzing several oligodeoxynucleotide pairs with different numbers of modified bases. The details of such calculations are included in Appendix 3. The end result is that the single parameter S may be used to characterize the effect of a base modification on either PCR product melting, as described above, or on primer–template melting:

$$\Delta t_m \approx \frac{N(t_m + 273.2)810000}{-\Delta H°(t_m^\infty + 273.2)} S$$

Many aspects of primer selection, such as avoidance of primer-dimer, are not unique to primers with modified bases. However, special consideration is warranted in two areas:

As a rule of thumb, the greatest specificity and yield of PCR is obtainable when the two primers have similar melting temperatures. The same balance is also important for long PCR (Cheng et al., 1994). If modified bases are incorporated into one primer of a primer pair, a decrease in specificity and/or yield may be the result of nonspecific priming by the higher melting primer. Compensatory changes in the second primer would allow adjustment of annealing and extension conditions to a new optimum.

The melting temperature of a PCR primer is but one thermodynamic variable in determining whether it will be a useful primer. The distribution of nearest-neighbor free energies within a primer may also be important. In particular, relatively higher 5'-stability and hence lower 3'-stability may prevent extension at sites with 3'- but lacking 5'-template complementarity. Thus not only the inclusion but also the positioning of modified bases in a primer may affect primer specificity.

Enzymology of Modified Nucleic Acids

The relative theoretical t_m values for unmodified and modified primer–template duplexes are more important than the absolute numbers. The third step of PCR, polymerization, is not separable from primer hybridization and in fact occurs simultaneously. The thermodynamics of primer extension have been addressed in detail by Petrushka et al. (1988). Because primer is continuously being extended to produce a primer–template duplex with a higher melting temperature, primer hybridization may be driven to completion at temperatures above the theoretical t_m, where the rate of dissociation of unextended primer–template duplex exceeds its rate of formation.

The ability to prime polymerization efficiently at temperatures above t_m depends upon having sufficient primer so that multiple hybridization events may take place. Using the value of 5×10^4 for the nucleation rate constant in PCR buffer, the half-time for hybridization, $t_{1/2}$ (seconds), may be calculated by (adapted from Wetmur, 1991):

$$t_{1/2} = 14/(C'' \times L^{0.5})$$

where C'' is the micromolar primer concentration and L is the primer length. Thus, for 1 μM primer of length 25, the half-time for hybridization is 2.8 sec. If the concentration is reduced 10-fold, the half-time is increased to 28 sec, with multiple hybridization events becoming less likely during any PCR cycle. In fact, too low a primer concentration may kinetically limit yields per cycle to less than 2-fold using otherwise optimum PCR conditions.

Modified Primers

Modified primers must not only hybridize to template and facilitate polymerization but must also be capable of acting as a template for subsequent cycles. Clearly, modifications at the 5' end are unlikely to affect template function. Reduced PCR yield with an internally modified primer may be the result of incomplete polymerization. This explanation may be verified by detection of truncated single strands in the PCR products. Increased extension times, especially at later cycles, may allow increased yield. Another possibility would be to test alternative thermostable DNA polymerases which may be more amenable to the primer modification.

Modified Deoxynucleoside Triphosphates

Polymerization using a modified deoxynucleoside triphosphate not only requires the polymerase to accept modified bases in the template but also to accept the modified deoxynucleoside triphosphate as a substrate. The modified template may have more or less secondary structure. Decreased secondary structure permits use of lower hybridization and extension temperatures, and a wider choice of length and composition for the PCR primer pairs. If the secondary structure is increased, PCR primer pairs may need to be lengthened to achieve optimum results. Modified deoxynucleotides which have been introduced by polymerization must have been recognized by the polymerase as substrates and must have been able to base pair. They are less likely to affect template function directly than the more complex modifications which can be introduced during chemical synthesis of primers. A functional PCR assay is used to measure relative incorporation of modified and unmodified deoxynucleoside triphosphate. Although PCR is a complex process, the results are routinely interpreted as measuring the relative affinity (or K_M) of the modified and unmodified deoxynucleoside triphosphates for the enzyme.

The functional PCR assay requires amplification using deoxynucleotide triphosphate mixtures containing various ratios of the modified and unmodified forms of one deoxynucleotide triphosphate. The simplest assay for incorporation would employ a radiolabeled modified base. Alternatively, modified deoxynucleotides often prevent recognition of a site by the corresponding restriction endonuclease (Innis, 1990). When investigating a new modification, it may be necessary to test several restriction endonucleases to determine which enzymes respond to an incorporation of a single modified base into the recognition site (e.g., Seela and Röling 1992). The best data are those for low levels of incorporation of the modified base, where the relationship between incorporation and resistance to cleavage by the restriction endonuclease is linear, and where the incorporation will have a minimal effect on the template secondary structure.

Modified Deoxynucleotides in PCR Primers

As outlined in the introduction, modified deoxynucleotides may be incorporated into PCR primer to label the PCR product for affinity purification (e.g., a biotin label) or detection, either directly (e.g., a fluorescent label) or indirectly (e.g., mediated by a digoxigenin label). These labels are typically added at the 5' end where their effect on template function and t_m is minimal. Modified deoxynucleotides may also be incorporated at other positions to alter the ability of the primers and/or the PCR product to serve as a substrate for various enzymes. Examples include incorporation of phosphorothioates at the 3' end to inhibit 3' → 5' exonuclease activity of certain DNA polymerases (de Noronha and Mullins, 1992) and incorporation of deoxyuridine triphosphate (dUTP) to permit subsequent treatment by uracil N-glycosylase (uracil DNA-glycosylase; ung) to expose cohesive ends and destroy primer dimer, thus facilitating cloning (Rashtchian et al., 1992).

Reduction of Primer Degeneracy

PCR may be used to amplify gene fragments using primer pairs with sequences derived from amino acid sequence data, either for the gene of interest or for related genes in other species. The simplest approach is to employ degenerate primers containing all of the possible codons.

Three important considerations come into play when using degenerate primers. First, at constant total primer concentration, the t_m for the correct primer in a mixture is reduced with increased degeneracy because its relative concentration is reduced. Thus the hybridization step must be carried out at lower temperatures where template secondary structure may interfere with priming. However, when nontemplate and nondegenerate deoxynucleotides are introduced at the 5' end of the primers, as is often the case to facilitate cloning, t_m will increase after a few cycles and the PCR program may be changed to compensate. Second, and perhaps more important, the increase in the half-time for forming a primer–template hybrid is directly proportional to the degeneracy. Thus available primer–template polymerase substrate may decrease with increasing degeneracy. Finally, the total PCR yield may be reduced if the primer concentration is limiting. Thus it is highly desirable to increase the primer concentrations to partially or completely compensate for their degeneracies. However, primer concentrations cannot be increased without limit without affecting PCR efficiency and/or specificity. One strategy to reduce the degeneracy of the primers is to rely on codon preferences for the particular organism, especially at the 5' end of the primers where mismatches may be less critical. A second strategy is to incorporate modified deoxynucleotides which may pair with more than one deoxynucleotide.

Some investigators have erroneously assumed that deoxyinosine (dI) could pair with all four deoxynucleosides. Base pairing by dI has been extensively investigated (Martin et al., 1985). Deoxyinosine base pairs with dC, but with lower affinity than dG with dC. Mismatch binding of dI with dG and deoxythymidine (dT) is weaker than mismatch binding of dG with dG or dT, which is already very weak. However, dI binds better to deoxyadenosine (dA) than does dG. Thus, dI may be used to pair with dC or dA and has even been used at the 3' end of a PCR primer to enhance evolutionary PCR (Batzer et al., 1991). Nevertheless, although dI does have an additional base-pairing capability, dI is not a neutral deoxynucleoside substitution in a PCR primer. Similarly, some investigators have erroneously concluded that because G · U base pairs in RNA are relatively stable mismatches, dG · dT mismatches in DNA would be similar and dG could act as a neutral purine deoxynucleoside. However, a dG · dT mismatch is just as destabilizing as a dG · dA mismatch (Batzer et al., 1991; Patel et al., 1984).

Brown and Lin (1991) have synthesized two modified deoxynucleosides, N^6-methoxy-2'-deoxyadenosine (dZ) and 2-amino-9-(2'-deoxy-β-ribofuranosyl)-6-methoxyaminopurine (dK) and have shown that dZ · dT and dZ · dC have similar stability, as have dK · dT and dK · dC, making these deoxynucleosides truly neutral purine deoxynucleosides. Habener et al. (1988) have found that 5-fluorodeoxyuridine pairs with dG and dA with affinities approximately equal to those of dG · dC base pairs. Similarly, Anand et al. (1987) have shown that N^4-methoxycytosine may act as a neutral pyrimidine deoxynucleoside. Most recently, Nichols et al. (1994) have shown that a 3-nitropyrrole deoxynucleoside has low but approximately equal affinity for the four natural deoxynucleosides. The melting temperature of an oligodeoxynucleotide of length L containing N of these universal nucleosides is approximately that expected for a primer of length $L - N$.

Modified Deoxynucleotides in the Sequence

As outlined in the introduction, one of the three purposes for incorporating modified bases has been to label the PCR product for detection or affinity purification. Although labels have usually been incorporated into the PCR primers, they have also been incorporated into DNA by polymerization using modified deoxynucleoside triphosphates. For example, biotin-11-dUTP, digoxigenin-11-dUTP, DPN-11-dUTP and fluorescein-11-dUTP have been polymerized into DNA to produce a set of *in situ* hybridization probes (Ried *et al.*, 1992). Many of these modified deoxynucleoside triphosphates have been used in PCR, including biotin (Lanzillo, 1990) and digoxigenin (Lo *et al.*, 1988). In all of these labeling reactions, the corresponding unmodified deoxynucleoside triphosphate must be included in the reaction because complete replacement blocks polymerization. 5-Bromodeoxyuridine (5-BrdU) has also been used as an indirect label. In this case deoxythymidine triphosphate (dTTP) was completely replaced by 5-BrdUTP in the PCR reaction (Tabibzadeh *et al.*, 1991).

Another purpose for introducing modifications has been to alter the thermodynamic properties of the product, raising or lowering the melting temperature, t_m^∞. Complete substitution of dC by 5-MedC has been used to raise t_m^∞ (D. M. Wong *et al.*, 1991; Wong and McClelland, 1991). Other modified bases which raise t_m^∞ but have not been systematically studied in PCR include 5-bromodeoxycytosine for dC (D. M. Wong *et al.*, 1991) and 2,6-diaminopurine for adenine (Cheong *et al.*, 1988). Partial substitution of c^7dG (McConlogue *et al.*, 1988) or dI (Wong and McClelland, 1991) for dG has been used to lower t_m^∞. Complete substitution of c^7dG for dG is usually avoided because the PCR product cannot be detected by fluorescence of bound ethidium bromide in agarose gels. Complete substitution of deoxyinosine triphosphate (dITP) for deoxyguanosine triphosphate (dGTP) has been employed in the Applied Biosystems cycle sequencing kit. The dITP concentration must be increased severalfold above the normal dGTP concentration because of the lower affinity of dITP for the *Taq* polymerase.

The third purpose for introducing modifications has been to alter the substrate properties of the PCR product. Substitution of a deoxynucleoside phosphorothioate for the corresponding deoxynucleotide has been employed to provide exonuclease resistance (Olsen and Eckstein, 1989). Complete replacement of dC by 5-MedC completely blocks template cleavage by a variety of restriction endonucleases, including those sensitive to dC methylation at dCdG sequences in mammalian DNA. In many cases, complete substitution has proved to be impossible, as with the deaza-deoxynucleotides c^7dA and c^7dI (Seela and Röling, 1992), and the effects of substitution on restriction enzyme recognition sites have been studied with partially substituted PCR products. For restriction endonucleases with palindromic sites, an alternative to complete substitution in both strands is complete substitution in one strand. For example, asymmetric PCR has been used to produce one strand with dA completely replaced by 2-chloro-dA (Hentosh et al., 1992). A very important modification which causes the PCR product to become a substrate, rather than the reverse, is complete substitution of deoxyuridine for dT (Longo et al., 1990). This substitution creates a substrate for uracil DNA glycosylase, allowing PCR products to be degraded and preventing contamination of PCR templates by carryover PCR products.

Appendix 1: Calculation of the Melting Temperature of Long, Unmodified DNA

The melting temperature (°C) of long, duplex DNA is given by:

$$t_m^x = 81.5 + 16.6 \cdot \log_{10}\left[\frac{(SALT)}{1.0 + 0.7(SALT)}\right] + 0.41(\% \text{ G} + \text{C}) - \frac{500}{L}$$

where

$$(SALT) = (K^+) + 4(Mg^{+2})^{0.5}$$

and sodium or other monovalent ions may be substituted for potassium.

Appendix 2: Calculation of the Melting Temperature of Unmodified Primer–Template Hybrids

The melting temperature (°C) of a primer–template hybrid is given by:

$$t_m = \frac{T°\Delta H°}{\Delta H° - \Delta G° + RT°\ln(C)} + 16.6 \cdot \log_{10}\left[\frac{(SALT)}{1.0 + 0.7(SALT)}\right] - 269.3$$

where enthalpy

$$\Delta H° = \Sigma_{nn} (N_{nn} \times \Delta H°_{nn}) + \Delta H°_e$$

and free energy

$$\Delta G° = \Sigma_{nn} (N_{nn} \times \Delta G°_{nn}) + \Delta G°_e + \Delta G°_i$$

and

$$T° = 298.2°K$$

and

$$R = 1.99 \text{ cal/mol°K}$$

The upper case T refers to temperatures in °K. The symbols i, nn, and e indicate contributions from initiation of base-pair formation, nearest-neighbor interactions, and dangling-end interactions. For the nearest-neighbor values in the table below, $\Delta G°_i = +2200$ cal/mol. An estimated average contribution of a single dangling end is $\Delta H°_e = -5000$ cal/mol and $\Delta G°_e = -1000$ cal/mol. Average values are used because the complete set of thermodynamic data is unavailable. Based on the enthalpy measurements of Breslauer et al. (1986) and melting data from many sources, one table of nearest-neighbor interactions is (Quartin and Wetmur, 1989):

or	AA TT	AT	TA	CA TG	GT AC	CT AG	GA TC	CG	GC	GG CC
$-\Delta H°_{nn}$	9.1	8.6	6.0	5.8	6.5	7.8	5.6	11.9	11.1	11.0
$-\Delta G°_{nn}$	1.55	1.25	0.85	1.15	1.40	1.45	1.15	3.05	2.70	2.30

For example, for the purposes of demonstrating a calculation, consider the short sequence ATGCAGCTAAGTCA with two dangling ends. PCR primers are much longer and the relative contribution of the dangling end is less. There is one each nearest-neighbor interaction of the types AA/TT, AT, TA, GT/AC, and GA/TC; two each of the type GC; and three each of the types CA/TG and CT/AG. Thus:

$$-\Delta H°_{nn} = (9.1 + 8.6 + 6.0 + 6.5 + 5.6) + 2(11.1) + 3(5.8 + 7.8)$$
$$+ 2(5.0) = 108.8 \text{ kcal/mol}$$
$$-\Delta G°_{nn} = (1.55 + 1.25 + 0.85 + 1.4 + 1.15) + 2(2.7)$$
$$+ 3(1.15 + 1.45) + 2(1.0) - 2.2 = 19.2 \text{ kcal/mol}$$

Continuing the example, for (C) of 0.2 μM and PCR buffer with 0.05 M K$^+$ and 0.0015 M Mg^{2+},

$$t_m = \frac{-298.2(108800)}{-108800 + 19200 + 1.99(298.2)\ln(2 \times 10^{-7})}$$
$$+ 16.6 \cdot \log_{10}\left[\frac{(0.205)}{1.0 + 0.7(0.205)}\right] - 269.3 = 46.8°C$$

Appendix 3: Calculation of the Melting Temperature of Modified Primer–Template Hybrids

The best value for the change in the free energy due to a modification, $\Delta\Delta G°$, may be determined by measuring Δt_m for several oligodeoxynucleotide pairs with normal and N modified bases and applying the first formula below. Spectrophotometric determinations of oligonucleotide melting are best accomplished when the complementary oligonucleotides have the same concentration. In this case, the (C) term in the equation in Appendix 2 becomes $(C/4)$ where C is the total oligonucleotide concentration.

$$\Delta\Delta G° \approx \frac{\Delta H°}{N \times T_m} \Delta t_m$$

or

$$\Delta t_m \approx \frac{N \times T_m}{\Delta H°} \Delta\Delta G°$$

Once $\Delta\Delta G°$ has been determined, the second formula above may be used to determine the change in the melting temperature for any modified PCR primer–template hybrid. Similarly, the characteristic PCR product melting parameter for this modification, S, for, S, is interconvertible with $\Delta\Delta G°$ or Δt_m using an average base pair $\Delta H° = -8100$ cal/mol and the formulas below. Thus, a single parameter may be used to characterize the effect of a base modification on either primer–template melting or PCR product melting.

$$S = \frac{-T_m^x(\Delta\Delta G°)}{100(8100)}$$

or

$$\Delta\Delta G° = \frac{100(8100)S}{-T_m^\infty}$$

or

$$\Delta t_m \approx \frac{N \cdot T_m \cdot 100 \cdot 8100}{-\Delta H°(T_m^\infty)} S$$

For example, substitution of three 5-MedC residues in either complementary strand of the sequence above, ATGCAGCTAAGTCA, or in both strands, results in $\Delta t_m \approx 3°C$ or $6°C$, respectively ($t_m \approx 60°C$)

$$\Delta\Delta G° \approx \frac{-98800}{3(333.2)} \times 3 = -0.3 \text{ kcal/mol}$$

or

$$S = 0.13,$$

which is in good agreement with the value of 0.12 from melting the naturally substituted DNA.

Literature Cited

Anand, N. N., D. M. Brown, and S. A. Salisbury. 1987. The stability of oligodeoxyribonucleotide duplexes containing degenerate bases. *Nucleic Acids Res.* **15**:8167–8176.

Batzer, M. A., J. E. Carlton, and P. L. Deininger. 1991. Enhanced evolutionary PCR using oligonucleotides with inosine at the 3'-terminus. *Nucleic Acids Res.* **19**:5081.

Breslauer, K. J., R. Frank, H. Blöcker, and L. A. Marky. 1986. Predicting DNA duplex stability from the base sequence. *Proc. Natl. Acad. Sci. U.S.A.* **83**:3746–3750.

Brown, D. M., and P. K. Lin. 1991. Synthesis and duplex stability of oligonucleotides containing adenine-guanine analogues. *Carbohydrate Res.* **216**:129–139.

Cheng, S., C. Folkler, W. M. Barnes, and R. Higuchi. 1994. Effective amplification of long targets from cloned inserts and human genomic DNA. *Proc. Natl. Acad. Sci. U.S.A.* **91**:5695–5699.

Cheong, C., I. Tinoco, Jr., and A. Chollet. 1988. Thermodynamic studies of base pairing involving 2,6-diaminopurine. *Nucleic Acids Res.* **16**:5115–5122.

de Noronha, C. M., and J. I. Mullins. 1992. Amplimers with 3'-terminal phosphorothioate linkages resist degradation by vent polymerase and reduce *Taq* polymerase mispriming. *PCR Meth. Appl.* **2**:131–136.

Ehrlich, M., K. Ehrlich, and J. A. Mayo. 1975. Unusual properties of the DNA from *Xanthomonas* phage XP-12 in which 5-methylcytosine completely replaces cytosine, *Biochim. Biophys. Acta* **395**:109–119.

Habener, J. F., C. D. Vo, D. B. Le, G. P. Gryan, L. Ercolani, and A. H. Wang. 1988. 5-Fluorodeoxyuridine as an alternative to the synthesis of mixed hybridization probes for the detection of specific gene sequences. *Proc. Natl. Acad. Sci. U.S.A.* **85:**1735–1739.

Hentosh, P., J. C. McCastlain, P. Grippo, and B. Y. Bugg. 1992. Polymerase chain reaction amplification of single-stranded DNA containing the base analog, 2-chloroadenine. *Anal. Biochem.* **201:**277–281.

Innis, M. A., *PCR Protocols: A guide to methods and applications* (1st ed.), Academic Press, San Diego, 1990, pp. 54–59.

Lanzillo, J. J. 1990. Preparation of digoxigenin-labeled probes by the polymerase chain reaction. *Biotechniques* **8:**620–622.

Lo, Y. M. D., W. Z. Mehal, and K. A. Fleming. 1988. Production of vector-free biotinylated probes using the polymerase chain reaction. *Nucleic Acids Res.* **16:**8719.

Longo, M. C., M. S. Berninger, and J. L. Hartley. 1990. Use of uracil DNA glycosylase to control carry-over contamination in polymerase chain reactions. *Gene* **93:**125–128.

Martin, F. H., M. M. Castro, F. Aboul-ele, and I. Tinoco, Jr. 1985. Base pairing involving deoxyinosine: implications for probe design. *Nucleic Acids Res.* **13:**8927–8938.

McConlogue, L., M. A. D. Brow, and M. A. Innis. 1988. Structure-independent DNA amplification by PCR using 7-deaza-2'-deoxyguanosine. *Nucleic Acids Res.* **16:**9869.

Nichols, R., P. C. Andrews, P. Zhang, and D. E. Bergstrom. 1994. A universal nucleoside for use at ambiguous sites in DNA primers. *Nature* **369:**492–493.

Olsen, D. B., and F. Eckstein. 1989. Incomplete primer extension during *in vitro* DNA amplification catalyzed by *Taq* polymerase: exploitation for DNA sequencing. *Nucleic Acids Res.* **17:**9613–9620.

Patel, D. J., S. A. Kozlowski, S. Ikuta, and K. Itakura. 1984. Dynamics of DNA duplexes containing internal G -T, G -A, A -C and T -C pairs: hydrogen exchange at and adjacent to mismatch sites. *Fed Proc.* **43:**2663–2670.

Petruska, J., M. F. Goodman, M. S. Boosalis, L. C. Sowers, C. Cheong, and I. Tinoco, Jr. 1988. Comparison between DNA melting thermodynamics and DNA polymerase fidelity. *Proc. Natl. Acad. Sci. U.S.A.* **85:**6252–6256.

Quartin, R. S., and J. G. Wetmur. 1989. Effect of ionic strength on the hybridization of oligodeoxynucleotides with reduced charge due to methylphosphonate linkages to unmodified oligodeoxynucleotides containing the complementary sequence. *Biochemistry* **28:**1040–1047.

Ried, T., A. Baldini, T. C. Rand, and D. C. Ward. 1992. Simultaneous visualization of seven different DNA probes by *in situ* hybridization using combinatorial fluorescence and digital imaging microscopy. *Proc. Natl. Acad. Sci. U.S.A.* **89:**1388–1392.

Rashtchian, A., G. W. Buchman, D. M. Schuster, and M. S. Berninger. 1992. Uracil DNA glycosylase-mediated cloning of polymerase chain reaction-amplified DNA: application to genomic and cDNA cloning. *Anal. Biochem.* **206:**91–97.

Seela, F., and A. Röling. 1992. 7-Deazapurine containing DNA: efficiency of c^7G_dTP, c^7A_dTP and c^7I_dTP incorporation during PCR-amplification and protection from endodeoxyribonuclease hydrolysis. *Nucleic Acids Res.* **20:**55–61.

Tabibzadeh, S., U. G. Bhat, and X. Sun. 1991. Generation of nonradioactive bromodeoxyuridine-labeled DNA probes by polymerase chain reaction. *Nucleic Acids Res.* **19:**2783.

Wetmur, J. G. 1991. DNA probes: applications of the principles of nucleic acid hybridization. *Crit. Rev. Biochem. Mol. Biol.* **26**:227–259.

Wong, D. M., P. H. Weinstock, and J. G. Wetmur. 1991. Branch capture reactions: displacers derived from asymmetric PCR. *Nucleic Acids Res.* **19**:2251–2259.

Wong, K. K., and McClelland, M. 1991. PCR with 5-methyl-dCTP replacing dCTP. *Nucleic Acids Res.* **19**:1081–1085.

7

Practical Considerations for the Design of Quantitative PCR Assays

Robert Diaco

As a result of the relative simplicity of reproducing as little as single target nucleic acid sequence into nanogram quantities of faithfully amplified products, which can range in size from 50 bp to more than 10 kb, polymerase chain reaction (PCR) technology is rapidly becoming one of the most powerful methods available for nucleic acid analysis. This technology has been shown to have wide utility for researchers interested in cloning and sequencing; for forensics analysis; paternity, genetics, and oncology testing; diagnostics applications; for monitoring therapeutic treatments; and for myriad other research and clinical applications. As PCR technology has gained more users, its applications have continuously been refined and exciting new techniques have been developed. One particular application, the use of PCR technology to quantitate nucleic acid target molecules, appears to be of tremendous interest to a number of PCR users, and has spawned a number of new strategies and techniques. Quantitation of nucleic acid molecules through the use of PCR technology provides the researcher and clinician with a powerful tool for analyzing nucleic acids.

Information Required to Start Quantitative Assay Development

The polymerase chain reaction is an enzymatic reaction which follows relatively simple, predictable, and well-understood mathematical principles. A basic understanding of the principles governing the production of PCR amplification products will aid in designing meaningful and reliable quantitative PCR assays.

Understanding PCR Efficiency

Most researchers view PCR as a process which yields exponential amplification of input target nucleic acid molecules at a rate which can be calculated by using the following formula:

$$product = target*2^n$$

where *product* is the number of PCR product molecules formed, *target* refers to the number of input target nucleic acid molecules, and *n* is the number of PCR amplification cycles used in the process.

Of course, no biological process is 100% efficient, so one must consider the relative efficiency of the specific PCR amplification process to be used in the quantitative assay. The term "relative PCR efficiency" is used to describe the overall efficiency of the amplification process averaged across the total number of PCR cycles. It is reasonable to predict that the actual PCR efficiency of each cycle will vary significantly throughout the process of amplification. During the first few cycles, where the number of input target molecules is very low relative to the total population of nucleic acids in the mixture, PCR amplification efficiency is predicted to be low. Similary, in the last few cycles, where the number of PCR product molecules is very high and is approaching the concentration of unreacted primers, PCR amplification is also predicted to operate at a low efficiency. PCR amplification is most efficient during the middle cycles of the process, where input target molecules, newly synthesized PCR product molecules, and unreacted primers are all present in abundance.

For the purposes of designing quantitative PCR assays, it is sufficient to measure only relative PCR amplification efficiency, which is relatively easy, rather than attempt the more cumbersome and challenging task of characterizing the actual PCR amplification efficiency for each cycle in the PCR amplification process. Table 1 illustrates the significance of understanding the effect of relative PCR efficiency on the rate of PCR product accumulation. As shown in this example, a 10% reduction in the relative efficiency of a PCR reaction results in a greater than 95% reduction in PCR product accumulation. Achieving an equivalent amount of product from a PCR reaction operating at 90% efficiency would require an increase of more than 2250% in the number of input target molecules. Similarly, a 20% increase in cycle number (36 instead of 30 in this example) is required to accumulate an equivalent amount of PCR product from a PCR amplification process operating at only 90% relative efficiency. A thorough understanding of the relationships which exist between relative PCR efficiency, cycle number, and the rate of product accumulation will aid the researcher in developing quantitative PCR applications.

To estimate the relative amplification efficiency of a specific PCR process, one must be able to quantitate both the number of input target molecules added into the reaction and the number of product molecules generated from the amplification. This is possible using a number of standard analytical techniques. Since the accuracy of the PCR amplification efficiency estimate directly depends upon the reliability of the measurements for input target molecules and ouput product, steps must be taken to ensure that these measurements are performed carefully. In addition, one must be careful not to perform these measurements on PCR reactions which have entered into the plateau phase (described later), or the quantity of product may be significantly underestimated. Several methods are available to help

Table 1

Effect of PCR Efficiency and Cycle Number on Product Accumulation

Input target molecules	Cycle number	Relative PCR efficiency	Accumulated product molecules	Reduction in accumulated product
100	30	100	1.07×10^{11}	
100	30	90	4.55×10^{9}	95.8%
100	30	80	1.33×10^{8}	99.9%
2350	30	90	1.07×10^{11}	Not applicable
100	36	90	1.55×10^{11}	Not applicable

the researcher with these analytical measurements. Some of the most common ones include (1) the use of fluorescent dye-binding assays, such as the one marketed by Hoechst AG, Frankfurt, Germany; (2) spectrophotometric estimates of nucleic acid content (Maniatis et al., 1982); and (3) estimation of nucleic acid quantity from an agarose gel (Piatak et al., 1993).

From estimates of the number of input target molecules and the number of product molecules, it is possible to calculate the relative efficiency (η) of the PCR process. Once a reliable estimate for relative PCR efficiency has been obtained, it is easy to calculate (1) the quantity of PCR product molecules formed from a specific target quantity; (2) the number of PCR cycles required to produce a desired quantity of product when amplifying a specific range of target molecules; and (3) the number of target molecules that were present in the original sample. These calculations can all be made by applying the following general formula, which calculates product accumulation from a PCR reaction with a known relative efficiency:

$$product = targets*(1 + \eta)^n$$

In this formula, *target* refers to the number of input target molecules, *product* is the number of product molecules formed in the PCR, *n* refers to the number of PCR cycles used in the process, and η is the estimate of relative PCR efficiency (averaged across all cycles). This formula is particularly useful for developing and characterizing quantitative PCR assays.

PCR Plateau

The point at which PCR enters the plateau phase is usually exhibited as an inflection point on a curve relating the number of target molecules and the quantity of product molecules produced. It is important to know where a PCR process enters into a plateau phase because this point defines the upper range of the linear quantitative response, and as such defines the upper limit for a reliable quantitative PCR assay. The relative efficiency of a PCR reaction determines when the process enters into the plateau phase. Well-developed PCR amplification reactions, which use standard concentrations of *Taq* polymerase and primers, and amplify low concentrations of input target molecules, usually achieve an average relative efficiency ranging from 70 to 95%. The relative PCR efficiency for a specific amplification is influenced by the cleanliness of the extracted sample, the length and secondary structure of the target nucleic acid molecules, and a wide range of other factors.

A simple experimental approach for measuring the point at which PCR enters the plateau phase involves measuring the number of product molecules produced as increasing quantities of input target molecules are added into the reaction. Representative results from an experiment of this type are shown in Fig. 1, which illustrates PCR product accumulation as a function of input target molecules for amplification of HIV-1 target sequences. In this 35-cycle PCR amplification, the plateau is reached when fewer than 10,000 molecules of target are input into the amplification reaction. Although the addition of even higher numbers of input target molecules leads to more PCR product, the increase in product accumulation switches from an exponential log–log relationship to a less efficient log–linear relationship.

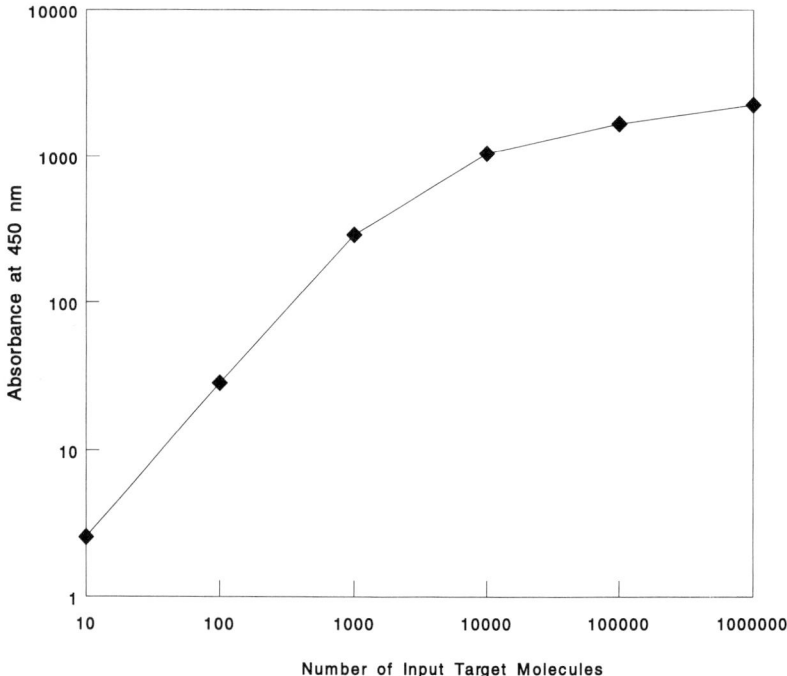

Figure 1 PCR product accumulation as a function of input target number from a 35-cycle amplification of HIV-1 sequences.

Know the Range of Target Molecules to Be Encountered

To develop a useful quantitative PCR procedure, one must know the range of target molecules present in the typical biological sample. This information is critical for adjusting the amplification and detection portions of the quantitative assay for maximum utility. This information can be obtained by using a number of standard analytical techniques (Maniatis et al., 1982; Landgraf et al., 1991; Ferre et al., 1992; Piatak et al., 1993).

Working Range of the Detection Assay

The most common methods for routinely measuring PCR product accumulation are solid-phase-dependent DNA-probe hybridization assays (Keller et al., 1991; Casareale et al., 1992, Winters et al., 1992; Siebert and Larrick, 1993). These methods are particularly useful because the specificity inherent in DNA-probe hybridization reactions ensures that the detection assay measures a faithful PCR product, and because these solid-phase reactions can be configured to occur quickly, have very high binding efficiencies, and usually exhibit low backgrounds. Most rapid DNA-probe assays use signal amplification technologies, such as an enzymatic reporter system, to achieve the high sensitivities and short reaction times required to detect the relatively low concentrations of nucleic acid normally measured. While assays of this type often provide fast results and high sensitivity, the high slope sensitivity of these assays limits the dynamic range for quantitative applications.

The Roche Diagnostic Systems Amplicor microwell plate (MWP) format (Casareale et al., 1992; Loeffelholz et al., 1992), which is similar in concept and performance to many DNA–probe assays that are dependent on a solid phase, will be used as an example for most of the discussions that follow.

Relationship of Product Molecules to Signal

Since numerous methods exist for determining the concentration of product molecules formed during a PCR reaction, it is relatively easy to generate a standard curve relating the number of product molecules to the signal obtained from the detection assay. A typical standard curve from an early prototype of the Amplicor *Chlamydia* assay is shown in Fig. 2. In this example, the linear range for the detection assay is approximately 1×10^8 to 1×10^{10} product molecules, which were amplified from an input of between 1 and 20 target molecules. Although this relatively narrow working range is appropriate for a high-sensitivity qualitative assay, it has limited applications for quantitative results.

Minimum Sensitivity

The limit of sensitivity of any assay is defined as the minimum concentration, or quantity, of analyte that can be reliably measured over background noise. For DNA-probe assays, the minimum concentration of nucleic acid that can be detected varies widely, depending on the length and sequence of the molecule being analyzed, the duration, stringency, and efficiency of the hybridization reaction,

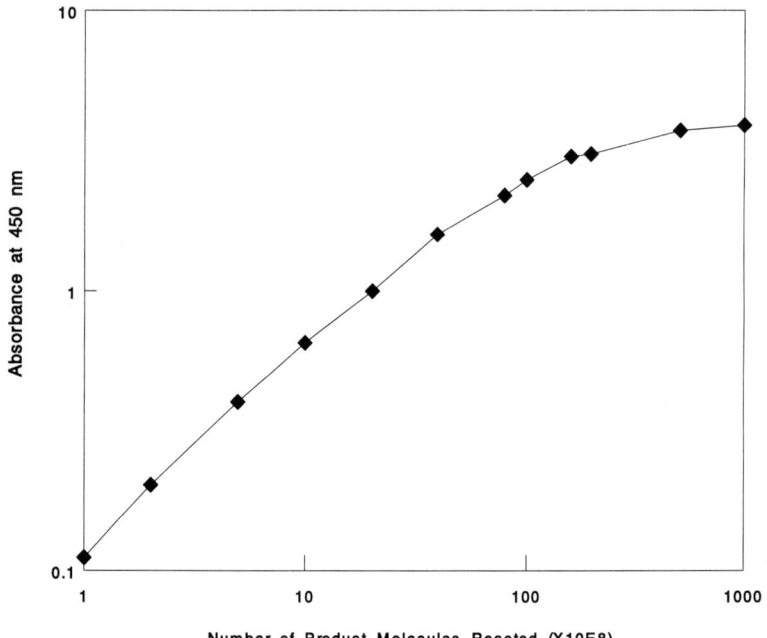

Figure 2 Standard curve from an early prototype of the Roche Diagnostic Systems Amplicor *Chylamydia* assay.

the choice of reporter molecule, and a number of other factors. The sensitivity limit of the detection assay differs significantly from the sensitivity of the overall PCR assay. Typically, DNA-probe detection assays can reliably detect only nanogram to picogram quantities of nucleic acids; with PCR amplification it is possible to generate reliably detectable (nanogram to picogram quantities) quantities of PCR product molecules from amplification of a single input target molecule. Although it is possible to adjust the lower limit of sensitivity to detect almost any number of input target molecules, it is useful to know the actual limit of sensitivity of the detection assay for product molecules when designing quantitative PCR assays. Adjusting the quantity of product molecules produced when amplifying the required low end of the target range to coincide with the minimum sensitivity of the detection assay will ensure that the dynamic range of the quantitative PCR assay will be as wide as possible.

Overcoming the Upper Limit of Detection in a Conventional DNA-Probe Assay

The high slope sensitivity and limited dynamic range of most hybridization assays places considerable constraints on the development of quantitative PCR assays. Several approaches for modifying existing qualitative assays for PCR quantitation have been proposed and extensively evaluated (Becker-André and Hahlbrock, 1989; Chelly et al., 1990; Gilliland et al., 1990; Kellog et al., 1990; Lundeberg et al., 1991; Cone et al., 1992; Ferre et al., 1992; Siebert and Larrick, 1993). These strategies, when used independently or in combination, have been shown to help overcome many of the factors which limit the usefulness of hybridization assays for quantitative applications. Included in these strategies are (1) diluting the input target number (sample dilution), (2) limiting the number of PCR cycles, (3) diluting the PCR product, and (4) inclusion of an internal control to monitor the efficiency of the PCR reaction. Each of these strategies provides individual as well as incremental value for improving the quantitative capabilities of PCR assays. The choice of which strategy, or combination of strategies to use, is dictated largely by the requirements for quantitation for the assay under development.

Dilution of Input Target Number

This strategy (often referred to as sample dilution) is frequently used to overcome the limited working range of the hybridization assay and to extend the range before the PCR plateau. Diluting the sample before the PCR reduces the number of product molecules formed during the reaction and thereby extends the upper range of the assay. This phenomenon is illustrated theoretically in Fig. 3 and experimentally in Fig. 4.

Figure 3 shows the theoretical number of product molecules formed as a function of input target number for a 35-cycle PCR reaction operating at an average efficiency of 78%. By marking both the detection assay working range and the PCR plateau range, it can be seen that this PCR system is capable of performing quantitation in the range of approximately 1 to 20 input target molecules (this is limited mostly by the working range of the detection assay), although product molecules continue to be formed at a linear rate (log–log) until the amplification reaction reaches the PCR plateau at approximately 2000 input target molecules (this is limited by the PCR

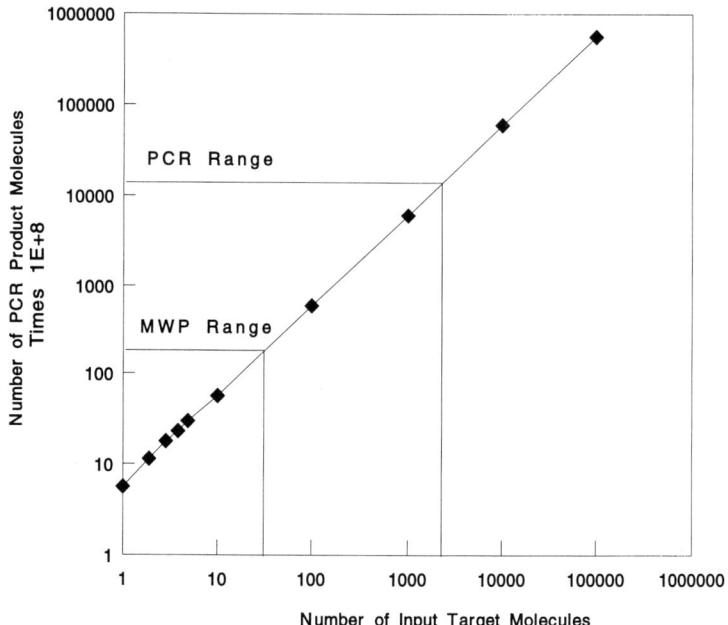

Figure 3 Estimated quantity of product formed as a function of input target number for a 35-cycle PCR reaction operating at an average efficiency of 78%.

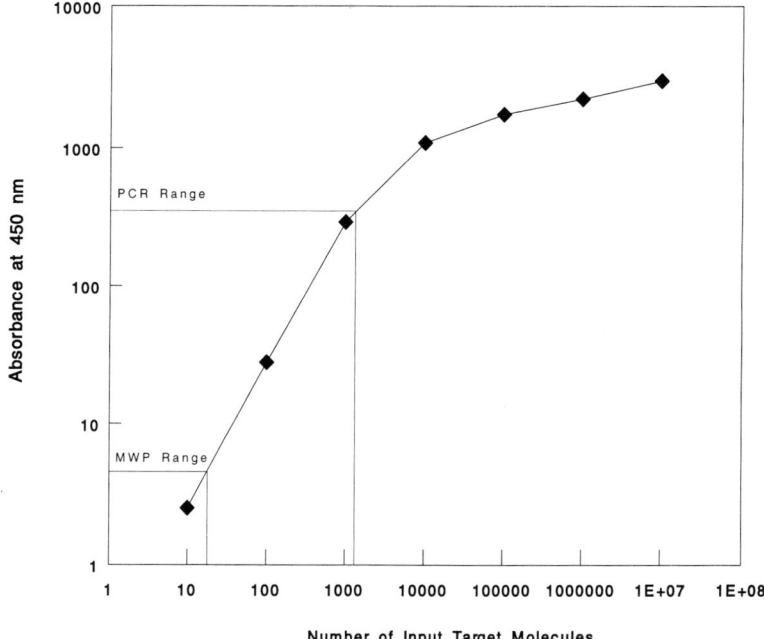

Figure 4 Standard curve from an early prototype of the Roche Diagnostic Systems Amplicor HIV-1 assay. The efficiency of this PCR reaction is approximately 78%.

reaction). Even though the PCR range is approximately 1–2000 input target molecules, the quantitative range of the overall assay is only approximately 1–20 input target molecules. An early prototype of the Roche Molecular Systems HIV-1 assay, shown in Fig. 4, exhibits a dynamic range and PCR plateau consistent for an assay operating with a relative efficiency of approximately 78%. Because the PCR amplification working range (1–2000 copies) is much greater than that for the detection assay (1–50 copies), diluting the input sample so that the number of PCR product molecules formed falls within the range of the detection assay (approximately 5×10^8 to 5×10^{10} PCR product molecules) can easily extend the quantitative range of this assay manyfold.

Limiting the Number of PCR Amplification Cycles

Reducing the number of PCR amplification cycles will limit the number of product molecules formed in the reaction. Careful choice of the number of PCR amplification cycles allows effective control of the working range of the assay (this strategy can be effectively used to adjust the working range, and can also be used to extend the point of inflection into the PCR plateau phase). This is shown clearly in Fig. 5, which displays the theoretical relationship between the number of input target molecules and the number of PCR product molecules formed when the number of PCR amplification cycles is varied (all calculated for amplifications with an average efficiency of 80%). Clearly, the working range of a PCR assay can be effectively modulated by controlling the number of PCR cycles. This is experimentally demonstrated in Fig. 6, which shows the impact of reducing the number of PCR amplification cycles in the HIV-1 qualitative system from the standard 35 to only 25 cycles. In this example, the apparent detection assay working range was shifted from approximately 1–20 input target molecules for 35 cycles to approximately 750–50,000 for 25 cycles, and the inflection point for the PCR plateau

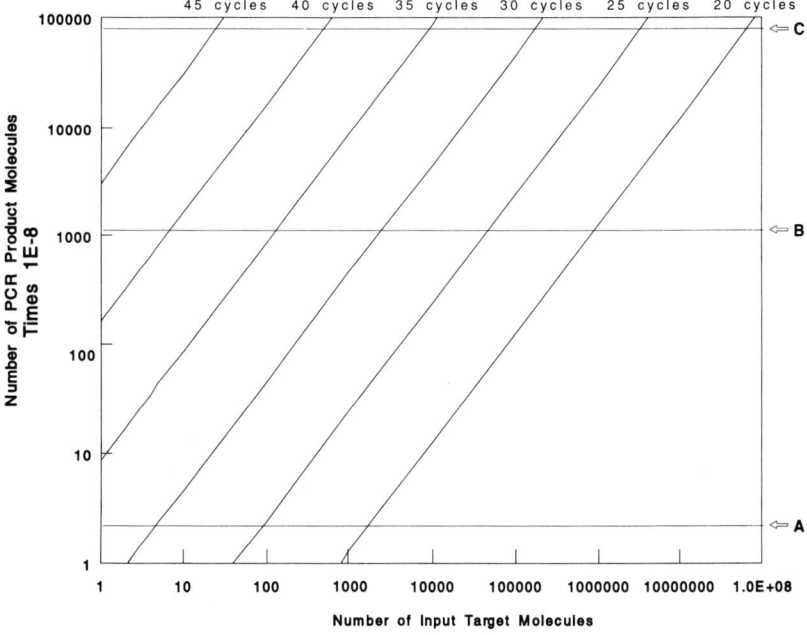

Figure 5 Relationship between the concentration of input target molecules and the quantity of PCR product formed as a function of the number of PCR amplification cycles (theoretical estimates calculated for amplifications with an average efficiency of 80%).

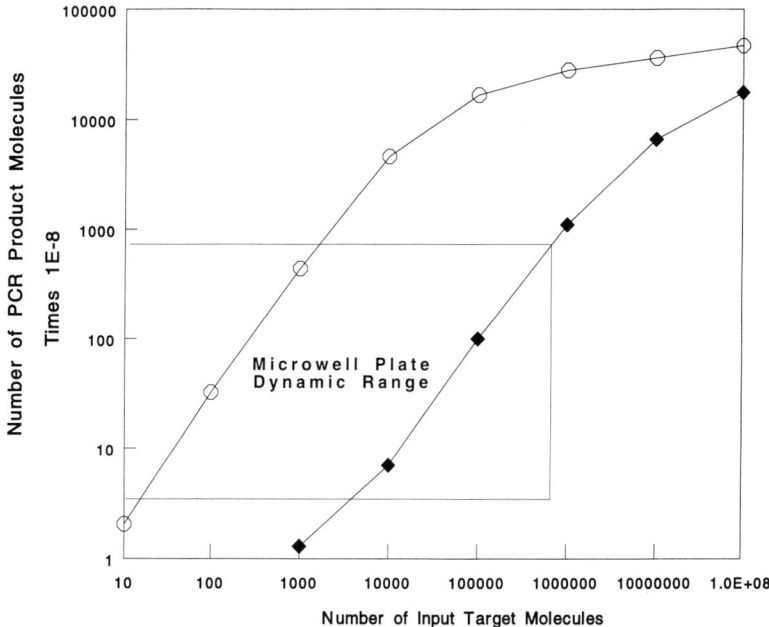

Figure 6 Product accumulation from 25-cycle and 35-cycle amplifications of HIV-1 sequences.

was extended from approximately 2000 input target molecules for 35 cycles to approximately 5,000,000 when using only 25 PCR amplification cycles.

Dilution of the PCR Product

This strategy is a very effective tool for extending the linear working (dynamic) range of the detection assay. As shown in the examples described previously, the working range of the detection assay is often much narrower than that of the PCR reactions (defined, for our purposes, as extending from the lower limit of assay sensitivity to the point of inflection into the PCR plateau phase). The effectiveness of this strategy is best illustrated by the experimental data shown in Fig. 4 and Fig. 7; in these examples, the working quantitative range of the detection assay is estimated to be less than 1–50 input target molecules, while the working quantitative range for the PCR is approximately 1–2000 input target molecules. By serially diluting the PCR product prior to performing the detection assay, it

96   Part One. Key Concepts for PCR

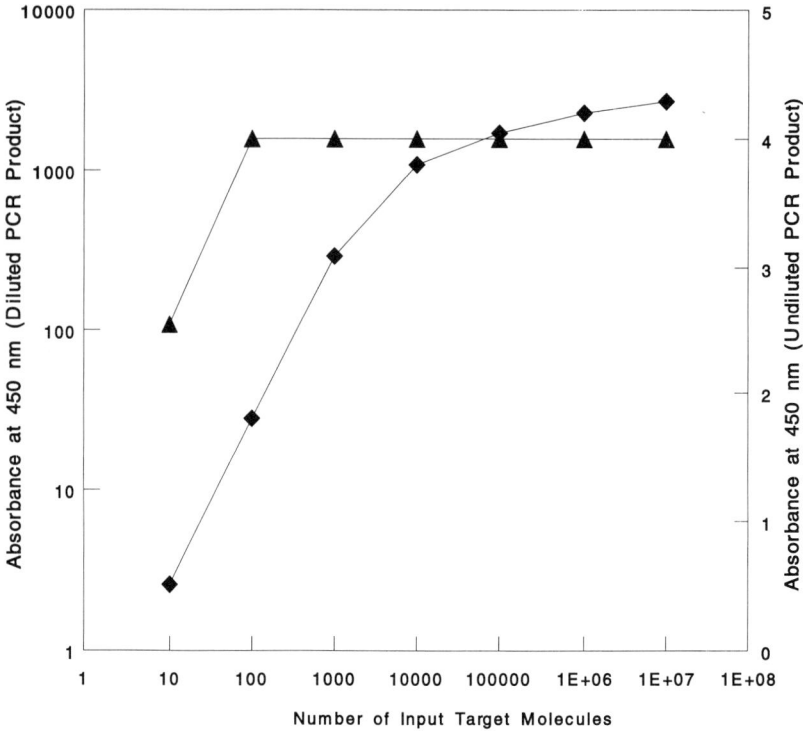

Figure 7 Extending the linear range of quantitative measurement by diluting PCR product prior to performing the detection assay.

is possible to extend the working quantitative range by approximately two orders of magnitude.

Internal Controls for the PCR Reaction

As described previously, the effectiveness of sample preparation procedures in reproducibly recovering target molecules which are free of PCR inhibitors will have a significant impact on the reliability and accuracy of quantitative PCR analyses. Recall that PCR product formation exhibits a log–log relationship with the number of input target molecules, and a log–linear relationship with the efficiency of the PCR reaction (of these two possible sources of variation, the latter, which is directly affected by PCR inhibitors, is more critical). Including internal controls, which monitor the efficiency of the PCR amplification reaction, will improve the reliability of the quantitative results by providing a means to correct for the efficiency of the PCR reaction. Internal controls also help control for variations in thermal cycler performance, and reagent formulations. Of course, this strategy will only work if the efficiency of both PCR reactions (the sample under analysis and the internal control) is equally affected by the nature and quantity of inhibitors in the reaction.

Several approaches for using internal controls to improve the quantitative resolution of PCR have been described (Becker-André and Hahlbrock, 1989; Gilliland et al., 1990; Chelly et al., 1990; Lundeberg et al., 1991; Ferre et al., 1992; Piatak et al., 1993; Siebert and Larrick, 1993). A common method calls for normalizing the estimated number of unknown targets against estimates of the quantity of genomic DNA targets contained within the sample (Lundeberg et al., 1991; Ferre et al., 1992). While these methods provide a means to correct for variation in PCR efficiency, they have the disadvantage of relying on different primer sequences, which might be affected differentially by PCR inhibitors. Another strategy relies on coamplification of a modified target using the same primer pairs (Becker-André and Hahlbrock, 1989; Gilliland et al., 1990; Chelly et al., 1990; Lundeberg et al., 1991; Piatek et al., 1993). The latter approach is less subject to individual variations in amplification efficiency, provided the modified target is similar in length and composition (nucleotide content) to the desired target, and the amplifications do not interfere with each other. These approaches were recently reviewed (Ferre, 1992), and mathematical models have been described (Nedelman et al., 1992) which demonstrate the utility of internal controls in quantitative PCR analyses.

Combining Strategies for Quantitative PCR

When used alone, none of the previously described strategies can accommodate all requirements for quantitative PCR analyses. By appropriately combining these strategies, however, PCR quantitation can be achieved over a wide range of input target molecules. The best combinations for quantitative PCR are those which combine strategies to extend the working ranges of both the PCR reaction and the detection assay. An example of a good combination strategy would be to control the number of PCR cycles to generate an appropriate range of products from the target concentrations normally encountered (to accomodate the low-end and high-end sensitivity requirements for the assay), and then to dilute the PCR products so that they fall within the working range of the detection assay.

Factors Affecting the Practical Resolution of a Quantitative PCR Assay

A number of factors will affect the practical resolution of a quantitative PCR assay. They include, but are not limited to (1) variability in the number of target molecules added into the quantitative PCR reaction; (2) the efficiency of the PCR amplification process; and (3) variability in the results from the detection assay. Although each of these factors contributes to the overall variability of the quantitative result, some have a more significant potential impact on the final results.

Variability in the number of target molecules added to the quantitative PCR reaction, which is caused by sampling error, comes from two major sources. The first error occurs because the efficiency of recovery of target molecules from the biological sample is almost never 100%. The reproducibility of target molecule recovery is dictated largely by the efficiency of the sample preparation procedure. As a result, as the efficiency of a sample preparation process increases, there is less variability in the amount of target molecules recovered. In addition, sample preparation procedures for quantitative applications must also efficiently separate target molecules from inhibitors of the PCR reaction. Because PCR inhibitors affect the efficiency of the amplification process, small variations in the amount and type of inhibitors that are not removed can lead to large variations in PCR product accumulation. Thus, it is critical to have reproducible and efficient sample preparation methods in order to obtain precise quantitative measurements.

The second source of sampling error is caused by variability in the distribution of target molecules in the biological sample. This situation is best described by Poisson distribution, which can be used to predict the normal distribution of members of a population. Sampling error which occurs as a consequence of target distribution is almost impossible to control, and is most pronounced at low input target concentrations where a small change in imput target molecules induces a relatively large difference in the number of product molecules produced. It is important, however, to understand the existence of this source of variability and to consider its effect on the practical resolution of the quantitative method.

Use of the Poisson distribution to correct for sampling error is best achieved through the following general formula:

$$P_N \frac{C^N}{N! e^C}$$

where C is the average number of target molecules in the solution, P_N is the probability of picking up N copies, and e is the natural constant. Further simplification of this formula makes it possible to calculate the impact of normal target distribution on assay results.

Fortunately, since there is a log–log relationship between the number of input target molecules and the quantity of product molecules produced during PCR (this is illustrated in Fig. 8 and Fig. 9), variability resulting from sampling error affects the practical analytical resolution of a quantitative PCR assay less severely than some of the other factors to be described.

A more significant source of variability in the quantitative response from a given sample is the efficiency of the PCR reaction. Since there is a linear–log relationship between PCR efficiency and the number of product molecules generated from a PCR reaction

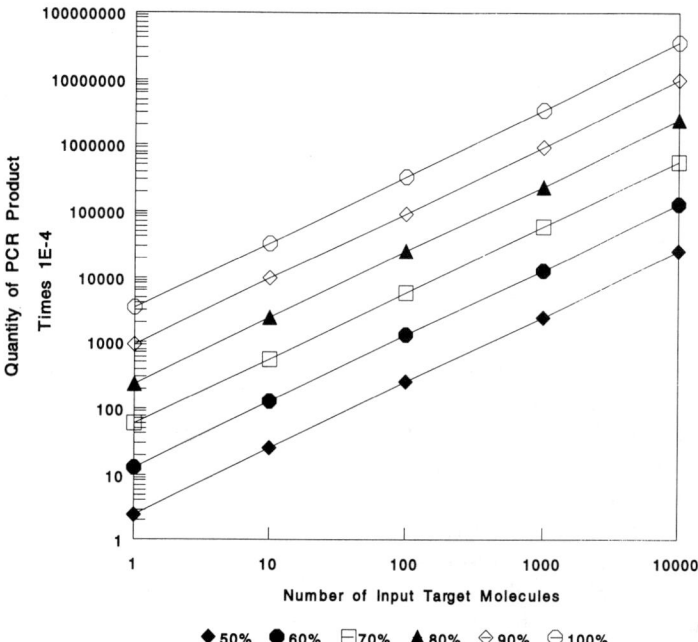

Figure 8 Estimates of product formation as a function of input target number for 25-cycle PCR reactions with average amplification efficiencies of 50, 60, 70, 80, 90, and 100%.

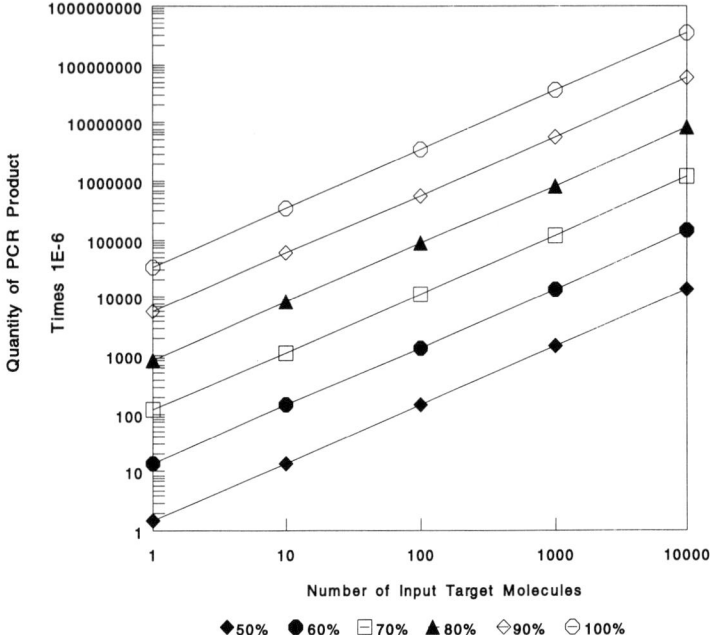

Figure 9 Estimates of product formation as a function of input target number for 35-cycle PCR reactions with average amplification efficiencies of 50, 60, 70, 80, 90, and 100%.

(see Fig. 10, Fig. 11, and Table 2), a small change in PCR efficiency will have a large effect on the number of product molecules generated. This is illustrated in Table 2 (e.g., with 35 cycles of PCR, a 10% change in PCR efficiency, from 50 to 60% leads to an approximately 95% change in the amount of product formed). This suggests that if PCR inhibition, which directly affects the efficiency of the PCR reaction, is variable from sample to sample, or sample aliquot to sample aliquot, there will be large differences in the amount of PCR product formed in the quantitative reactions. As a result, the resolution of a quantitative PCR method will be directly related to the ability to consistently and efficiently remove PCR inhibitors.

Because most solid-phase dependent DNA-probe hybridization assays exhibit a high slope sensitivity (there is a log–log relationship between the number of input target molecules and the response in most DNA-probe hybridization assays), relatively small variations in the results of the detection assay can cause significant variations in the quantitative resolution of the assay. The high slope sensitivity

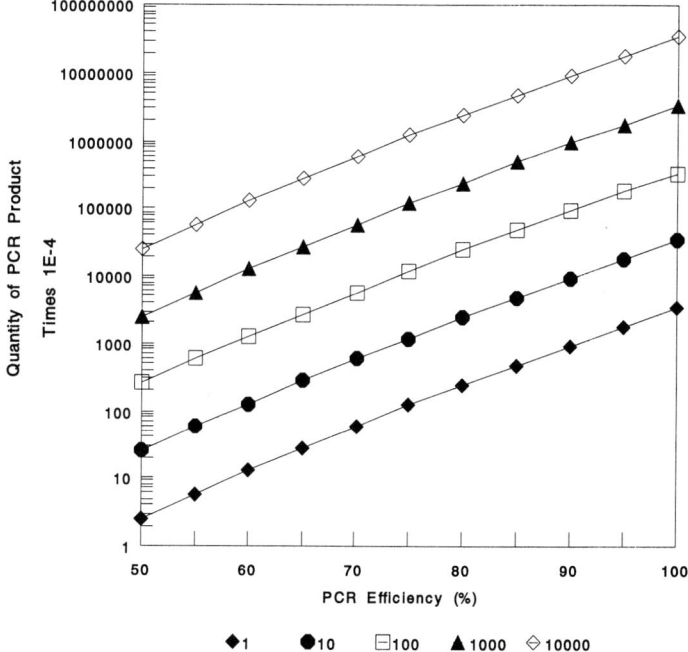

Figure 10 Estimates of product formation as a function of PCR efficiency for 25-cycle PCR reactions with inputs of 1, 10, 100, 1000, and 10,000 target molecules.

for these types of assays is illustrated in Fig. 1, which experimentally confirms the classic log–log relationship for this function. As a result of this high-slope sensitivity, quantitative resolution is also dependent on the precision of the assay.

Design of a Quantitative PCR Assay

Many groups have reported excellent success in developing quantitative PCR assays (see review by Ferre, 1992). A properly designed quantitative PCR assay is one which makes maximum use of the working range of both PCR amplification and DNA-probe detection, without sacrificing the analytical resolution of the results. As mentioned previously, a combination of the strategies described here should provide an appropriate linear working range, with the required

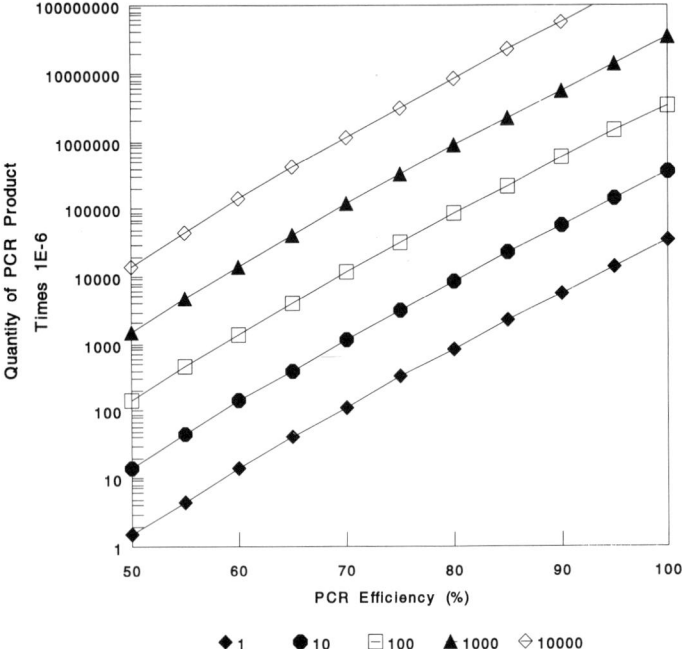

Figure 11 Estimates of product formation as a function of PCR efficiency for 35-cycle PCR reactions with inputs of 1, 10, 100, 1000, and 10,000 target molecules.

level of precision, to obtain reliable estimates of the quantities of targets in biological samples. Key points to consider when designing the assay will be (1) the minimum sensitivity required; (2) the range of target molecules in the samples; (3) the desired analytical resolution; and (4) the reliability (required confidence interval) of the results.

By carefully balancing the requirements for minimum assay sensitivity against the required dynamic range, it is possible to achieve reasonable analytical resolution over several orders of magnitude. There are, however, several factors which will significantly affect the ultimate resolution of the quantitative assay; these include, but are not limited to (1) the reproducibility of target recovery, (2) the efficiency of removing PCR inhibitors, (3) the concentration of targets placed into the PCR reaction, (4) the efficiency of the PCR reaction, (5) the numbers of PCR cycles, (6) the efficiency of hybridizing to a DNA-probe, and (7) the reproducibility (precision) of the DNA-probe assay. The potential impacts of many of these factors

Table 2
Relationship between PCR Efficiency and Product Accumulation in a 35-Cycle PCR

Relative PCR efficiency	Input target molecules					
	1	5	10	100	1000	10,000
50%	1.46×10^6	7.28×10^6	1.46×10^7	1.46×10^8	1.46×10^9	1.46×10^{10}
60%	1.39×10^7	6.97×10^7	1.39×10^8	1.39×10^9	1.39×10^{10}	1.39×10^{11}
70%	1.16×10^8	5.82×10^8	1.16×10^9	1.16×10^{10}	1.16×10^{11}	1.16×10^{12}
80%	8.60×10^8	4.30×10^9	8.60×10^9	8.60×10^{10}	8.60×10^{11}	8.60×10^{12}
90%	5.71×10^9	2.85×10^{10}	5.71×10^{10}	5.71×10^{11}	5.71×10^{12}	5.71×10^{13}
100%	3.44×10^{10}	1.72×10^{11}	3.44×10^{11}	3.44×10^{12}	3.44×10^{13}	3.44×10^{14}

have been described previously; some additional considerations are discussed in the following paragraphs.

When designing a quantitative assay, it is important to also consider that the accuracy of the quantitative analysis will depend upon the confidence interval required for the results and the range of input target molecules to be measured. As shown in Table 3, when attempting to quantitate low levels of normally distributed target molecules (a range of 10 to 100 input target molecules is used in this example) with a relatively high degree of confidence (confidence interval of 99.9%), the minimum error of the estimates is quite large. The situation described earlier is encountered when samples are diluted prior to amplification in order to generate product concentrations which conform to the relatively narrow working range of standard high-sensitivity DNA-probe detection assays. If quantitative assays are designed to measure aliquots containing larger concentrations of target molecules (e.g., by adjusting the number of PCR cycles to obtain an appropriate working range from the detection assay), the resolution of the results will be less dependent on the distribution of targets in the sample.

Another danger associated with dilution of samples prior to PCR is the increased probability of encountering false negative PCR results. PCR false negatives are sometimes referred to as dropouts. Dropouts can occur when (1) inhibitors reduce the efficiency of the PCR efficiency, thereby preventing the targets from being amplified to detectable levels; (2) PCR efficiency is reduced because of problems with either the thermal cycler or the PCR reagents; or (3) no targets are present in the sampled aliquot owing to normal distribution of targets in solution. Unlike PCR dropouts which occur as a result of reduced amplification efficiency, Table 4 estimates the frequency of dropouts caused by the normal distribution (Poisson estimate) of targets in a sample. As predicted by the log–log relationship between the number of input target molecules and product accumulation, the frequency of PCR dropouts diminishes rapidly as the number of targets in solution increases. It is important, therefore, to design sample preparation and amplification procedures which yield concentrations of target molecules which can be reliably amplified, to avoid the potential for PCR dropouts. If the sample preparation procedure efficiently removes PCR inhibitors, it is safer to amplify higher concentrations of target molecules, using reduced cycles of PCR, than to amplify very low concentrations of input target molecules.

Table 3
Quantitative Resolution When Measuring Low Concentrations of Target Molecules

Number of normally distributed target molecules per sample	Minimum estimate at 99.9% confidence ($p \leq 1/1000$)	Maximum estimate at 99.9% confidence ($p \leq 1/1000$)	Minimum error of estimate
10 copies	1	21	±110%
20 copies	7	35	±75%
40 copies	22	60	±50%
60 copies	39	83	±38%
80 copies	56	106	±33%
100 copies	73	128	±28%

$$\text{Minimum error} = \frac{\text{Maximum error} - \text{target number}}{\text{target number}} \cdot 100\%$$

Table 4

Relationship between Target Concentration and the Frequency of PCR Dropouts

Concentration of target molecules	Dropouts (%)	Frequency of occurrence
1	36.788	1 in 2.78
5	0.6738	1 in 148
10	0.0045	1 in 22,222
20	2.06×10^{-7}	1 in 4.85×10^8
50	1.93×10^{-20}	1 in 5.18×10^{21}
100	3.72×10^{-42}	1 in 2.67×10^{43}

As discussed previously, the reproducibility of the DNA-probe assay can also affect the quantitative resolution of a PCR assay. If either the efficiency of hybridization to homologous products is poor, or the stringency is insufficient to prevent hybridization with heterologous products, the accuracy and precision of the detection assay results will be diminished and there will be a marked degradation in the analytical resolution of the quantitative assay. For these reasons it is important to design an efficient solid-phase dependent DNA-probe assay which can differentiate the desired PCR products from primer-dimers and incorrectly amplified sequences.

Conclusions

It is possible to design quantitative PCR assays which can be used to reliably measure target molecules over many orders of magnitudes. Well-designed quantitative assays make use of sample preparation procedures which reliably liberate large quantities of targets from inhibitors of the PCR process, efficient amplification procedures, and precise DNA-probe detection assays. The accuracy and precision of the measurements that are obtained by these assays are dependent on the range of targets to be encountered, the efficiency and reliability of the sample preparation methods, the number of PCR cycles, the relative efficiency of the PCR amplification process, and the reliability of the detection assay.

Quantitative PCR will enhance the research capabilities for a number of important applications, and has led to the development of new approaches for detecting PCR products (Langraf *et al.*, 1991; Warren *et al.*, 1991; Glazer and Rye, 1992; Porcher *et al.*, 1992; Lu *et al.*, 1994). As the use of these techniques expands, additional refinements and approaches will undoubtedly be developed to further enhance the analytical resolution of quantitative PCR. A thorough understanding of the features of a well-designed quantitative PCR assay and the theoretical factors which contribute to analytical performance will simplify the process of designing and interpreting quantitative PCR assays.

Acknowledgments

The author wishes to thank Mr. Steven Soviero, Ms. Rita Sun, and Mr. Zhuang Wang for contributing valuable data. I am also grateful to Dr. John Sninsky for helpful comments and critical evaluation of the manuscript.

Literature Cited

Becker-André, M., and K, Hahlbrock. 1989. *Nuc. Acids Res.* **17**:9437–9446.
Casareale, D., R. Pottathil, and R. Diaco. 1992. *PCR Meth. Appli.* **2**:149–153.
Chelly, J., D. Montarras, C. Pincet, Y. Berwald-Netter, J.-C. Kaplan, and A. Kahn. 1990. *Nature* **333**:858–860.
Cone, R. W., A. C. Hobson, and M. L. W. Huang. 1992. *J. Clin. Microbiol.* **30**:3185–3189.
Ferre, F. 1992. *PCR Meth. Appli.* **2**:1–9.
Ferre, F., A. Marchese, P. C. Duffy, D. E. Lewis, M. R. Wallace, H. J. Beecham, K. G. Burnett, F. C. Jensen, and D. J. Carlo. 1992. *AIDS Res. Hum. Retrovir.* **8**:269–275.
Gilliland, G., S. Perrin, K. Blanchard, and H. F. Bunn. 1990. *Proc. Nat. Acad. Sci. U. S. A.* **87**:2725–2729.
Glazer, A., and H. S. Rye. 1992. *Nature* **359**:859–861.
Keller, H. H., D. P. Huang, and M. M. Manak. 1991. *J. Clin. Microbiol.* **29**:638–641.
Kellog, D. E., J. J. Sninsky, and S. Kwok. 1990. *Anal. Biochem.* **189**:202–208.
Landgraf, A., B. Reckmann and A. Pingoud. 1991. *Anal. Biochem.* **193**:231–235.
Loeffelholz, M. J., C. A. Lewinski, A. P. Purohit, S. A. Herman, D. B. Buonogurio, and E. A. Dragon. 1992. *J. Clin. Microbiol.* **30**:2847–2851.
Lu, W., D.-S. Han, Y. Ju, and J. M. Audrieu. 1994. *Nature* **368**:269–271.
Lundeberg, J., J. Wahlberg, and M. Uhlen. 1991. *BioTechniques* **10**:68–75.
Maniatis, T., E. F. Fritsch, and J. Sambrook. eds. (1982). In: *Molecular cloning. A laboratory manual.* Cold Spring Harbor Laboratory, New York.
Nedelman J., P. Heagerty, and C. Lawrence. 1992. *CABIOS.* **8**:65–70.
Piatak, M., K.-C. Luk, B. Williams, and J. D. Lifson. 1993. *BioTechniques* **14**:70–78.

Porcher, C., M.-C. Malinge, C. Picat, and B. Grandchamp. 1992. *BioTechniques* **13:**106–113.
Siebert, P. D., and J. W. Larrick. 1993. *BioTechniques* **14:**244–249.
Warren, W., T. Wheat, and P. Knudsen. 1991. *BioTechniques* **11:**250–255.
Winters, M. A., M. Holodniy, D. A. Katzenstein, and T. C. Merigan. 1992. *PCR Meth. Appli.* **1:**257–262.

Part Two

ANALYSIS OF PCR PRODUCTS

8

CARRIER DETECTION OF CYSTIC FIBROSIS MUTATIONS USING PCR-AMPLIFIED DNA AND A MISMATCH-BINDING PROTEIN, MutS

Alla Lishanski and Jasper Rine

The rapid rate of discovery of new disease genes has increased the need for high-throughput methods to detect mutations and allow many individuals to be screened at several genetic loci. It has now become possible to detect point mutations by a variety of techniques which have been reviewed (Cotton, 1993; Rossiter and Caskey, 1990). Two recent papers reflect a growing interest in DNA repair enzymes as potential tools for screening genomes for polymorphisms and disease-causing mutations (Lu and Hsu, 1992; Nelson et al; 1993).

We used the ability of MutS, a member of the family of *Escherichia coli* methyl-directed mismatch repair enzymes, to recognize and bind mismatches in order to develop a technique that would test for heterozygosity of a locus in PCR-amplified DNA from an individual. *In vivo*, MutS initiates a series of events leading to a correction of replication errors (Grilley et al., 1990; Au et al., 1992). This protein is able to bind various single-base mismatches in short synthetic oligonucleotides *in vitro* (Jiricny et al., 1988). We extended this finding to PCR products from two regions of the human cystic fibrosis (CFTR) gene that harbor the most frequently occurring mutations within this gene. The choice of the CFTR gene was based on the availability of a collection of well-characterized mutations and the importance of rapid screening methods to detect carriers. We have established that MutS preferentially bound mismatched heteroduplexes formed upon reannealing of PCR product from genomic DNA that was heterozygous for certain frameshift and point mutations within the CFTR gene, and that this ability is retained on PCR fragments as large as ~ 500 bp in length.

Rationale

A schematic representation of our experimental design is shown in Fig. 1. A DNA region of interest is PCR amplified from genomic DNA samples extracted from peripheral blood. The PCR products are heat denatured and reannealed. If the target DNA is heterozygous at the A locus (A/a), reannealing of a PCR product yields four different DNA species: two homoduplexes, AA and aa, and two heteroduplexes, Aa and aA. In the case of a homozygous allele, only homoduplexes are generated upon reannealing of the PCR product. Depending upon the nature of the difference between the two alleles, the resulting heteroduplexes contain various single-base pair mismatches or small bulges in one of the strands. MutS protein binds to mismatched base pairs, and formation of the complex is detected by a gel mobility-shift assay. MutS binds to homoduplex DNA poorly and the difference in binding of heteroduplex vs homoduplex is the signal on which this technique is based.

Protocols

1. Choice of PCR Primers

PCR primers are listed in Table 1. They were designed using the Primer software, except the 5/6 and 10/11 pairs, which were chosen according to Zielinski *et al.* (1991). Oligonucleotides were purchased from Genosys Biotechnologies Inc. (The Woodlands, TX).

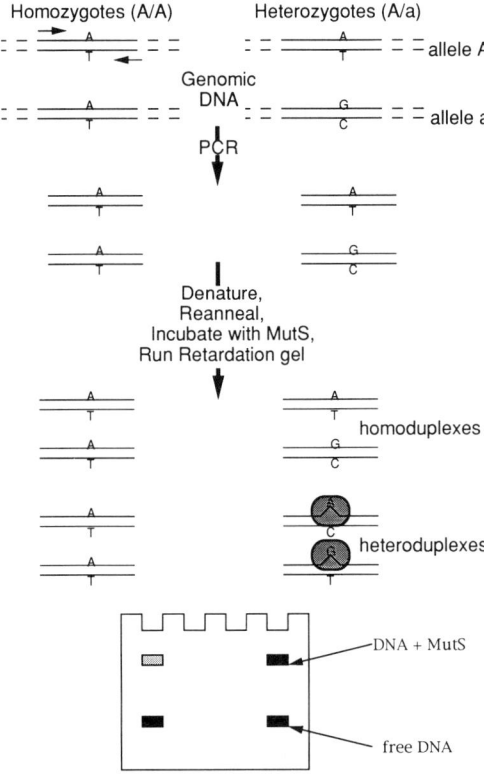

Figure 1 Experimental strategy (schematic representation). PCR primers are indicated by arrows; MutS protein is indicated by shaded ovals.

2. End Labeling of Primers

Primers were 5'end labeled using T4 polynucleotide kinase (Boehringer-Mannheim) and [γ-^{32}P]ATP (10 mCi/ml, >5000 Ci/mmole, Amersham) according to standard procedures.

3. PCR Amplification and Reannealing of PCR Products

Amplifications were carried out in the MiniCycler (M. J. Research, Cambridge, MA) or in the Perkin–Elmer 9600 Thermocycler using 0.2-ml thin-walled tubes. The hot start technique with AmpliWax PCR Gems was performed according to the procedure recommended by the manufacturer (Perkin–Elmer, Norwalk, CT); 250 ng of genomic DNA were used as a target. A total of 27 PCR cycles were performed

Table 1
Mutations within the CFTR Gene and Respective PCR Products[a]

Mutation	Location	PCR Primers (5' → 3')	PCR Product Size (bp)	Lesion	Mismatches Formed Upon Reannealing of PCR Product
ΔF508	exon 10	(1) CTCAGTTTCCTGGATATGCC (2) TGGCATGCTTTGATGACG	100	3 bp deletion (Phe deletion)	CTT + GAA bulges
		(1) CTCAGTTTCCTGGATATGCC (3) CTAACCGATTGAATATGGAGCC	200		
		(4) CAAGTGAATCCTGAGCGTGA (3) CTAACCGATTGAATATGGAGCC	340		
		(5) GCAGAGTACCTGAAACAGGA (6) CATTCACAGTAGCTTACCCA	491		
G542X	exon 11	(7) GCCTTTCAAATTCAGATTGAGC (8) TGCTCGTTGACCTCCACTC	141	G → T (Gly$_{542}$ → Stop)	G/A + T/C single-base mismatches
R553X	exon 11	(7) GCCTTTCAAATTCAGATTGAGC (9) GACATTTACAGCAAATGCTTGC	203	C → T (Arg$_{553}$ → Stop)	C/A + T/G single-base mismatches
		(10) CAACTGTGGTTAAAGCAATAGTGT (11) GCACAGATTCTGAGTAACCATAAT	425		
G551D	exon 11	(7) GCCTTTCAAATTCAGATTGAGC (9) GACATTTACAGCAAATGCTTGC	203	G → A (Gly$_{551}$ → Asp)	G/T + A/C single-base mismatches
		(10) CAACTGTGGTTAAAGCAATAGTGT (11) GCACAGATTCTGAGTAACCATAAT	425		

[a] Reproduced with permission from Lishanski, A., E. A. Ostrander, and J. Rine. 1994. Mutation detection by mismatch binding protein, MutS, in amplified DNA: Application to the Cystic Fibrosis Gene. *Proc. Natl. Acad. Sci. U.S.A.* **91**: 2674–2678.

and consisted of a 1-min denaturation step at 94° C, 1-min reannealing step at 58° C (55° C for the 5/6 and 52° C for the 10/11 primer sets), and a 1-min extension step at 74° C. The last cycle was followed by a 7-min final extension. Amplification was followed by denaturation (95° C, 3 min) and reannealing (67° C, 1 hr) of the PCR product. The last two steps were incorporated into the same program and were followed by storage at 4° C. A final volume of 100 μl was generated; 2.5 U of *Taq* polymerase (Perkin–Elmer) were used per reaction.

4. MutS Protein Binding and Mobility Shift Assays

Two to 6 μl of the reannealed PCR product were mixed with 1–4 μl (~ 4 pmoles) MutS and supplemented with the assay buffer (Jiricny *et al.*, 1988) so that the total volume never exceeded 10 μl. In some experiments, 1 μg bovine serum albumin (BSA) was added to each reaction. The mixtures were incubated on ice or at room temperature for 10–20 min, mixed with a dye solution, and loaded on polyacrylamide gels. Run conditions for each experiment are indicated in the figure legends. Wet gels were sealed in polyethylene and exposed to X-ray film at $-70°$ C.

Example 1: ΔF508

The carrier frequency for cystic fibrosis is about 1 in 25 North American Caucasians. One individual in 2500 inherits two altered copies and is thus affected with the disease. The most frequent mutant allele of the CFTR gene, ΔF508, is a three-base pair deletion, which results in a loss of a phenylalanine residue at amino acid position 508 in the coding region of the gene (Table 1). This mutation accounts for about 75% of CF chromosomes in this population.

Genomic DNA samples extracted from peripheral blood were amplified using the ^{32}P-5'end-labeled primers that bracket the site of the ΔF508 deletion to generate PCR products 100, 200, 340, and 491 bp in length. Heteroduplexes with this deletion in one of the strands contain a CTT or a GAA bulge in the complementary strand. After being reannealed and incubated with MutS protein, the reaction mixture was run on a 6% polyacrylamide gel (Fig. 2) or on a 4–15% gradient gel (Fig. 3a). MutS binding, which was detected as a shifted band, was consistently stronger in the amplified DNA of individuals who were heterozygous for ΔF508 than in the amplified

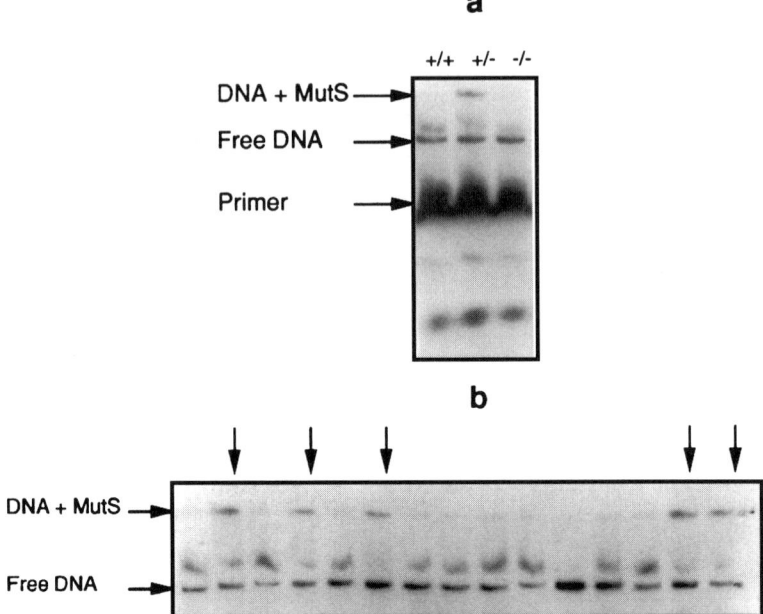

Figure 2 Detection of heterozygosity for ΔF508 by mobility shift assay. (a) View of the typical retardation gel. (b) Blind experiment (only the essential area of the gel is shown); 250 ng genomic DNA were amplified with *Taq* polymerase to generate fragments 100 bp in length; 27 cycles were performed. A mobility shift assay was performed in a 20 × 20 × 0.15-cm 6% polyacrylamide gel (PAG) (AA : Bis = 19.1) in 0.2X TBE. The gels were run at 4° C for 15 min at 500 V. +, normal allele; −, mutant allele; +/+, homozygous healthy individuals; −/−, homozygous affected patients; +/−, asymptomatic heterozygous carriers. Reproduced with permission from Lishanski, A., E. A. Ostrander, and J. Rine. 1994. Mutation detection by mismatch binding protein, MutS, in amplified DNA: Application to the Cystic Fibrosis Gene. *Proc. Natl. Acad. Sci. U. S. A.* **91**:2674–2678.

DNA of homozygotes (both affected and normal individuals). In a blind experiment, all 5 heterozygous carriers of the ΔF508 mutation were correctly detected among 15 individuals (Fig. 2b).

Example 2: Point Mutations

In order to determine the ability of this method to detect point mutations, we tested the most frequently occurring non-ΔF508 mutations in CFTR which are found in exon 11 (Table 1). Figure 3b shows

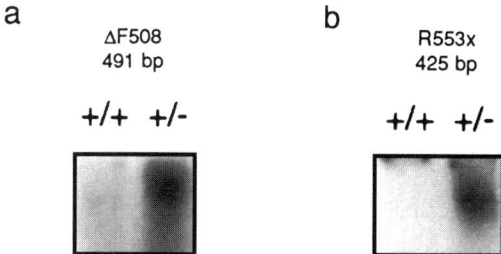

Figure 3 Detection of ΔF508 and R553X. Genomic DNA was amplified with *Taq* polymerase; 27 cycles were performed with primers 5/6 (a) or with primers 10/11 (b) to generate fragments 491 bp and 425 bp in length, respectively. Mobility shift assays were performed in a 4–15% precast gradient minigel (Bio-Rad) in the discontinuous Laemmli buffer system (Laemmli, 1970) without SDS. The gels were run at 200 V for 45 min (a) and for 40 min (b). Only the area of the gel containing shifted bands is shown. Reproduced with permission from Lishanski, A., E. A. Ostrander, and J. Rine. 1994. Mutation detection by mismatch binding protein, MutS, in amplified DNA: Application to the Cystic Fibrosis Gene. *Proc. Natl. Acad. Sci. U. S. A.* **91:**2674–2678.

the results obtained for the R553X C → T transition using a PCR product 425 bp in length. Reannealed PCR products contain G/A and A/C mismatches. MutS distinguished between heterozygotes for this point mutation and homozygotes. Positive results were also obtained for G542X and G551D point mutations (not shown).

Problems

1. One of the potential problems of this method is a low fidelity for all thermostable DNA polymerases. When the amplified DNA is reannealed, any polymerase errors that accumulated during amplification would result in artificially introduced mismatches. MutS binding to such erroneous mismatches could account for the noise in our experiments. The number of error-caused mismatches would increase with the length of PCR product and depend upon the fidelity of the polymerase used for amplification (Reiss *et al;* 1990). We examined the influence of these two factors on the signal-to-noise ratio in our mobility shift experiments. We tested the ability of MutS to select three-base pair bulges and single-base mismatches in different-sized DNA fragments (Table 1) and found that it was retained in PCR products as large as ~ 500 bp in length. We also carried out amplifications with either *Taq* or *Pfu* polymerase (the

latter has a 12-fold greater fidelity; Lundberg et al., 1991) and have not found any substantial difference between them in our mobility shift experiments.

2. One of the problems we encountered in the course of this study was the low stability of MutS–DNA complexes, resulting in their dissociation and, after long electrophoretic runs, loss of signal. On the other hand, the wavy bands that represent DNA dissociated from the complex with MutS were seen above free DNA in all the lanes in Fig. 2b, suggesting that MutS initially bound not only heteroduplexes but homoduplexes as well. Apparently, the latter nonspecific complexes had a shorter half-life than complexes of MutS bound to mismatched bases. Therefore, the choice of the duration of a run was a compromise between these two considerations and was rather short (15 min in Fig. 2). Omission of EDTA from 0.2 × TBE buffer (Tris–borate, boric acid, EDTA) did not appear to improve the stability of the complexes. In the discontinuous Laemmli buffer system (Laemmli, 1970), used in the experiments in Fig. 3, the MutS–DNA complexes remained stable for a longer time (30–45 min), but shifted bands appeared as smudges, revealing the inherent heterogeneity of the complexes. Stability of the complexes also appeared to increase with DNA length.

Future Developments

In its current version, this method does not differentiate between homozygotes with both wild-type or both mutant alleles, but it can be modified to do so by reannealing PCR products with an excess of a probe that corresponds to one of the strands of the wild-type allele. The ratio of the signals after MutS binding would be 0 : 1 : 2 for +/+, +/− and −/− DNA, respectively.

Preliminary experiments indicate that MutS protein immobilized onto a solid support retains its binding activity and specificity. Thus, it may be possible to develop a nongel form of the technique described here, which would offer a greater potential for automation.

Heterozygosity for ΔF508 is easily detectable by a variety of different methods, the simplest of which is polyacrylamide gel electrophoresis, which reveals a slowly moving heteroduplex band. It would be interesting to use the MutS-based approach to find other disease-causing mutations for which detection methods are not yet so advanced.

Conclusions

We have described a PCR-based technique for detecting mutations in genomic DNA by using a mismatch binding protein, MutS. This technique made use of preferential affinity of MutS for DNA heteroduplexes containing mismatches caused by mutations. Binding is detected by a band shift assay. This approach was tested on a three-base pair deletion and several single-base substitutions within the CFTR gene by using PCR products up to ~500 bp long. The method appears to have a potential for diagnostic applications.

Acknowledgments

The authors thank Paul Modrich for providing us with MutS protein, Patricia Laurenson and Elaine Ostrander for early advice, and Alex Glazer and Cristian Orrego for stimulating discussions. This work was supported by the Director, Office of Energy Research, Office of Health and Environmental Research, Human Genome Program, of the U.S. Department of Energy under Contract No. DE-AC03-76SF00098 and National Institute of Environmental Health Studies (NIEHS) Mutagenesis Center Grant P30 ES-01896-16.

Literature Cited

Au, K. G., K. Welsh, and P. Modrich. 1992. Initiation of methyl-directed mismatch repair. *J. Biol. Chem.* **267**(17):12142–12148.
Cotton RG. 1993. Current methods of mutation detection. *Mutat. Res.* **285**(1):125–144.
Grilley, M., J. Holmes, B. Yashar, and P. Modrich, 1990. Mechanisms of DNA-mismatch correction. *Mut. Res.* **236**(2–3):253–267.
Jiricny, J., S. S. Su, S. G. Wood, P. Modrich, 1988. Mismatch-containing oligonucleotide duplexes bound by the *E. coli* mutS-encoded protein. *Nucleic Acids Res.* **16**(16):7843–7853.
Laemmli, U. K. 1970. Cleavage of structural proteins during the assembly of the head of bacteriophage T4. *Nature*, **227**(259):680–685.
Lu, A. L., and I. C. Hsu. 1992. Detection of single DNA base mutations with mismatch repair enzymes. *Genomics* **14**(2):249–255.

Lundberg, K. S., D. D. Shoemaker, M. W. Adams, J. M. Short, J. A. Sorge, and E. J. Mathur. 1991. High-fidelity amplification using a thermostable DNA polymerase isolated from *Pyrococcus furiosus*. *Gene* **108**(1):1–6.

Nelson, S. F., J. H. McCusker, M. A. Sander, Y. Kee, P. Modrich, and P. O. Brown. 1993. Genomic mismatch scanning: a new approach to genetic linkage mapping *Nat. Genet.* **4**(1):11–18.

Reiss, J., M. Krawczak, M. Schloesser, M. Wagner, and D. N. Cooper. 1990. The effect of replication errors on the mismatch analysis of PCR-amplified DNA. *Nucleic Acids Res.* **18**(4):973–978.

Rossiter, B. J., and C. T. Caskey. 1990. Molecular scanning methods of mutation detection. *J. Biol. Chem.* **265**(22):12753–12756.

Zielenski, J., R. Rozmahel, D. Bozon, B. Kerem, Z. Grzelczak, J. R. Riordan, J. Rommens, and L. C. Tsui. 1991. Genomic DNA sequence of the cystic fibrosis transmembrane conductance regulator (CFTR) gene. *Genomics* **10**(1):214–228.

9

SINGLE-STRANDED CONFORMATIONAL POLYMORPHISMS

Anne L. Bailey

The rapid growth in the identification of new disease genes has increased the need to accurately characterize disease-causing mutations and identify DNA polymorphisms. Since a large portion of sequence variation in the human genome is caused by single base changes, any method used to detect mutations or polymorphisms must be capable of detecting a single-base substitution. Single-base genetic variations can easily be detected by DNA sequencing. However, with the advent of polymerase chain reaction (PCR) technology, rapid, new methods for detecting mutations and polymorphisms have emerged, negating the need for expensive and time-consuming sequencing.

Some of the recent methods described for detecting point mutations include RNase cleavage (Myers et al., 1985), chemical cleavage (Cotton et al., 1988), denaturing gradient gel electrophoresis (DGGE) (Fischer and Lerman, 1980; Myers et al., 1987; Sheffield et al., 1989; 1992) heteroduplex analysis (Keen et al., 1991; White et al., 1992), single-stranded conformational polymorphism (SSCP) analysis (Orita et al., 1989a,b) and modified SSCP techniques such as RNA-SSCP (Sarkar et al., 1992a) or dideoxy fingerprinting (Sarker et al., 1992b). SSCP analysis has gained wide acceptance because of its simplicity, and has been used to detect sequence changes in various contexts. For example, the SSCP technique has been effective in detecting polymorphisms in ALU repeats (Orita et al., 1990) and defining mutations in human diseases (Mashiyama et al., 1990, 1991; Demers et al., 1990; Dean et al., 1990; Ainsworth et al., 1991; Murakami et al., 1991a,b; Soto and Sukumar, 1992; Michaud et al., 1992; Pan et al., 1992).

The SSCP method is based on the principle that single-stranded DNA, in a nondenaturing condition, has a folded structure that is determined by intramolecular interactions (Fig. 1). These sequence-based secondary structures (conformers) affect the mobility of the DNA during electrophoresis on a nondenaturing polyacrylamide gel. A DNA molecule containing a mutation, even a single-base substitution, will have a different secondary structure than the wild type, resulting in a different mobility shift during electrophoresis than that of the wild type.

In the original method (Orita et al., 1989b), the DNA sequence of interest was amplified and simultaneously radiolabeled by PCR. The PCR sample was denatured prior to being separated on a sequencing size polyacrylamide gel and visualizing the bands by autoradiography.

Subsequent improvements in method have diminished the gel size and eliminated the radiolabel by visualizing the bands with silver staining (Ainsworth et al., 1991). Automation has been achieved by using fluorescent-labeled primers in the PCR reaction and separating them on an automated DNA sequencer (Makino et al., 1992). Both of these improvements have further simplified the SSCP method by eliminating the need for radioactivity, but to date the most widely accepted SSCP techniques use radiolabels and autoradiography for band detection.

The sensitivity (ability to detect mutations) of SSCP analysis is widely disputed in the literature, with reports ranging from 35% (Sarkar et al., 1992a) to nearly 100% (Orita et al., 1989b). The one factor that has the greatest effect on SSCP sensitivity is the size of the DNA fragment (Hayashi, 1991, 1992; Sheffield et al., 1993). As the DNA fragment increases in length, the sensitivity of the SSCP analysis decreases. An optimal size of 200 bp or less is the most sensitive for single-base substitutions. The electrophoretic mobility of the single-stranded conformer may also be affected by temperature, ionic strength (Spinardi et al., 1991), gel additives (glycerol), acrylamide concentration (Savov et al., 1992), high acrylamide to cross-linker ratio (Mashiyama et al., 1990), sequence composition, position of substitution, and type of mutation. Because of the wide variety of parameters that may affect the mobility of the single-stranded conformers, the SSCP technique is usually optimized for maximum sensitivity by each laboratory (Dean and Gerrard, 1991).

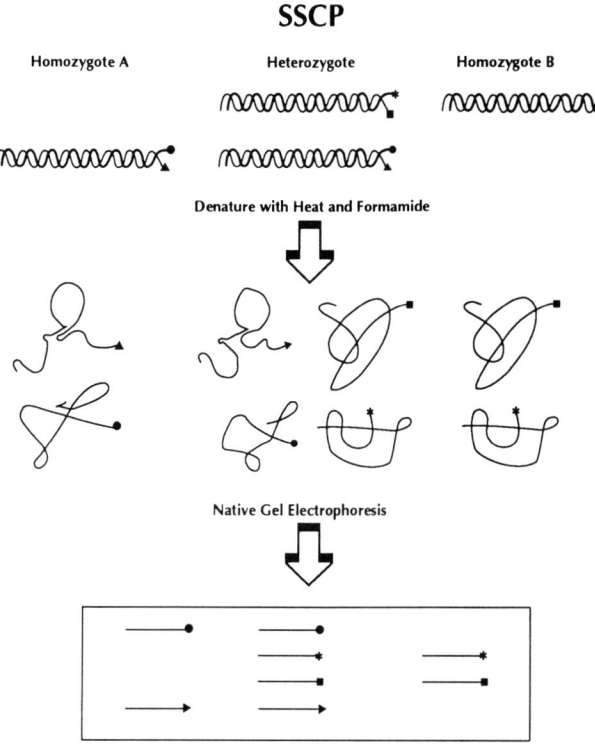

Figure 1 To test for a mutant gene or polymorphism, PCR products from control and test samples are prepared by heat denaturation in the presence of formamide. The double-stranded character of the samples is eliminated, and folded single-stranded secondary structures emerge during cooling. The three-dimensional structures are unique to the primary sequence, and move at different rates through a nondenaturing MDE gel. The sensitivity of mutation detection is high because single-base changes may radically alter the migration of the nucleic acid.

Therefore the following protocol should be used as a starting method; it may require additional optimization by each laboratory.

Protocols

DNA Amplification

If a radiolabel is to be incorporated during the PCR reaction, either end-labeled primers or deoxynucleotides may be used. Since the amount of radiolabeled product required for detection is less than that needed for staining, the volume of the reaction and concentration of substrates may be reduced. For radiolabeled PCR-SSCP, the optimum reaction volume is 5 to 20 µl with 25 to 100 ng of genomic DNA. Other reactants and buffers are reduced accordingly. DNA fragment size for SSCP analysis is optimum at 200 bp or less. If larger fragments are amplified, they should be digested with an appropriate restriction enzyme prior to being run on an SSCP gel to obtain the optimal detection sensitivity. If a fluorescent label is used for detection, the same reduced reaction volume and concentration of substrates should be used.

DNA Denaturation

Complete denaturation of the amplified sample prior to loading it onto the gel for electrophoresis is essential for achieving optimal sensitivity. To ensure complete denaturation, the amplified sample can be diluted in a stop solution (95% formamide, 10 mM NaOH, 0.1% bromophenol blue, 0.1% xylene cyanol). Since end-labeled primers have a higher specific activity than radioactive deoxynucleotides, 1µl of the reaction mixture is usually mixed with 50 to 100 µl of the stop solution. For radioactive deoxynucleotides, 1.5–2.5 µl of reaction mixture are usually mixed with 10 µl of stop solution. Denaturation is accomplished by heating the diluted samples at 90–94°C for 2–5 min. Samples are immediately placed on ice and 1–3 µl loaded onto the electrophoresis gel. If larger fragments are amplified, they should be placed in restriction enzyme digest buffer and digested before being diluted in stop solution. A recent paper (Weghorst and Buzzard, 1993) suggests the use of methyl mercury hydroxide as an additional denaturant for enhanced strand separation. However, owing to the extreme toxicity of this chemical, additional data are needed before it can be routinely included in the stop solution.

Electrophoresis

As previously discussed, there are many parameters which affect the separation of the single-stranded conformers. There is universal agreement in the literature that DNA fragment size has a significant impact on SSCP sensitivity. Therefore for optimum detection of mutations, a fragment size of 200 bp or less is recommended. In a DNA fragment of 200–400 bp, the sensitivity of SSCP analysis decreases from 95 to 100% to 65 to 75% (Sheffield et al., 1993 and Hayashi, 1991). The minimum fragment size has not been fully addressed in the literature, but several groups (Sheffield et al., 1993 and Spinardi et al., 1991) show good results with fragments of 60 to 115 bp.

Other parameters which affect the sensitivity of SSCP analysis are related to the gel composition and electrophoresis conditions. Recent reports have shown that a reduced concentration of crosslinker to acrylamide (49:1 to 99:1 acrylamide: N, N^1-methylene bisacrylamide) improves separation of single-stranded DNA conformers. The mutation detection enhancement (MDE) gel matrix (AT Biochem, Malvern, PA) has been optimized for resolving conformational differences and has been shown (Soto and Sukumar, 1992) to improve the sensitivity of SSCP analysis. Other electrophoretic conditions that have been explored in the literature are temperature, gel additives (glycerol), acrylamide concentration, and ionic strength. Listed below are the best conditions for gels run at room temperature and at 4°C. As discussed earlier, these are starting conditions which may require further optimization by the laboratory.

Room Temperature Gels

1. We recommend that the laboratory evaluate a range of acrylamide concentrations (5–10%) at an acrylamide to bis ratio of 49:1. The gels should be prepared in 0.5X TBE (Tris Base, Boric Acid, and EDTA). Alternatively, the MDE gel matrix supplied as a 2X concentrate may also be used in a range of 0.3 to 0.7X, with a recommended starting concentration of 0.5X. For some DNA sequences, the addition of 5–10% glycerol to the gel has increased the sensitivity of the SSCP analysis. Once the optimal gel concentration is determined, the effects of glycerol on sensitivity can be evaluated.
2. For the best heat dissipation, thin gels of 0.4 mm or less should be used. Sequencing-size plates and apparatuses are usually used for SSCP analysis. A constant temperature of 20–25°C is essential for obtaining reproducible SSCP analysis, and sequencing apparatuses with aluminum plates, water-jacketed cooling, or buffer backs are best for heat dissipation.
3. Run the SSCP gel in 0.5X TBE (pH 8.3).
4. Prerun the gel for 10 min at 8 to 20 W (see step 5).
5. Load 1–3 µl of the denatured samples onto the gel. For sharp

bands, use a shark's-tooth comb or well-forming comb with 5 mm teeth.
6. The power supply should be set at a constant wattage to maintain even heat. In an apparatus with good heat dissipation, 8 to 20 W (0.5 W/cm) will sustain a constant temperature of 20–25°C. The run time for a 200-bp fragment at 20 W is approximately 2.5–3 hr. For larger fragments, run times should be increased to 5 hr. If a lower wattage is used (8 W), the run time should be increased to 14–16 hr for a 200-bp fragment. Marker dyes in the stop solution can be used to visualize and monitor movement of the DNA fragments.
7. Once electrophoresis is complete, transfer the gels to 3 MM Whatman paper, dry, and expose to X-ray film for 2–3 hr if ^{32}P is used and longer for lower specific activity labels.

4°C Gels

1. If gels can be conveniently run in the cold (4–10°C), glycerol may be omitted. For the researcher who wishes to evaluate gels at 4°C with and without glycerol, the run time with glycerol must be increased, since glycerol at cold temperatures reduces mobility. The dyes in the stop solution can be used as visual markers to monitor run time.
2. To prepare and run gels, the researcher should follow steps 1–7 for room-temperature gels. The wattage can be increased from 20 to 50 for gels 40 cm in length. As noted above, dyes in the stop solution make excellent visual markers for monitoring run times.

Discussion

Owing to simplicity of performance, the PCR–SSCP method is becoming widely used for detecting mutations. For optimal sensitivity, the DNA fragment size should be kept small (200 bp or less) and the cross-linker concentration reduced. Further enhancements of the sensitivity (ability to detect mutations) will increase the acceptance of this technique. In addition, the elimination of the radioactive label by using silver staining and smaller gels shows great promise for developing the PCR–SSCP method as a robust clinical tool. However, additional data are needed to assess the sensitivity limits of the method.

In conclusion, PCR–SSCP for single-base substitutions is an effective research method when the DNA fragment size is small. Larger DNA fragments can be amplified and cut by restriction enzymes prior to SSCP analysis. For polymorphisms, larger fragments can easily be used in the PCR–SSCP analysis if the detection of each base substitution is not critical.

Literature Cited

Ainsworth, P. J., L. C. Surth, and M. B. Couler-Mackie. 1991. Diagnostic single strand conformational polymorphism, (SSCP): A simplified non-radioisotopic method as applied to a Tay-Sachs B1 variant. *Nucleic Acids Res.* **19**:405–406.

Cotton, R. G. H., N. R. Rodriquez, and R. D. Campbell. 1988. Reactivity of cytosine and thymine in single-base-pair mismatches with hydroxylamine and osmium tetroxide and its application to the study of mutations. *Proc. Natl. Acad. Sci. U.S.A.* **85**:4397–4401.

Dean, M., and B. Gerrard. 1991. Helpful hints for the detection of single-stranded conformation polymorphisms. *BioTechniques* **10**:332–333.

Dean, M., M. D. White, J. Amos, B. Gerrard, C. Stewart, K.-T. Khaw, and M. Leppert. 1990. Multiple mutations in highly conserved residues are found in mildly affected cystic fibrosis patients. *Cell* **61**:863–870.

Demers, D. B., S. J. Odelberg, and L. M. Fisher. 1990. Identification of a factor IX point mutation using SSCP analysis and direct sequencing. *Nucleic Acids Res.* **18**:5575.

Fischer, S. G., and L. S. Lerman. 1980. Separation of random fragments of DNA according to properties of their sequences. *Proc. Natl. Acad. Sci. U.S.A.* **77**: 4420–4424.

Keen, J., D. Lester, C. Inglehearn, A. Curtis, and S. Bhattacharya. 1991. Rapid detection of single base mismatches as heteroduplexes on hydrolink gels. *Trends Genet.* **7**:5.

Hayashi, K. 1991. PCR-SSCP: A simple and sensitive method for detection of mutations in the genomic DNA. *PCR Meth. Appl.* **1**:34–38.

Hayashi, K. 1992. PCR-SSCP: A method for detection of mutations. *GATA* **9**:73–79.

Makino, R., H. Yazyu, Y. Kishimoto, T. Sekiya, and K. Hayashi. 1992. F-SSCP: Fluorescence-based polymerase chain reaction-single-strand conformation polymorphism (PCR-SSCP) analysis. *PCR Methods App.* **2**:10–13.

Mashiyama, S., T. Sekiya, and K. Hayashi. 1990. Screening of multiple DNA samples for detection of sequence changes. *Technique* **2**:304–306.

Mashiyama, S., Y. Murakami, T. Yoshimoto, and T. Sekiya. 1991. Detection of p53 gene mutations in human brain tumors by single-strand conformation polymorphism analysis of polymerase chain reaction products. *Oncogene* **6**:1313–1318.

Michaud, J., L. C. Brody, G. Steel, G. Fontaine, L. S. Martin, D. Valle, and G. Mitchell.

1992. Strand-separating conformational polymorphism analysis: Efficacy of detection of point mutations in the human ornithine Δ-aminotransferase gene. *Genomics* **13**:389–394.

Murakami, Y., K. Hayashi, and T. Sekiya. 1991a. Detection of aberrations of the p53 alleles and the gene transcript in human tumor cell lines by single-strand conformation polymorphism analysis. *Cancer Res.* **51**:3356–3361.

Murakami, Y., M. Katahira, R. Makino, K. Hayashi, S. Hirohashi, and T. Sekiya. 1991b. Inactivation of the retinoblastoma gene in a human lung carcinoma cell line detected by single-strand conformation polymorphism analysis of the polymerase chain reaction product of cDNA. *Oncogene* **6**:37–42.

Myers, R. M., Z. Larin, and T. Maniatis, 1985. Detection of single base substitutions by ribonuclease cleavage at mismatches in RNA: DNA duplexes. *Science* **230** (4731):1242–1246.

Myers, R. M., T. Maniatis, and L. S. Lerman. 1987. Detection and localization of single base changes by denaturing gradient gel electrophoresis. *Methods Enzymol.* **155**:501–527.

Orita, M., H. Iwahara, H. Kanazawa, K. Hayashi, and T. Sekiya, 1989a. Detection of polymorphisms of human DNA by gel electrophoresis as single-strand conformation polymorphisms. *Proc. Natl. Acad. Sci. U.S.A.* **86**:2766–2770.

Orita, M. Y., Y. Suzuki, T. Sekiya, and K. Hayashi, 1989b. Rapid and sensitive detection of point mutations and DNA polymorphisms using the polymerase chain reaction. *Genomics* **5**:874–879.

Orita, M., T. Sekiya, and K. Hayashi. 1990. DNA sequence polymorphisms in *Alu* repeats. *Genomics* **8**:271–278.

Pan, Y., A. Metzenberg, S. Das, B. Jing, and J. Gitschier. 1992. Mutations in the V2 vasopressin receptor gene are associated with X-linked nephrogenic diabetes insipidus. *Nature Genetics* **2**:103–106.

Sarkar, G., H.-S. Yoon, and S. S. Sommer, 1992a. Screening for mutations by RNA single-strand conformation polymorphism (rSSCP): comparison with DNA-SSCP. *Nucleic Acids Res.* **20**:871–878.

Sarkar, G., H.-S. Yoon, and S. S. Sommer. 1992b. Dideoxy fingerprinting (ddF): A rapid and efficient screen for the presence of mutations. *Genomics* **13**:441–443.

Savov, A., D. Angelicheva, A. Jordanova, A. Eigel, and L. Kalaydijieva. 1992. High percentage acrylamide gels improve resolution in SSCP analysis. *Nucleic Acids Res.* **20**:6741–6742.

Sheffield, V. C., D. R. Cox, and R. M. Myers. 1989. Attachment of a 40-base-pair G-C rich sequence to genomic DNA fragments by the polymerase chain reaction results in improved detection of single-base changes. *Proc. Natl. Acad. Sci. U.S.A.* **86**:232–236.

Sheffield, V. C., J. S. Beck, A. Lidral, B. Nichols, A. Cousineau, and E. M. Stone. 1992. Detection of polymorphisms within gene sequences by GC-clamped denaturing gradient gel electrophoresis. *Am. J. Hum. Genet.* **50**:567–575.

Sheffield, V. C., J. S. Beck, A. E. Kwitek, D. W. Sandstrom, and E. M. Stone. 1993. The sensitivity of single-strand conformation polymorphism analysis for the detection of single base substitutions. *Genomics* **16**:325–332.

Soto, D., and S. Sukumar. 1992. Improved detection of mutations in the p53 gene in human tumors as single-stranded conformation polymorphs and double-stranded heteroduplex DNA. *PCR Methods Appl.* **2**:96–98.

Spinardi, L., R. Mazars, and C. Theillet. 1991. Protocols for an improved detection of points mutations by SSCP. *Nucleic Acids Res.* **19**:4009.

Weghorst, C. M., and G. S. Buzzard. 1993. Enhanced single-strand conformation polymorphism (SSCP) detection of point mutations utilizing mehylmercury hydroxide. *Biotechniques* **15**:397–400.

White, M. B., M. Carvalho, D. Derse, S. J. O'Brien, and M. Dean. 1992. Detecting single base substitutions as heteroduplex polymorphisms. *Genomics* **12**:301–306.

10

ANALYSIS OF PCR PRODUCTS BY COVALENT REVERSE DOT BLOT HYBRIDIZATION

Farid F. Chehab, Jeff Wall, and Shi-ping Cai

Detection strategies for the analysis of polymerase chain reaction (PCR) products have been numerous since PCR found a home in many laboratories. In the research laboratory, radioactivity and conventional methods such as dot blot hybridization, denaturing gradient gel electrophoresis (DGGE), single-stranded conformational polymorphism (SSCP), and DNA sequencing have not yielded to the more thoughtful assays devised for the clinical molecular diagnostics laboratory. In this new environment, PCR strategies have focused on turnaround time and cost, two major considerations that have shaped the techniques utilized in the clinical laboratory.

In genetic and infectious diagnoses as well as in tissue typing, a clinical laboratory's role is to determine by molecular techniques if a specific DNA sequence is present in an individual. Typically in genetic diseases and tissue typing, many potential mutation or polymorphic sites within a specific gene or gene cluster have to be determined. A similar situation exists for some infectious diseases, such as those caused by mycobacteria, where a sample from a single individual must be typed for many etiologic agents. In these clinical situations, a universal method for investigating one sample for a multitude of specific DNA sequences is required.

In our laboratory, we have adopted a method that utilizes the immobilization of a multitude of oligonucleotide probes onto a solid support and subsequent hybridization to biotin-amplified DNA from an individual. The hybrids are then detected via a streptavidin–horseradish peroxidase (HRP) conjugate and chromogenic substrate. This method, known as the reverse dot blot, (RDB) was initially devised by Saiki et al. (1989), modified by Zhang et al. (1991), and implemented by us for cystic fibrosis (Chehab and Wall, 1992), β-thalassemia (Maggio et al., 1993; Cai et al., 1994), and mycobacteria typing (Fiss et al., 1993). We describe in this chapter our general laboratory protocol, which has evolved over the past 4 years to become a robust and clinically reliable technique for detecting subtle known mutations in amplified DNA. This protocol should allow research or clinical laboratories to screen for mutations within a particular gene of interest. Once the technique is established, a change of PCR primers and oligonucleotide probes allows switching from one gene to another or from one type of organism to the next.

Protocol

The protocol described here is currently used to test for cystic fibrosis mutations in our laboratory. A list of PCR primers and probes can be obtained from Wall, Cai, and Chehab (1995).

Reagents

0.1 M HCl
10% [1-Ethyl-3-(3 dimethylaminopropyl)carbodiimiole-HCl (EDC) prepared just before use (Cat. No. 22980, from Pierce Co., Rockford, IL)
0.1 M NaOH
Sodium bicarbonate buffer [0.5 M NaHCO$_3$–Na$_2$CO$_3$ (pH 8.4)]
Sodium citrate buffer [1 M (pH 5.0)]
Working citrate buffer [0.1 M (pH 5.0)]
Tetramethylbenzidine (2 mg/ml in 100% ethanol). Stable at least 6 months at 5°C.
HRP chromogenic substrate. Add 1.0 ml of the stock tetramethylbenzidine solution and 10 μl of 3% H$_2$O$_2$ to 19 ml of 0.1 M Na citrate

(ph 5.0). Mix and protect from light. Must be prepared fresh just before use.

Preparation of Membrane Strips

The membrane (Biodyne C purchased from Mensco, Fremont, CA) used to bind the oligonucleotide probes bears a carboxyl group which when activated by EDC can covalently bind the amino group attached to the 5' end of the probe. For further information on this procedure, refer to the paper by Zhang *et al.* (1991).

1. Use a rubber stamp (10.5 cm × 7.8 cm) to print a square grid pattern on the sheet of membrane. Our stamp is divided into 21 rows. Each strip (7.8 cm × 1.5 cm) holds 3 rows, with 8 squares in each row, and the whole stamp can be cut into 7 individual strips. After printing the grid, cut the membrane around the outside of the grid patterns.
2. Place the membrane in a small plastic box or tray and rinse it several times with 0.1 N HCl. Drain off the HCl and rinse with water. Several membranes may be treated at once. Some of the ink from the stamp may bleed during the HCl washes.
3. Prepare 5 ml of 10% EDC to cover the piece of membrane. If more than 4 membranes are being treated at once, use 10 ml. As many as 12 membranes can be treated together in 20 ml of 10% EDC solution.
4. Place the box containing the membrane and EDC solution on a rocker and rock for 15 min. Be sure that the EDC is being distributed over all the membrane. If many sheets are being treated at once, increase the time to 1 hr.
5. Rinse each membrane briefly in water and allow to dry on an absorbent surface such as filter paper towels. The activated membrane is fairly stable in this condition and can be used to bind 5'-amino oliogonucleotides anytime during the next few days. Long-term stability has not been determined.

Binding of Oligonucleotide Probes onto the Membrane

1. Allele-specific oligonucleotide probes (ASOs) are usually not purified after synthesis. They must be dissolved in water, quantitated, and diluted for use. Dilute the stock solution of 5'

amino-oligonucleotides to 5 picomole/μl with 0.5 M NaHCO$_3$–Na$_2$CO$_3$ buffer (pH 8.4).
2. Using an 8-barrel multipipetter (obtained from Robbins Scientific, Sunnyvale, CA), apply 2 μl of each ASO to the membrane. Apply the normal probes in a row across the top of the strip and the mutant probes across the bottom of the strip. Center the spots in the marked squares and be extremely careful not to miss any spot or dot any probe in the wrong location. If the membrane is dry when the spots are applied, the spots will be faintly visible for several minutes until they also dry. This is helpful in keeping track of where the spots are.
3. Allow the spots to dry for 15 min and then rinse the membrane in 0.1 N NaOH for 1 min.
4. Rinse the membrane several times in water to completely remove the NaOH and place on an absorbent surface to dry. When dry, cut the membrane into the individual strips and, if necessary, store them in an envelope until needed. The strips are stable at least 6 months at room temperature.

PCR Amplification

It is essential that the PCR product carry a biotin tag that can be subsequently used in the detection system. We initially used biotin 16-deoxyuridine triphosphate (dUTP) in a 1 : 3 ratio with deoxythymidine triphosphate (dTTP) during amplification. However, the availability of biotin phosphoramidites to automated DNA synthesis has eased biotin conjugation onto the 5' end of PCR primers. Since then, we have been routinely using biotinylated PCR primers to amplify DNA products that require typing by the reverse dot blot.

PCR reactions are set up in 50 μl with a standard PCR buffer at 2.5 mM MgCl$_2$, 10 pmoles of each primer and 100 μM of deoxyribonucleoside triphosphates (dNTPs). In multiplexing PCR products, the Stoffel buffer with 6 mM MgCl$_2$ and Stoffel *Taq* DNA polymerase gave the best amplification efficiencies of all the PCR products.

Hybridization and Washing

1. Label one strip for each sample (patient and controls), using either pencil or a ballpoint pen.

2. Place each patient's strips in a 15-ml plastic tube with a screw top. Add 12 ml of 2X standard sodium citrate (SSC),0.1% sodium dodecyl sulfate (SDS). Add 20 μl of PCR reaction. Mix by inversion, and place each tube in a boiling water bath for 5–10 min.
3. Transfer the 15-ml screw-top tubes from the boiling water bath to the 42°C water bath. Hybridize for at least 30 min. Longer hybridization times, up to overnight, may be helpful if sensitivity is a problem. We usually hybridize overnight and perform the detection the following day.
4. Equilibrate a 50-ml screw-top tube containing about 40 ml of the washing buffer in a 42°C water bath. The strength of the buffer will vary with the wash stringency requirement of each set of ASOs. For our panel of cystic fibrosis mutations, we use 0.75X SSC, 0.1% SDS as the washing buffer.
5. Transfer the strips from the hybridization tubes to the 50-ml tube containing the washing buffer. All the strips can be put into one tube of washing buffer. Wash the strips for 10 min at 42°C.
6. Pour off the washing buffer, and add 20 ml of 2X SSC, 0.1% SDS, and 5 μl of streptavidin–HRP conjugate. Mix on the rocker at room temperature for 15 min. If there are more than 12 strips, use 40 ml of buffer and 10 μl of streptavidin-HRP and mix for 30 min.
7. Pour off the conjugate solution and wash the strips 3 times for 2 min each with 2X SSC, 0.1% SDS, and twice for 2 min with 0.1 M Na$^+$Citrate (pH 5.0).

Color Development

1. Add 20 ml of the chromogenic substrate of HRP to the tube containing the strips. Double the amount if there are more than 12 strips.
2. Place the tube on the rocker and mix while color develops. This takes about 10 min but will vary. Check the strips every few minutes to monitor the color and stop the reaction once all the spots are visible. Leaving the strips too long in the color development will produce excessive background and will bring out the faint cross-hybridization that occurs with some ASOs.
3. Rinse the strips several times with *distilled* water. The color is fairly stable in air once they have been rinsed well. To make a permanent record of the results, you may either save the dried

strip, photograph the strips on a light box, or simply photocopy them.

Results and Discussion

A critical component of this system during the development phase is the optimization of the temperature and salt conditions during the hybridization and washing steps. This is especially the case when one is attempting to combine a multitude of probes for many mutations, as for example, in cystic fibrosis or β-thalassemia. Our experience in developing these multiprobe reverse dot blot assays was guided by the availability of *mutant* genomic DNA controls for each *mutation*. PCR products from these controls are absolutely required for ensuring the specificity of the signal. As a guide, we often attempted to use a hybridization temperature 5° C below the T_m of a probe; this is calculated according to the formula: 2X (the number of A plus T residues) + 4X (the number of G plus C residues). Although this rule did not always hold, the only flexibility at hand, keeping the temperature and salt conditions constant, was the length of the probe. Thus, it is possible to shorten or lengthen a probe by one or two bases, depending on its specificity of hybridization to the controls. To achieve a specific hybridization pattern, in many instances we had to synthesize 3 to 5 oligonucleotide probes per mutation. In developing our CF mutational panel, we synthesized over 150 probes to cover 31 mutations. This initial investment in time and cost was justified, since it led to an assay which is currently a routine diagnostic test for screening CF mutations.

Once the probes are optimized, it is equally essential to ensure that the multiplexed PCR products are efficiently amplified and hybridized to the probes on the membranes. The faint hybridization signal of some mutations is indicative of a low efficiency during either amplification or hybridization. A low amplification efficiency arises from multiplexing targets over a wide range, for example from 100 to 1000 bp. We circumvented this problem by dividing our amplification reactions into two separate PCRs according to the size of the expected products. Thus for each individual to be typed, one PCR is set up for the long (410–560 bp) and one for the short (161–

410 bp) products, which are then combined for the hybridization step. In order to increase a low hybridization efficiency, which may result from a secondary structure in the DNA sequence of either the probe or the amplified product, one can attempt to amplify a smaller region that does not have a secondary structure. If, however, the mutation is embedded in a G+C-rich region, then a secondary structure in the probe may still result. This situation was encountered in devising probes for the promoter mutations of the β-globin gene in β-thalassemia, and was resolved by boiling the membrane strip containing the oligonucleotide probes for 2 min prior to hybridization. In some instances, we were unable to increase the hybridization efficiencies for some probes; however, these weak signals were consistently faint and were eventually included in the assay.

In the CF panel, three probes produced consistently faint cross-hybridization signals which could not be eliminated. However, these false signals do not represent a major problem of the assay because their true hybridization signal is fairly intense and also because a cross-hybridization pattern is obtained in a negative control PCR and hybridization. For this reason, we always run negative and positive controls in parallel, which ensures that the system is running adequately.

Figure 1 shows reverse dot blot strips for cystic fibrosis mutations routinely obtained from our laboratory. Each individual is represented by two strips. The most common mutations, 1–16, are dotted as normal and mutant probes whereas mutations 17–31 are represented by only the mutant probe. Should an individual show a hybridization signal with any rare mutant probe, zygosity at this site can be subsequently determined with a custom-made strip for this mutation, as shown in the figure by the R553X ministrip. The genotype of individuals is easily scored in this assay by the appearance of normal and mutant hybridization signals at different mutation sites. "Normal" individuals (N/N) do not show any mutation, heterozygotes (N/W1282X and N/R553X) and double heterozygotes (G542X/R334W, ΔF508/621+1 and ΔF508/3905insT) show a normal and

Figure 1 Reverse dot blot strips from individuals screened for 31 cystic fibrosis mutations. Each individual is represented by two strips with three rows of probes dotted onto each strip. The normal (N) probes for mutations 1–16 are displayed on the upper row, the mutant (M) probes for the same mutations are in the second row, and the mutant probes for the rare mutations (RM) 17–31 are on the bottom row. The genotype for each individual is shown next to the strips. See the text for explanations.

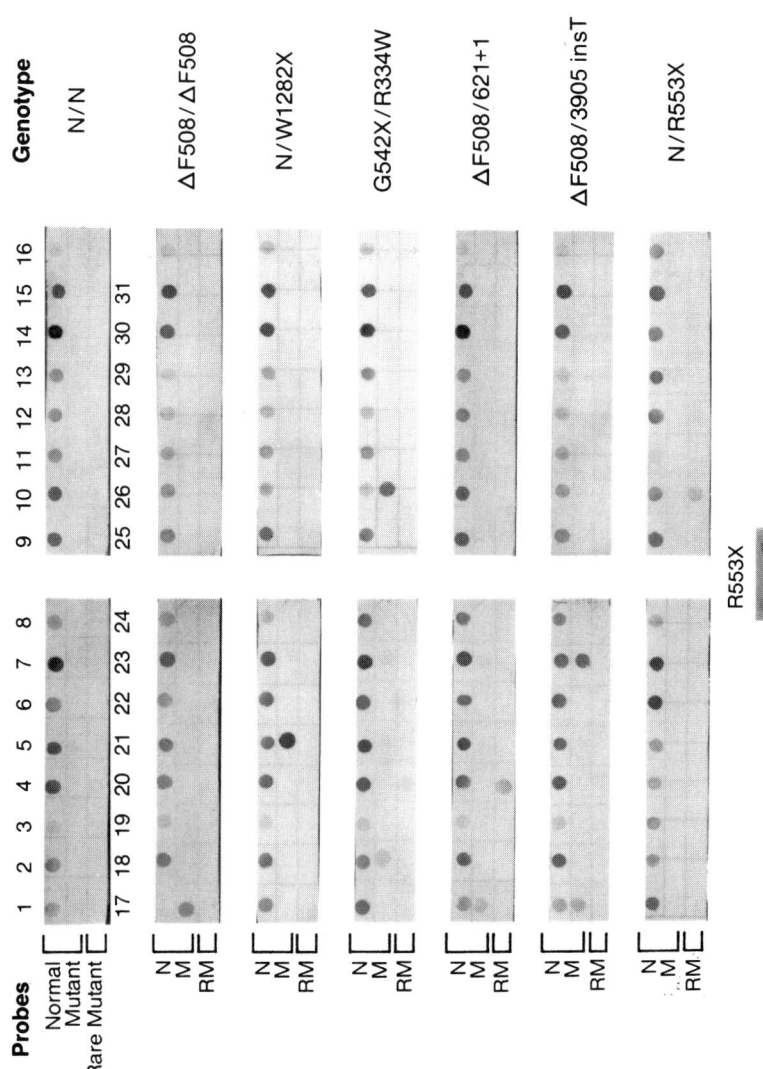

mutant signal at respectively, the same and different mutation sites; whereas homozygotes (ΔF508/ΔF508) hybridize to a single mutant probe and not its corresponding wild-type allele. It is noticeable that the signal intensity of all the probes is not uniform for the reasons explained above; however, a comparison of different strips ensures the reproducibility of this pattern. Mutation 4 (621+1) is an example of how a true signal supersedes its faint cross-hybridization signal.

The initial time invested in developing this system has now clearly paid off, as reverse dot blot hybridization for cystic fibrosis mutations is a routine test among our panel of molecular assays. We performed this assay clinically on approximately 700 samples that came to us for cystic fibrosis testing. Our experience has taught us that the assay is reliable, reproducible, and economical. However, caution needs to be exercised at various levels, for example, in ensuring the correct order of probes on the membranes, in assaying the strips before use, and in preventing PCR contamination. The use of positive and negative controls every time the assay is performed is a prerequisite we have imposed on this test. The positive controls are changed frequently and it is preferable to use as positive control DNA a sample homozygous for a single mutation. Such a control will lack a hybridization signal at the normal probe site and therefore PCR contamination from a previously amplified sample is likely to be normal at this site. It will thus show hybridization at the normal probe site, turning the homozygous control into a heterozygote.

Overall, we feel that this assay has now matured enough that the next generation should include an automated detection instrument that will couple hybridization and detection, possibly in a microtiter plate format. Such an instrument would facilitate screening programs that are in place or that are being contemplated.

Literature Cited

Cai, S. P., J. Wall, Y. W. Kan, and F. F. Chehab, 1994. Reverse dot blot probes for the screening of β-thalassemia mutations in Asians and American Blacks. *Hum. Mutat.* **3**:59–63.

Chehab, F. F., and J. Wall, 1992. Detection of multiple cystic fibrosis mutations by reverse dot blot hybridization: A technology for carrier screening. *Hum. Genet.* **89**:163–168.

Fiss E. H., F. F. Chehab, and G. F. Brooks, 1992. DNA amplificaiton and reverse dot blot hybridization for detection and identification of myocobacteria to the species level in the clinical laboratory. *J. Clin. Microbiol.* **30**:1220–1224.

Maggio, A., A. Giambona, S. P. Cai, J. Wall, Y. W. Kan, and F. F. Chehab, 1993. Rapid and simultaneous typing of Hb S, Hb C and seven Mediterranean β-thalassemia mutations by covalent reverse dot blot hybridization analysis: Application to prenatal diagnosis in Sicily. *Blood* **81:**239–242.

Saiki, R. K., P. S. Walsh, C. H. Levenson, and H. A. Erlich, 1989. Genetic analysis of amplified DNA with immobilized sequence-specific oligonucleotide probes. *Proc. Natl. Acad. Sci. U. S. A.* **86:**6230–6234.

Wall, J., S. P. Cai, and F. F. Chehab, (1995). A 31-mutation assay for cystic fibrosis testing in the clinical molecular diagnosis laboratory. *Hum. Mut.* (in press).

Zhang, Y., M. Y. Coyne, S. G. Will, C. H. Levenson, and E. S. Kawasaki, 1991. Single-base mutational analysis of cancer and genetic diseases using membrane bound modified oligonucleotides. *Nucleic Acids Res.* **19**(14):3929–3933.

11

HIGH-PERFORMANCE LIQUID CHROMATOGRAPHY ANALYSIS OF PCR PRODUCTS

John M. Wages, Xumei Zhao, and Elena D. Katz

Numerous techniques have been described for the analysis of PCR-amplified DNA. In contrast to methods involving probe hybridization or DNA sequencing, direct detection methods visualize the amplified DNA without any intervening hybridization or other analysis. In the methods which have been described to date, such as high-performance liquid chromatography (HPLC) (Katz and Dong, 1990), gel electrophoresis (Piatak et al., 1993), and capillary electrophoresis (Sunzeri et al., 1991), size fractionation is employed, but no sequence-specific identification is made.

Direct assays are potentially more quantitative than indirect assays, with many intervening steps, each of which contributes a characteristic fraction of the assay's total variation. The problem of misidentification of nonspecific PCR products has been significantly alleviated by the development of new techniques for performing hot start PCR (Chou et al., 1992), which can render synthesis so specific that in many cases no nonspecific products at all are detectable on gels or by HPLC. For such specific reactions, direct assays for amplified DNA may have advantages, both in terms of simplicity and of assay precision. High-performance liquid chromatography, a proven tool in the analytical laboratory, can be used to good advantage in analyzing PCR products.

HPLC Separation of Double-Stranded DNA

While most separations of nucleic acids are performed by gel electrophoresis, development of HPLC column packings and biocompatible instrumentation has seen a parallel development of methods for size fractionation of double-stranded DNA (dsDNA). Recently, nonporous packings with anion-exchange (Kato et al., 1989) and reversed-phase (Huber et al., 1993) functionalities have been employed for rapid separation of dsDNA. These techniques have also been used directly for rapid analysis of PCR products (Katz and Dong, 1990; Warren et al., 1991).

Comparison of Gel Electrophoresis and HPLC

Analysis of PCR products by HPLC has several advantages over other quantitative techniques. They include the following:

1. HPLC separation is rapid (10 min or less per product), compared with gel electrophoresis. A minimum of several hours is required to pour, load, run, stain, and document a gel. The advantage of speed is most significant when only a small number of products need to be quantitated. For large numbers of products, gel electrophoresis may still be faster, owing to the capacity for running multiple products per gel.
2. The precision which makes HPLC one of the principal analysis techniques in the environmental or analytical chemistry laboratory is also an advantage in nucleic acid analysis. Quantitation within about ± 4% is possible (Katz et al., 1990a,b; Davey et al., 1991), which would be difficult to attain reproducibly by gel-based methods.
3. On the scale typically required for DNA cloning and sequencing experiments, an analytical HPLC can easily be transformed into a preparative HPLC by using a fraction collector. The collected PCR product typically requires only concentration and desalting to be suitable for further manipulation. Preparative gel electrophoresis necessitates extraction of the DNA from the gel matrix, for example, by passive diffusion or electroelution. The preparative advantage is also useful in preparing large

quantities of amplified DNA for use as calibration standards (e.g., for hybridization-based assays).
4. HPLC separation avoids the use of the mutagen ethidium bromide, which is employed for staining in most gel electrophoresis applications. UV detection is nondestructive and does not involve the use of any radioisotopes or toxic chromogenic enzyme substrates; furthermore, it provides DNA concentration measurements in absolute units.
5. HPLC analysis also has the potential for multiplex PCR analysis. In the same way that gel electrophoresis can be used to resolve multiple amplicons from a single reaction product, HPLC can be used to accomplish the same separation, but with greater precision and higher sensitivity. Applications to internal standard-based quantitative PCR will be discussed later in this chapter.

Utility of HPLC in the PCR Laboratory

Quantitation of PCR products by HPLC can be highly reproducible. While this represents a distinct advantage for quantitation of specific sequences by PCR, it is not the only application of HPLC in the PCR laboratory. HPLC is also eminently suited for rapid PCR protocol optimization, and for preparative applications.

The need to optimize magnesium ion concentration [Mg^{2+}] for maximum PCR efficiency is well known (Williams, 1989). However, amplification reactions can also benefit from optimization of other reaction variables, including concentrations of deoxynucleoside triphosphate (dNTP), *Taq* DNA polymerase, and primer, in addition to thermal cycling parameters. Since most PCR assays involve some form of sample preparation prior to the PCR amplification itself, optimization of the nucleic acid extraction step may also be required. With many novel product analysis techniques from which to choose, it is advantageous to optimize these by reference to a proven high-precision analytical technique. The rapidity and precision of HPLC analysis can provide rapid turnaround in experimental results, yielding data with a high information content. Peak areas provide a more objective measure of the effect of changing PCR variables on PCR yield than the subjective evaluation of band intensities.

HPLC Procedures

System Configuration

All HPLC data presented in this chapter were collected with a Perkin–Elmer (Norwalk, CT) HPLC system equipped with a 410 Bio-LC biocompatible pump, ISS-200 autosampler, LC-95 UV/Vis detector, Model 1020 integrator, diethylaminoethyl nonporous (DEAE–NPR) anion-exchange column, a DEAE–NPR guard column, and 0.5 μm in-line column filter (Rheodyne-Cotati, CA). A column oven to maintain constant operating temperature helps ensure constancy of retention times, which can be affected by ambient temperature fluctuations. Buffers A [25 mM Tris–HCL (pH 9.0), 1.0 M NaCl, 1% acetonitrile] and B [25 mM Tris–HCl (pH 9.0), 1% acetonitrile] were obtained as premade solutions from Brand-Nu Labs (Meriden, CT). A mixture of 30% acetonitrile in water was used to flush the autosampler needle and valve between injections, although high salt buffers (buffer A or 1 M NaClO$_4$) can also be used. Following each use, the column was purged with water and stored in 30% acetonitrile in water. In order to separate PCR products by the method given here, a pump capable of binary gradient operation (and preferably of biocompatible design, for resistance to the 1 M NaCl in buffer A) is required.

Analysis of PCR Products

A rapid separation method was employed for relatively noncomplex product mixtures. A description of the procedure of quantitating PCR products by HPLC follows.

1. PCR is performed according to standard procedures. Preferably, hot start is employed to facilitate more specific synthesis and more rapid separation times.
2. For amplicons in the range of 100–300 base pairs, 20–30 μl of each PCR product are injected onto the DEAE–NPR column. For longer products, as little as 5–10 μl of product may be sufficient for quantitation. An automatic sample injector should be used for greater precision.

3. A rapid gradient, requiring approximately 10 min per run, is used to resolve specific and nonspecific PCR products. A suitable protocol is given below:
   a. Equilibrate column 5 min at 46% buffer A.
   b. Ramp linearly to 54% buffer A over 0.1 min following injection.
   c. Ramp linearly to 60% buffer A over 3.9 min.
   d. Ramp linearly to 75% buffer A over 1 min.
   e. Return to 46% buffer A over 0.1 min.

Buffer A 25 mM Tris–Cl (pH 9.0), 1.0 M NaCl
          1% acetonitrile
Buffer B 25 mM Tris–Cl (pH 9.0)
          1% acetonitrile

Higher resolution methods may also be defined, which can enable separation of longer amplicons. Resolution of a 4.1-kb PCR product has been reported (Katz et al., 1990). In order to increase the resolution of an HPLC method for PCR product analysis on DEAE–NPR columns, the rate of change in the NaCl concentration is reduced. Slight changes in rate may produce large changes in resolution.

4. Products are detected by UV absorbance at 260 nm.

An example of the separation which is typically achievable with the method presented here is provided in Fig 1. Human herpesvirus type 6 strain Z29 (HHV-6^{Z29}) DNA sequences were amplified in the presence of genomic DNA from approximately 150,000 cells per reaction. Amplification was performed in reactions containing 10 mM Tris–HCl (pH 8.3), 1.25 mM MgCl$_2$, 0.2 μM each of primers (Pruksananonda et al., 1992), 200 μM each of deoxyadenosine triphosphate (dATP), deoxycytidine triphosphate (dCTP), deoxyguanosine triphosphate (dGTP), thymidine triphosphate (TTP), and 2.5 U AmpliTaq DNA polymerase (Perkin–Elmer). Cycling was in the GeneAmp PCR System 9600 (Perkin–Elmer) as follows: 94°C for 5 min; then, 94°/56°/72°, 15 sec/60 sec/60 sec, for 10 cycles; 92°/56°/72°, 15 sec/60 sec/60 sec, for 30 cycles; hold at 72°C.

As shown by Fig. 1, adequate resolution of this 384-bp product from minor nonspecific products, primers, and dNTPs, is obtained within 9 min. Separation times could probably be further shortened when a hot start is employed to produce highly specific reaction products.

5. Specific peaks are identified based on characteristic retention times compared with molecular weight standards, PCR product from high copy-number reactions, or previously amplified prod-

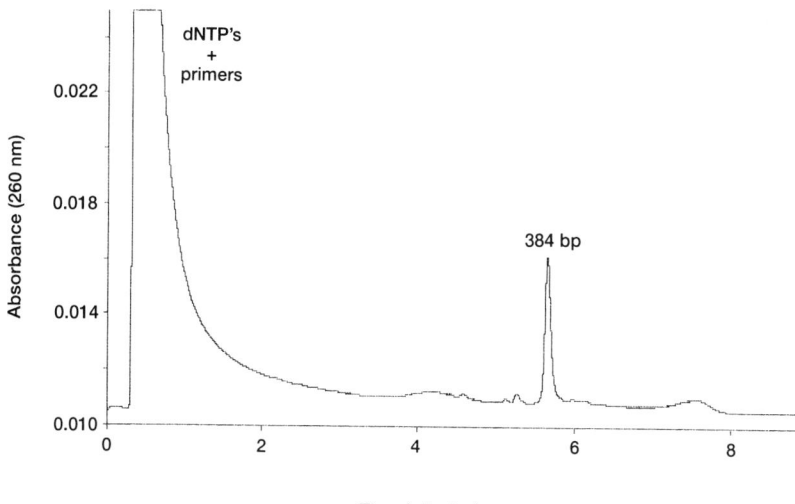

Figure 1 HPLC separation of HHV-6 PCR products. HHV-6^{z29} DNA sequences were amplified for 40 cycles as described in the text in the presence of genomic DNA from approximately 150,000 cells per reaction. Following amplification, a total of 30 µl of product were injected onto a DEAE–NPR column for resolution of specific product from genomic DNA, dNTPs, and PCR primers.

uct which has been identified by DNA sequencing or probe hybridization analysis. The injection of 5 µl of 250 µg/ml of the HaeIII digest of pBR322 provides molecular size markers covering the range of 100 to approximately 500 base pairs. It should be noted, however, that the elution of double-stranded DNA from DEAE–NPR does not strictly follow molecular size (Fack and Sarantoglou, 1991). A recently described technique is reported to remove the elution dependence on AT content of anion exchange on the DEAE matrix (Bloch, 1993).

6. To reflect the empirical nature of postexponential phase quantitation by PCR, data should generally be plotted as peak area vs log initial copy number. Linear regression analysis then permits copy number to be estimated in unknown samples. Using the techniques described here, a usable linear range is typically obtained over 30 to 30,000 initial copies of template.

A typical standard curve is shown in Fig. 2. Briefly, total DNA extracted from the ACH-2 cell line (Clouse et al., 1989) was diluted in total DNA extracted from non-HIV-infected cells, for a total of approximately 300,000 cells/reaction. A high concentration of uracil-

N-glycosylase (UNG) was employed to facilitate hot start PCR (Longo et al., 1990). Amplification reactions contained 10 mM Tris–HCl (pH 8.3), 50 mM KCl, 2.0 mM MgCl$_2$, 1.0 μM each primer SK 145 and SK 431, 2.0 U UNG, 200 μM each of dATP, dCTP, and dGTP, 300 μM dUTP, and 2.5 U AmpliTaq DNA polymerase. Thermal cycling was performed (GeneAmp PCR System 9600, Perkin–Elmer) using the following method: at 50°C for 2 min; 94°C for 5 min; then, 94°C for 15 sec, 60°C for 60 sec, 72°C for 60 sec, for 10 cycles; and then, 92°C for 15 sec, 60°C for 60 sec, 72°C for 60 sec, for 30 cycles; hold at 72°C.

Calibration of the HPLC System

If calibrated against an appropriate standard, the HPLC can be used to determine molar concentration of the PCR product. The molar

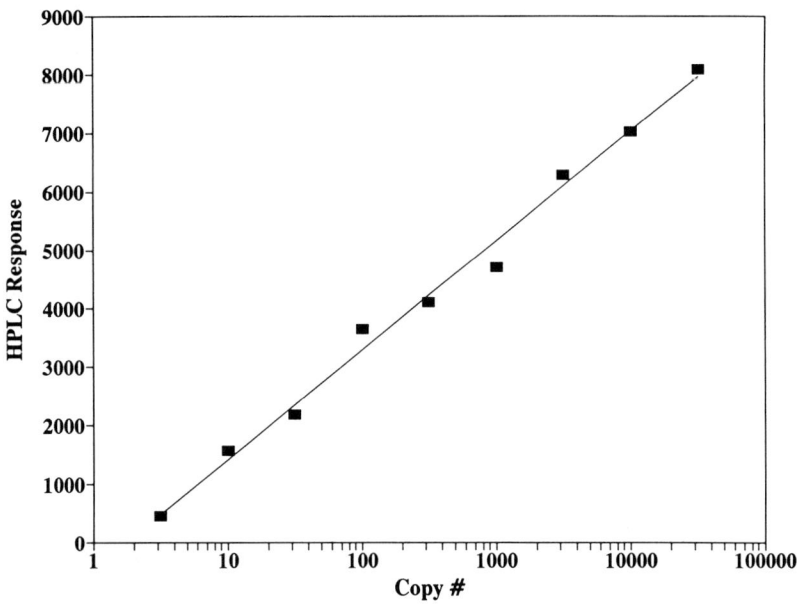

Figure 2 Quantitation of HIV-1 DNA by PCR followed by HPLC. Total DNA extracted from ACH-2 cells was diluted and amplified in the presence of total DNA from 300,000 cells/reaction. A high concentration of UNG was employed to facilitate hot start PCR, as described. PCR products were quantified by injection (30 μl) on a DEAE–NPR NPLC column.

concentration is useful when designing hybridization-based assays in which the probe concentration must be adjusted to be in large excess over the molar concentration of the double-stranded PCR product. In order to determine the analytical sensitivity of a PCR product quantitation technique, it is necessary to have an accurately quantitated standard PCR product. When the HPLC response is calibrated by injection of DNA molecular weight standards, calibration factors are derived which permit rapid estimation of concentration from the peak area. A procedure for calibration is as follows:

1. Purchase or prepare (digest to completion, verify digestion by gel electrophoresis, measure o.d. to determine total DNA concentration) an appropriate restriction digest containing fragments of length comparable to the PCR products to be quantitated. With 100- to 500-bp products, the *Hae*III digest of pBR322 can be used.
2. Inject the standard multiple times (e.g., three times) at each of several injection volumes.
3. Choose a peak comparable in size to the PCR product of interest (e.g., the 123-bp peak). This should be a DNA fragment which is fully resolved from the other fragments. Calculate a mean peak area per unit volume injected for this DNA fragment.
4. The DNA concentration in a PCR product can now be calculated, assuming the length of the amplicon is known and assuming equivalent extinction coefficients (i.e., approximately equal G + C content) between the amplicon and standard.

The HPLC system was calibrated with the *Hae*III digest of pBR322 (Sigma Chemical Co., St. Louis, MO; 288 μg/ml, concentration provided by the manufacturer), by injection of 10 μl a total of three times. An average peak area for the 123-bp fragment was measured as 1065.7 ± 11.1 standard deviation (sd). For confirmation, 20 μl and 40 μl were also injected, giving peak areas of 2083.2 ± 87.2 (sd) and 4233.8 ± 101.1 (sd), respectively. The value from 10 μl was used to derive a formula for calculating PCR product concentration. The concentration of each fragment in this digest is 94.3 fmoles/μl: (288 μg/ml) × (moles bp/700 g) × (moles DNA/4363 moles bp) = 94.3 fmoles/μl.

In order to convert peak areas into femtomoles of product, the following equation may be used:
(A) (10 μl/V) (123 bp/L) (943 fmoles/1065.7) = fmoles injected

where A = peak area obtained from injection of unknown DNA fragment
V = volume injected (μl)
L = length in base pairs of PCR product injected

It should be recognized that this method allows an estimate of concentration which may be adequate for most, but not all, applications. When standard PCR product concentration must be more accurately determined, HPLC provides a method of rapid purification, following which a bulk o.d. 260-nm measurement can be made.

HPLC Analytical Performance

HPLC Precision

The reproducibility of determination of PCR products by HPLC has been estimated as approximately ± 4% coefficient of variation (c.v.) (Katz et al., 1990a,b; Davey et al., 1991). Typically, less than ± 5% variability among replicate injections is observed. When HIV-1 gag PCR products (142 bp, SK 145/150 primers) from 1000 initial copies of DNA were pooled and injected a total of five times, approximately ± 2% variability was observed (1972 ± 41 relative peak area units, $N = 5$ injections).

The ability to precisely measure the concentration of PCR products makes it possible to assess the variability of a PCR amplification protocol itself, while minimizing the effect of variation at the post-PCR analysis step. To estimate interreaction variation in PCR product yield, a master mix was prepared with DNA from approximately 100 ACH-2 cells per reaction. Individual reactions were aliquotted from this single master mix to 20 GeneAmp PCR reaction tubes, which were overlaid with 70 μl of mineral oil and cycled in a DNA thermal cycler (Perkin–Elmer) as follows: hold at 94°C for 90 sec; then, 94°C for 75 sec, 55°C for 60 sec, 72°C for 60 sec, for 40 cycles; hold at 72°C. Products were analyzed by HPLC. Variation in peak area among the 20 replicate reactions was estimated to be within 8% (c.v.). This result indicates that the PCR is capable of highly precise quantitation and suggests that this result is frequently not realized because of other factors; these might include pipetting and other subtle errors which accumulate during the process of performing the relatively complex quantitative PCR assay. Access to a highly precise method for analysis of PCR products allows one to probe the assay for sources of variation, while having confidence in the precision of post-PCR analysis.

HPLC Linearity

An HPLC response is typically linear over a range of 2.4 to 240 ng PCR product for a 500-bp product (Katz et al., 1990a,b). Sensitivity is approximately the same or slightly higher than ethidium-bromide stained gels (Katz et al., 1990a,b), or about 300 pg DNA per peak (Katz et. al., 1990a,b). This level of sensitivity cannot approach currently available radioisotopic or chemiluminescence- and electrochemiluminescence-based detection methods (Wages et al., 1993; Van Houten et al., 1993). Trends in fluorescent and chemiluminescent HPLC detection may ultimately extend sensitivity considerably, and may make hybridization analyses by HPLC possible. To date, however, no such techniques have been described.

Internal-Standard Quantitative PCR Analysis by HPLC

In order to correct for inhibition or other effects which may be present in biological samples, quantitative PCR using internal standards as reference templates has been developed (Wang et al., 1989). Quantitative competitive PCR is a subset of internal standard methods which make use of the apparent competition observed under certain conditions (Piatak et al., 1993; Gilliland et al., 1990). Assuming the two templates are amplified with comparable efficiencies, the ratio of internal standard and wild-type sequence should remain the same following amplification as it was prior to PCR. Plotting this ratio as a function of copies of internal standard added to a series of reactions, all containing the same amount of specimen DNA, allows the initial copy number (Piatak et al., 1993) to be estimated.

Most internal-standard PCR assays have made use of gel electrophoresis (Piatak et al., 1993) or probe hybridization to separate and quantitate wild-type and internal-standard sequence in the PCR product (Jalava et al., 1993). Rapid separation and quantitation by

HPLC provides another method, with the advantage of high-precision analysis. Quantitation of differential PCR products by HPLC has been described recently (Zeilinger et al., 1993). Figure 3 shows the separation of a 136-bp HIV-1 PCR product from a 172-bp internal-standard fragment generated by PCR (Seibert and Larrick, 1993). Briefly, RNA PCR was performed by adding a constant amount (10,000 copies) of HIV-1 gag RNA transcript directly into the RT reaction mix (10 μl reaction mix per RT reaction). A dilution series of internal-standard transcript was then added (10 μl) to each tube containing RT reaction mix and gag RNA. Reverse transcription was allowed to proceed at 42°C for 15 min; then the reactions were heated to 99°C for 5 min, and quick cooled to 4°C. PCR was performed for 40 cycles in the GeneAmp PCR System 9600 (Perkin–Elmer) as follows: 94°C for 5 min; then, 94°/56°/72°, 15 sec/60 sec/60 sec, for 10 cycles; 92°/56°/72°, 15 sec/60 sec/60 sec, for 30 cycles; hold at 72°C. PCR products were analyzed by HPLC, and peak areas were converted to femtomoles of product by the calibration factors described earlier in this report.

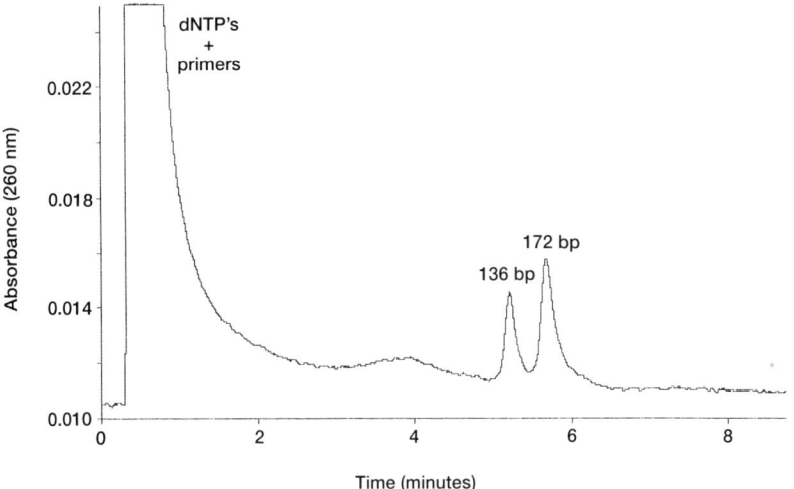

Figure 3 Resolution of wild-type and internal standard products by HPLC. Wild-type (136-base pair) HIV-1 and an internal standard (172 base pairs) were coreverse transcribed by MMLV-reverse transcriptase, then coamplified by PCR for 40 cycles. A total of 30 μl of product were injected onto the DEAE–NPR column for resolution and quantitation of the two products.

Potential Problems with HPLC

The use of column in-line and solvent filters to prevent particulates from clogging column inlet frits and the column matrix itself is good HPLC practice. In particular, when relatively crude samples are amplified and assayed by HPLC, the following steps may extend column lifetime: (1) use of a guard column of the same functionality as the analytical column, (2) insertion of a 0.5-μm in-line filter before the guard column, and (3) frequent cleaning by injections of strong solvent (i.e., the high-salt buffer), mixtures of 30% acetonitrile or 0.2 M NaOH in water, all of which help prevent genomic DNA buildup on the analytical column. In addition, the volume of PCR product injected should be kept to the minimum necessary for accurate and precise measurement. When these guidelines are followed, at least 1000 30-μl injections of PCR products, even those containing appreciable amounts of genomic DNA prepared by a crude cell lysate method (Higuchi, 1989), should be possible on one column.

Since slight changes in the salt concentration significantly influence retention time, it is important to take particular care in preparing the high-salt buffer used in this method. Crystalline NaCl should be dried in a vacuum oven and carefully weighed to ensure proper ionic strength in the buffer.

Conclusion

HPLC represents an advantageous technique for PCR product analysis. The rapidity of separation, the precision of quantitation, and the widespread availability of HPLC systems make it an attractive option for some PCR applications, and a generally useful tool in the PCR laboratory. Specific applications include optimization of PCR protocols, quantitation of PCR products as an end point in quantitative PCR assays, resolution and assay of internal standard product from specimen product in quantitative competitive PCR, the preparation of calibration standards of amplified product for use in developing new assays, and the purification of PCR products for DNA sequencing.

Literature Cited

Bloch, W. 1993. Anion-exchange separation of nucleic acids. Europ. Patent Appl. 92302930.0.

Chou, Q., M. Russell, D. E. Birch, J. Raymond, and W. Bloch. 1992. Prevention of pre-PCR mis-priming and primer dimerization improves low-copy-number amplifications. *Nucleic Acids Res.* **20**(7):1717–1723.

Clouse, K. A., D. Powell, I. Washington, G. Poli, K. Strebel, W. Farrar, P. Barstad, J. Kovacs, A. S. Fauci, and T. M. Folks. 1989. Monokine regulation of human immunodeficiency virus-1 expression in a chronically infected human T cell clone. *J. Immunol.* **142**:431–438.

Davey, M. P., M. Meyer, D. Munkirs, D. Badcock, M. P., Braun, J. B. Hayden, and A. C. Bakke. 1991. T-Cell receptor variable beta-genes show differential expression in CD4 and CD8 T cells. *Hum. Immunol.* **32**:194–202.

Fack, F., and V. Sarantoglou. 1991. Curved DNA fragments display retarded elution upon anion exchange HPLC. *Nucleic Acids Res.* **19**:4181–4188.

Gilliland, G., S. Perrin, and H. F. Bunn. 1990. Competetive PCR for quantitation of mRNA. In: *PCR protocols: A guide to methods and applications* (eds: M. A. Innis, D. H. Gefland, J. J. Sninsky, and T. J. White), pp. 60–69. Academic Press, San Diego, CA.

Higuchi, R. 1989. Simple and rapid preparation of sample for PCR. In: *PCR technology: Principles and applications for DNA amplification* (ed. H. A. Erlich), pp. 31–38. Stockton Press, New York.

Huber, C. G., P. J. Oefner, E. Preuss, and G. K Bonn. 1993. High-resolution liquid chromatography of DNA fragments on nonporous poly(styrene-divinylbenzene) particles. *Nucleic Acids Res.* **21**:1061–1066.

Jalava, T., P. Lehtovaara, A. Kallio, M. Ranki, and H. Soderlund. 1993. Quantitation of hepatitis B virus DNA by competitive amplification and hybridization on microplates. *BioTechniques* **15**, 134–139.

Kato, Y., Y. Yamasaki, A. Omaka, T. Kitamura, and T. Hashimoto. 1989. Separation of DNA restriction fragments by high-performance ion-exchange chromatography on a nonporous ion exchanger. *J. Chromatogr.* **478**:264–268.

Katz, E. D., and M. W. Dong. 1990. Rapid analysis and purification of polymerase chain reaction products by high-performance liquid chromatography. *BioTechniques* **8**(5):546–555.

Katz, E. D., W. Bloch, and J. Wages. 1990a. HPLC quantitation and identification of DNA amplified by the polymerase chain reaction (poster presented at V San Diego Conference, "Nucleic Acids: New Frontiers." Nov. 1990.).

Katz, E. D., L. A. Haff, and R. Eksteen. 1990b. Rapid separation, quantitation and purification of products of polymerase chain reaction by liquid chromatography. *J. Chromatogr.* **512**:433–444.

Longo, M., M. C. Beringer, and J. L. Hartley. 1990. Use of uracil DNA glycosylase to control carry-over contamination in polymerase chain reactions. *Gene* **93**:125–128.

Piatak, M., K.-C. Luk, B. Williams, and J. D. Lifson. 1993. Quantitative competitive polymerase chain reaction for accurate quantitation of HIV DNA and RNA species. *BioTechniques* **14**:70–81.

Pruksananonda, P. C. B. Hall, R. A. Insel, K. McIntyre, P. E. Pellett, C. E. Long, K. C. Schnabel, P. H. Pincus, F. R. Stamey, T. Dambaugh, and J. A. Stewart.

1992. Primary human herpesvirus 6 infection in young children. *N. Engl. J. Med.* **326**(22):1445–1450.

Siebert, P. D., and J. W. Larrick. 1993. Competitive DNA fragments for use as internal standards in quantitative PCR. *BioTechniques* **14**:244–249.

Sunzeri, F. J., T. H. Lee, R. G. Brownlee, and M. P. Busch. 1991. Rapid simultaneous detection of multiple retroviral DNA sequences using the polymerase chain reaction and capillary DNA chromatography. *Blood* **77**(4):879–886.

Van Houten, B. D. Chandrasekhar, W. Huang, and E. D. Katz. 1993. Mapping DNA lesions at the gene level using quantitative PCR methodology. *Amplifications* **10**:10–17.

Wages, J. M., Jr., L. Dolenga, and A. K. Fowler. 1993. Electrochemiluminescent detection and quantitation of PCR-amplified DNA. *Amplifications* **10**:1–6.

Wang, A. M., M. V. Doyle, and D. F. Mark. 1989. Quantitation of mRNA by the polymerase chain reaction. *Proc. Natl. Acad. Sci. U.S.A.* **86**:9717–9721.

Warren, W., T. Wheat, and P. Knudsen. 1991. Rapid analysis and quantitation of PCR products by high-performance liquid chromatography. *BioTechniques* **11**:250–255.

Williams, J. F. (1989). Optimization strategies for the polymerase chain reaction. *BioTechniques* **7**(7):762–769.

Zeilinger, M., C. Schneeberger, P. Speizer, and F. Kury. 1993. Rapid quantitative analysis of differential PCR products by high-performance liquid chromatography. *BioTechniques* **15**:89–95.

12

HETERODUPLEX MOBILITY ASSAYS FOR PHYLOGENETIC ANALYSIS

Eric L. Delwart, Eugene G. Shpaer, and James I. Mullins

The detection of genetic strain differences in bacterial and viral pathogens such as human immunodeficiency virus (HIV) has been revolutionized in recent years by the polymerase chain reaction (PCR), which permits large quantities of subgenomic fragments of pathogen DNA or RNA (following reverse transcription) to be generated in hours. Sequence variation among strains is increasingly being recognized and sometimes correlated with particular virulence, phenotypes, or antigenic traits. Variation among amplified fragments can be determined by DNA sequencing of derived clones or the PCR product itself. However, simpler, alternative methods to DNA sequencing have been developed that provide limited genetic information that is useful for a variety of applications. Applied to the study of HIV, they include PCR- (Simmonds et al., 1990) and restriction fragment-length polymorphism (PCR-FLP and RFLP, respectively) analysis, RNase A mismatch cleavage (Lopez-Galindez et al., 1991) and anchored primer PCR typing (McCutchan et al., 1991), primer mismatch-sensitive PCR (Larder et al., 1991), single-stranded conformational polymorphism analysis (SSCP) (Fujita et al., 1992), and denaturing gradient gel electrophoresis (DGGE) (Andersson et al., 1993). Because of their relative simplicity and speed, these methods can be applied to a larger number of samples than is currently feasible by DNA sequencing. Here we describe the use of a new method for genetic screening, referred to as heteroduplex mobility analysis or HMA (Delwart *et al.*, 1993), which not only can distinguish between individual strains but can also provide reliable inferences of phylogenetic relationships among strains of microorganisms. So far it has been used only to study HIV and related retroviruses, but the method can also be used to study other infectious agents and cellular genes.

Heteroduplex Mobility Analysis

HMA is based on the observation that the structural deformations in double-stranded DNA that result from mismatches and nucleotide insertions or deletions cause a reduction in the electrophoretic mobility of these fragments in polyacrylamide gels, but not in standard agarose gels. Heteroduplexes formed from molecules with mismatched nucleotides but without gaps can be detected using nondenaturing conditions of electrophoresis when the degree of divergence exceeds 1–2%, and generally increases with the degree of mismatch (Delwart *et al.*, 1993). These mismatches are thought to result in "bubbles" in the heteroduplex which retard the mobility of the fragment through the polyacrylamide matrix (Fig. 1). Heteroduplexes up to at least 1–3 kb in length containing internal gaps as small as 1–3 bases, respectively, all display mobility shifts, even in a genetic background of <1–2% mismatches. Mobility shifts due to insertions or deletions generally increase with the size of the gap. The greater effect of gaps relative to base substitutions on heteroduplex mobility is thought to be due to the "bend" of the DNA double helix required to accommodate the extra bases (Bhattacharyya and Lilley, 1989; Wang and Griffith, 1991) (Fig. 1). Most often, each heteroduplex that is formed migrates with a distinct mobility, indicating that the strand-specific composition of mismatched and unpaired nucleotides affects their mobility.

Heteroduplexes are formed during late stages of PCR reactions when high levels of products, such as in nested PCR (N-PCR) are generated, and other reactants such as primers and functional polymerase, are depleted. They are also formed by the deliberate mixing of separately amplified reactions. In the simplest application of HMA,

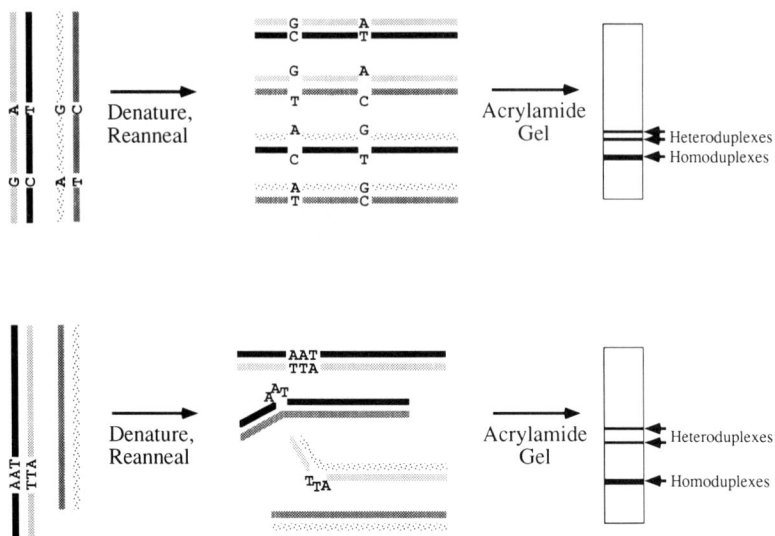

Figure 1 Schematic representation of heteroduplex mobility assay. For simplicity, the positions of only two point mutations (top) or one three-base gap (bottom) are shown. To be detected in ethidium bromide-stained gels, as depicted on the right, mismatches need to amount to at least 1–2% of the base pairs in the heteroduplex, whereas three nucleotide gaps can be detected in molecules at least as large as 3 kb.

heteroduplexes are formed by denaturing and reannealing mixtures of PCR-amplified DNA fragments from divergent but related genes. When these products are separated on polyacrylamide gels, nearly comigrating homoduplex bands plus two additional slower migrating heteroduplex bands are observed. The mobilities of the heteroduplexes were found to be generally proportional to the degree of genetic distance between the two strands (Delwart et al., 1993).

Heteroduplexes also form spontaneously when a complex mixture of sequences is PCR amplified from an individual infected with a rapidly evolving microorganism, such as HIV. In such cases, PCR products reanneal to form heteroduplexes at later heating and cooling cycles of the amplification process if product levels are sufficiently high. The overall degree of complexity of the genetic variant pool can thus be assessed by the display of heteroduplex bands following electrophoresis through an acrylamide matrix (Delwart et al., 1993).

Procedure Outline

1. Procedures for isolating DNA from clinical specimens that result in DNA of sufficient purity for PCR are all that is required for HMA. Our laboratory uses IsoQuick kits (MicroProbe Corp., Garden Grove, CA), which typically result in clean DNA product. However, it takes approximately 3–4 hr to prepare DNA from a group of twelve specimens and faster procedures, such as those producing cell lysates, are suitable.

2. Because of the low levels of HIV present in typical clinical specimens, the PCR preparation for HMA typically requires a nested series of reactions (Mullis and Faloona, 1987), with 1–5% of the first-round reaction products diluted into a second round of PCR with internally annealing primers.

    Note: The gene segment to be amplified should contain constant as well as variable regions. Large gaps in variable regions can lead to very large mobility shifts and diminished ability to quantitate relationships among genes. In an analysis of HIV surface envelope gene variability, a more predictable relationship between heteroduplex mobility and genetic distance was obtained throughout the range of sequence divergence examined when 1.3-kb fragments, rather than 0.7 kb fragments (Delwart et al., 1993) were used.

    Notes: The distribution of mismatches also affects heteroduplex mobility, with larger shifts resulting from distributed rather than clustered mismatches (M. Bachmann and J. I. Mullins, unpublished observations).

    Because of the complex mixture of variants often present in microorganism template preparations, and the efficiency of product formation afforded by nested PCR, heteroduplexes are often formed during the latter rounds of PCR amplification. These heteroduplexes can be resolved into homoduplexes by transferring 10 μl of the second-round PCR reaction to 90 μl of fresh standard PCR reaction mix, followed by a single round of denaturation, primer annealing, and extension.

    When evaluating variant pool complexity, estimates of template abundance in the specimen are essential to ensure that multiple templates are being coamplified. Several methods for quantitating the template have been described (Piatak et al., 1993; Mulder et al., 1994; Delwart et al., 1993). Ignorance of the template load in a specimen could result in the erroneous conclusion that little or no genetic diversity exists within the microorganism population when actually only one or a few templates from a larger mixture were amplified and evaluated.

3. To form heteroduplexes, 4.5 µl from two second-round PCR reactions are combined and 1 µl of 10× annealing buffer is added [1 M NaCl, 100 mM Tris–HCl (pH 7.8) 20 mM EDTA]. The DNA mixture is then denatured at 94°C for 2 min and annealed by rapid cooling in a thermal cycler or on ice. Samples are stored on ice until they are loaded on gels.

Note: The efficiency of heteroduplex formation, compared with homoduplexes, is proportional to the genetic relationship between the strands. As a consequence, rapid cooling (on ice) is essential to form appreciable amounts of heteroduplexes between distantly related sequences (>15% divergence), whereas the slower cooling that occurs within the thermal cycler is used for more closely related sequences.

4. Fragment mixtures are separated in 5% polyacrylamide gels (30 : 0.8 acrylamide : bis) in TBE buffer (0.088 M Tris–borate, 0.089 M boric acid, 0.002 M EDTA). A variety of conditions have been employed, including 250 V for 3 hr (for 0.7-kb fragments), and 70 mA for 1100 Volt-hr or 200 V for 6 hr (for 1.3-kb fragments), all in a BRL V16 vertical gel apparatus. In general, a more linear relationship between heteroduplex mobility and DNA distance is achieved when distantly related molecules are examined under lower voltage conditions.

Note: Heteroduplex mobilities are strongly affected by the high gel temperature reached (approximately 60°C) during electrophoresis at 70 mA. Consequently, strict reproducibility of run conditions is essential for quantitative measures (see following discussion).

5. Genetic relationships between two unknown sequences or between a known and an unknown sequence can be estimated from a standard curve generated by plotting relative heteroduplex mobility against known DNA distances. DNA distances are calculated by the standard method of counting only mismatches between aligned sequences and, discounting unpaired bases due to insertions and deletions (Felsenstein, 1988; Kusumi et al., 1992). Curves approximating this

relationship can then be generated, for example, by the least-squares method, and used to calculate genetic distances.

Note: In our analysis of HIV envelope genes, estimated genetic distance values above 30% were not reliable and were kept at that maximum limit (Delwart et al., 1993).

6. To generate phylogenetic trees, a triangle matrix is generated from inferred DNA distances of all or a subset of the possible combinations of heteroduplex pairings. Phylogenetic trees are then constructed using the program FITCH (Felsenstein, 1989) with global rearrangements to improve branching order.

Note: Because of the requirement for a DNA distance matrix to analyze heteroduplex mobility data (versus aligned sequences when DNA sequences are known), we generate DNA distance-based trees for comparison using a parallel method (Felsenstein, 1989). DNADIST and FITCH is used to generate 100 trees and CONSENSE is used to find the consensus tree and the frequencies of branch clusters. Branch lengths are calculated for the consensus tree using FITCH.

Acknowledgments

We thank Dr. Michael Bachmann for helpful discussions. This work was supported by Public Health Service Grant AI32885 and a grant from the Stanford Program in Molecular and Genetic Medicine.

Literature Cited

Andersson, B., J.-H. Ying, D. E. Lewis, and R. A. Gibbs. 1993. Rapid characterization of HIV-1 sequence diversity using denaturing gradient gel electrophoresis and direct automated DNA sequencing of PCR products. *PCR Methods Appl.* **22**:293–300.

Bhattacharyya, A., and D. M. J. Lilley. 1989. The contrasting structures of mismatched DNA sequences containing looped-out bases (bulges) and multiple mismatches (bubbles). *Nucleic Acids Res.* **17**:6821–6841.

Delwart, E. L., H. W. Sheppard, B. D. Walker, J. Goudsmit, and J. I. Mullins. 1994. Tracking HIV evolution *in vivo* using a DNA heteroduplex mobility assay. *J. Virol.* **68**:6672–6683.

Delwart, E. L., E. G. Shpaer, F. E. McCutchan, J. Louwagie, M. Grez, H. Rübsamen-Waigmann, and J. I. Mullins. 1993. Genetic relationships determined by a heteroduplex mobility assay: Analysis of HIV *env* genes. *Science* **262**:1257–1261.

Felsenstein, J. 1988. Phylogenies from molecular sequences: Inference and reliability. *Annu. Rev. Genet.* **22**:521–565.

Felsenstein, J. 1989. PHYLIP—Phylogeny inference package. *Cladistics* **5**:164–166.

Fujita, K., J. Silver, and K. Peden. 1992. Changes in both gp120 and gp41 can account for increased growth potential and expanded host range of human immunodeficiency virus type 1. *J. Virol.* **66:**4445–4451.

Kusumi, K., B. Conway, S. Cunningham, A. Berson, C. Evans, A. K. N. Iversen, D. Colvin, M. V. Gallo, S. Coutre, E. G. Shpaer, D. V. Faulkner, A. DeRonde, S. Volkman, C. Williams, M. S. Hirsch, and J. I. Mullins. 1992. HIV1 envelope gene structure and diversity *in vivo* and following co-cultivation *in vitro*. *J. Virol.* **66:**875–885.

Larder, B. A., K. E. Coates, and S. D. Kemp. 1991. Zidovudine-resistant human immunodeficiency virus selected by passage in cell culture. *J. Virol.* **65:**5232–5236.

Lopez-Galindez, C., J. M. Rojas, R. Najera, D. D. Richman, and M. Perucho. 1991. Characterization of genetic variation and 3′-azido-3′-deoxythymidine- resistance mutations of human immunodeficiency virus by the RNase A mismatch cleavage method. *Proc. Natl. Acad. Sci. U.S.A.* **88:**4280–4284.

McCutchan, F. E., E. Sanders-Buell, C. W. Oster, R. R. Redfield, S. K. Hira, P. L. Perine, B. L. Ungar, and D. S. Burke. 1991. Genetic comparison of human immunodeficiency virus (HIV-1) isolates by polymerase chain reaction. *J. Acquir. Immun. Defic. Syndr.* **4:**1241–1250.

Mulder, J., N. McKinney, C. Christopherson, J. Sninsky, L. Greenfield, and S. Kwok 1994. Rapid and simple PCR assay for quantitation of human immunodeficiency virus type 1 RNA in plasma: Application to acute retroviral infection. *J. Clin. Microbiol.* **32:**292–300.

Mullis, K. B., and F. A. Faloona. 1987. Specific synthesis of DNA *in vitro* via a polymerase-catalyzed chain reaction. *Methods Enzymol.* **155:**335–350.

Piatak, M., M. S. Saag, L. C. Yang, S. J. Clark, J. C. Kappes, K.-C. Luk, B. H. Hahn, G. M. Shaw, and J. D. Lifson. 1993. High levels of HIV-1 in plasma during all stages of infection determined by competitive PCR. *Science* **359:**1749–1754.

Simmonds, P., P. Balfe, C. A. Ludlam, J. O. Bishop, and A. J. L. Brown. 1990. Analysis of sequence diversity in hypervariable regions of the external glycoprotein of human immunodeficiency virus type 1. *J. Virol.* **64:**5840–5850.

Wang, Y.-H., and J. Griffith. 1991. Effects of bulge composition and flanking sequence on the kinking of DNA by bulged bases. *Biochemistry* **30:**1358–1363.

13

PCR AMPLIFICATION OF VNTRs

S. Scharf

The human genome contains large amounts of repetitive DNA sequences, some of which are arrayed as tandem repeat units. Polymorphic tandem repeats occur in two general families, variable number tandem repeats (VNTRs) or short tandem repeats (STRs). VNTRs, or minisatellites, contain repeat units that range from 8 to 50 base pairs in length (Nakamura et al., 1987). STRs, or microsatellites, are genetically similar to VNTRs at the molecular level, with the distinction that the core repeat unit of STRs ranges from 2 to 6 bases (Sharma and Litt, 1992). Both types of markers typically show some degree of heterogeneity in the core repeat unit itself, e.g., base substitutions, insertions, or deletions.

Historically, these markers have been used for constructing chromosomal linkage maps and for mapping genetic disease genes to their chromosomal loci (Nakamura et al., 1987). More recently they have been used for forensics and testing paternity. Many transplantation units and laboratories are now using these markers for monitoring bone marrow or solid organ transplant engraftment, since they provide a simple means to distinguish donors from recipients (Lawler et al., 1991; Negrin et al., 1991). These individuals cannot be distinguished by other polymorphic markers, such as the HLA loci (Kimura and Sasazuki, 1991), because they are specifically chosen to be matched at these loci to minimize the risk of rejection or graft-versus-host disease.

Using primers complementary to conserved sequences flanking the tandem repeat regions and PCR to amplify the VNTR or STR produces a product whose length is directly proportional to the number of repeat units present. These amplification produts can be resolved as individual alleles by gel electrophoresis. Thus a person can have at most two different alleles, one from each chromosome, and the variable number of repeat units constitute a fragment-length polymorphism. In determining the optimum PCR amplification protocols for VNTR markers, it is important to ascertain the reaction conditions which maximize yield and specificity of the desired product, so that limited amounts of intact starting DNA, or inhibitors that may be present in the sample, do not prove refractory to amplification. This is particularly important if the DNA sample is partially degraded (as is often the case with forensic samples). To minimize the possibility that the larger allele of the sample is not amplified because the DNA is degraded, the reaction should be robust to maximize any traces of intact DNA being amplified. When amplifying VNTR or STR-length polymorphisms in a heterozygous sample, the smaller allele is often more efficiently amplified than the larger allele. This "preferential amplification" can also potentially result in a heterozygous sample being typed as homozygous.

When optimizing a PCR amplification protocol for a particular VNTR or STR, it is particularly important to vary the concentration of the amplification constituents to enhance the desired aspects (e.g., specificity, yield, or robustness) of the reaction. For example, optimizing enzyme deoxyribonucleoside triphosphate (dNTP) Mg^{2+}, and primer concentrations can significantly improve performance for the desired amplification product (Saiki et al., 1988) while minimizing undesired effects, e.g., preferential amplification or the formation of primer-dimer (Walsh et al., 1992). Likewise, varying the thermocycling conditions such as cycle number or annealing temperature can similarly improve performance. While it is not within the scope of this chapter to describe the optimal reaction conditions for all of the PCR-based VNTR and STR loci, we can illustrate how optimizing a marker of interest can prove beneficial.

The improvements that can be realized can be exemplified by the D1S80 VNTR marker. The D1S80 locus RFLP was originally described and characterized by Nakamura et al. (Nakamura et al., 1988) using the probe pMCT118. Kasai et al. (1990) and Budowle et *al.* (1991) have described the PCR amplification of this marker, and identified 16 alleles and 37 phenotypes (Budowle *et al.*, 1991). To date, 29 alleles have been identified. This marker has been rapidly accepted by the forensic community as a very good PCR VNTR marker for its robust PCR performance and high polymorphic information content. Kasai *et al.* (1990) determined that the repeat consists, on average, of 16 base pairs. D1S80 alleles segregate as a codominant autosomal marker, so that individuals inherit one allele from each parent (Kasai *et al.*, 1990). The primers described by Kasai *et al.*, (the Cetus primers CRX51 and CRX66—see Fig. 1) produce amplification fragments which vary in length, depending on the number of repeat units present in each allele. The PCR products representative of the alleles characterized to date range from approximately 400 to 800 base pairs in length. For the purposes of illustration, the optimization of enzyme concentration, primer concentration, and cycle number is described. Additional benefits can be obtained by examining the Mg^{2+} and temperature cycling conditions.

Oligonucleotide Probes and Southern Blotting

To fully optimize any PCR amplification system, it is often useful to use a hybridization probe to analyze the reaction products. Southern blot analysis often reveals information that may not be apparent when simply examining ethidium bromide (EtBr)–stained gels. Using a unique sequence probe internal to the amplification primers produces a signal that is dependent on the quantity of the DNA fragment present. For example, many samples heterozygous for a VNTR marker exhibit preferential allele amplification, where the larger allele amplifies less efficiently than the smaller one. We developed a horseradish peroxidase (HRP)-conjugated oligonucleotide probe (CRX73) for this purpose (Levenson and Chang, 1990). The two alleles of a heterozygous sample may appear to be equally amplified on EtBr-stained gels, but may prove in fact to be preferentially amplified when probed. The probe provides rapid, sensitive, chemiluminescent detection of the specific amplification products, and eliminates the need to prepare radiolabeled probes every few weeks.

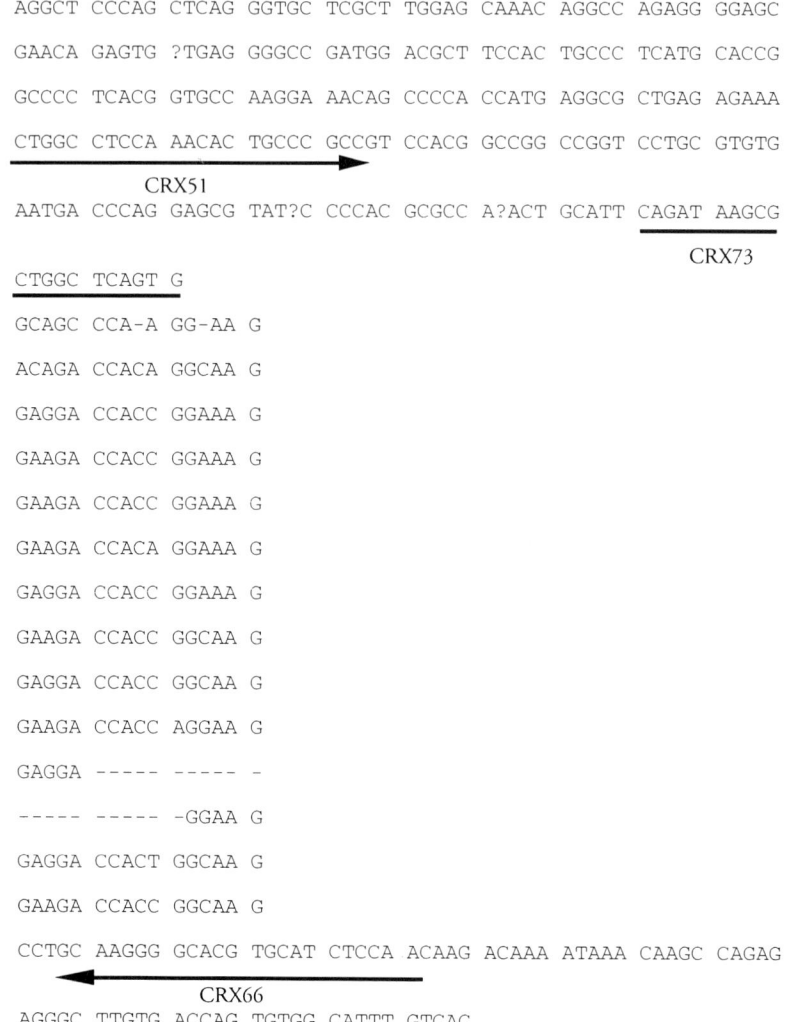

Figure 1 Sequence of a D1S80 locus VNTR allele published by Kasai et al. (1990). Arrows indicate the sequences corresponding to the PCR primers CRX51 and CRX66. The sequence corresponding to the HRP-conjugated oligonucleotide probe CRX73 is shown underlined next to the primer CRX51.

Protocol

Enzyme Concentration

The amount of *Taq* DNA polymerase was varied from 1 to 5 units/ 100 μl reaction volume to determine if there was an optimal enzyme concentration for amplification of the D1S80 alleles (Fig. 2). Though the samples amplified with 1 and 2 units of *Taq* polymerase show slightly less product than those with 3, 4, or 5 units, enzyme concentration has little effect on the efficiency and specificity of the reaction. Interestingly, there is a monomorphic band in lane B of the 3-, 4-,

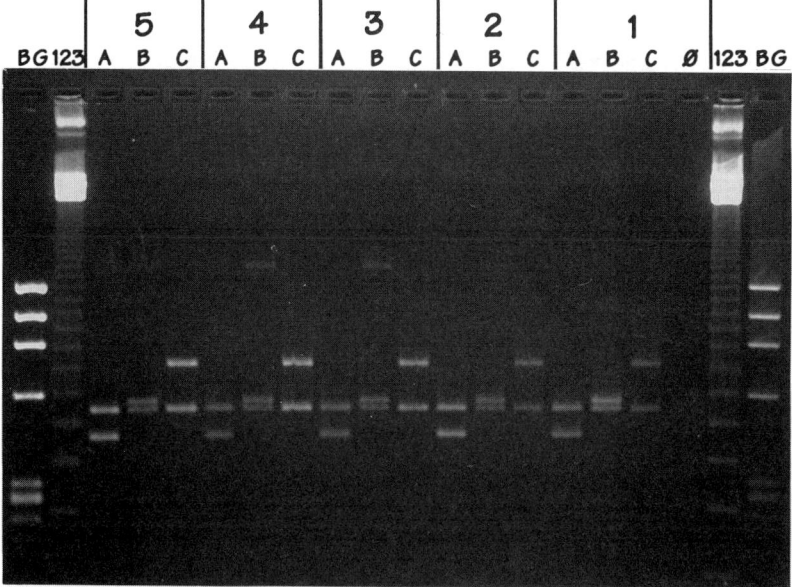

Figure 2 Amplification of the D1S80 VNTR with varying *Taq* DNA polymerase concentrations. One hundred nanograms of three DNA samples (A, B, and C) from unrelated individuals were amplified as described in the Methods except that the reactions contained 1.5 mM MgCl$_2$ and 800 mM total dNTPs. Samples were amplified for 28 cycles with the following 45-sec incubations: 94° C. for denaturation, 55° C. for annealing, and 72° C. for extension. 123 is the 123-base pair repeat ladder, used initially for estimating the sizes of the PCR products. BG is a ladder of PCR products amplified from the human β-globin gene used for keying the 123-base pair repeat ladder (S. Scharf, unpublished results). Lanes 1, 2, 3, 4, and 5 represent number of units of *Taq* polymerase added to each set of samples, respectively. φ is a negative control reaction with water added instead of DNA.

and 5-unit samples that migrates more slowly than the alleles. This monomorphic band was not detected when the gel was Southern blotted and probed with CRX73, indicating that this was a nonspecific PCR product (data not shown). In contrast to other genetic loci, efficient and specific amplification of the D1S80 VNTR is possible across a range of enzyme concentrations, and does not appear to be dependent on a specific concentration of enzyme.

Primer Concentration

Primer concentrations were compared using either 2.5 or 1.25 units of *Taq* polymerase in a 50-μl reaction with 26 cycles of amplification. Fifty and 25 pmoles of each primer gave identical results with respect to efficiency and specificity of product when used, but 12.5 pmoles of primer showed a slight decrease in efficiency. In a separate but similar experiment using 1.25 units of *Taq* polymerase in a 50-μl reaction, primer concentration was varied from 50 pmoles to 3.125 pmoles each primer. A decrease in efficiency was noted at 12.5 pmoles or less for each primer. As a result of these experiments, 25 pmoles of each primer per 50-μl reaction was chosen as the optimal concentration for this marker (data not shown).

Cycle Number

To examine the reaction at various cycles, we amplified 150 ng of two different DNA samples and removed a 10-μl aliquot at 20, 22, 24, 26, 28, and 30 cycles. The allelic products become clearly visible on an EtBr-stained gel at 26 cycles and increase in intensity up to 30 cycles of amplification (Fig. 3A). A 1-min exposure of a Southern blot of the cycle number experiment gel shows that the bands are detectable at 26 cycles, and appear to increase in intensity with increasing cycle number (Fig. 3B). In a separate experiment starting with 100 ng of DNA, amplification to 35 and 40 cycles led to the

---

Figure 3 Amplification of the D1S80 VNTR with varying cycle number. One hundred fifty nanograms of two DNA samples (A and B) from unrelated individuals were amplified as described in the Methods except that the reactions contained 1.5 mM MgCl$_2$ and 800 mM total dNTPs. Samples were amplified for 28 cycles with the following 45-sec incubations: 94° C. for denaturation, 55° C. for annealing, and 72° C. for extension. Samples were amplified in a 100-μl volume for 20 cycles; a 10-μl aliquot was collected at 20 cycles and every two cycles thereafter. After the 30-cycle samples were collected, the 10-μl aliquots were loaded onto a 1.5% agarose gel for electrophoresis. Panel A depicts the ethidium bromide-stained gel. Lanes 123 are the 123-base pair repeat ladders, used initially for estimating the sizes of the PCR products. BG is

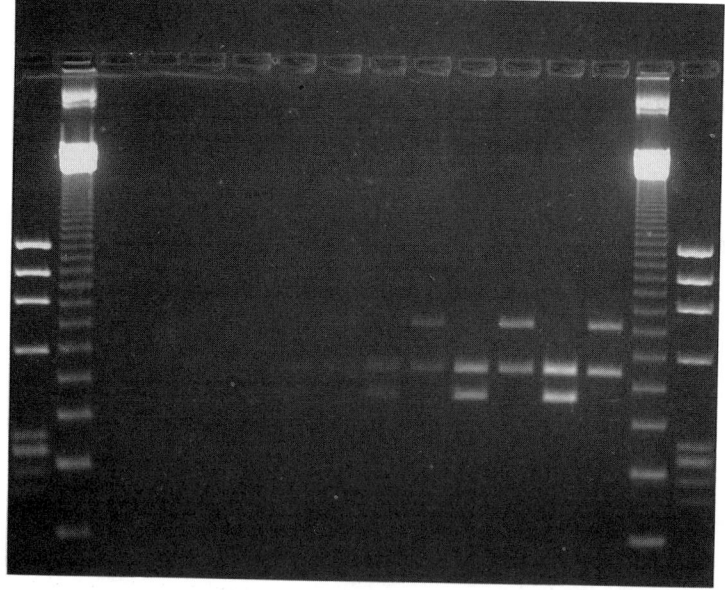

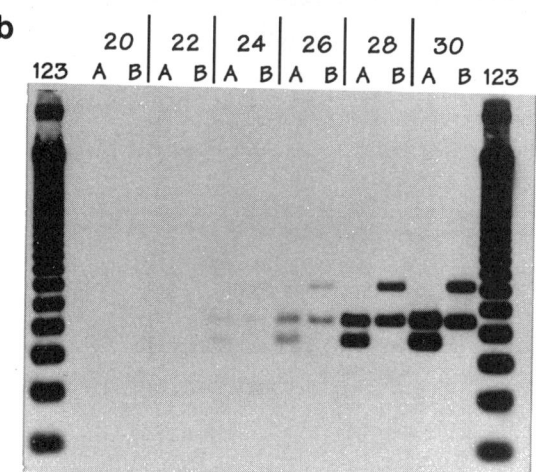

a ladder of PCR products amplified from the human β-globin gene used for keying the 123-base pairs repeat ladder (S. Scharf unpublished results). The cycle number for each set of samples collected is indicated at the top of the photo. Panel B: Southern blot of the gel depicted in panel A. In addition to CRX73, the filter was hybridized with the HRP-conjugated probe CRX69, which hybridizes specifically to the 123-base pair ladder. CRX69 (5'—GGCCTATCCTGAAGCCAAAG—3') was added at a concentration of 2 pmoles/5ml hybridization solution, and hybridized simultaneously with CRX73. After immersion in the ECL solution, the filter was exposed to Kodak XAR-5 autoradiography film for 1 min.

production of nonspecific PCR product and concomitant loss of the alleles (data not shown).

Based on the above experiments, and optimization of the other parameters of the reaction, we have developed an amplification protocol for this marker which is rapid and robust, and which has less tendency to produce preferential amplification than the protocol originally described.

Methods

The standard reaction volume for all experiments described is 50 μl.

Basic Reaction

50-μl reaction volume

25 pmoles CRX51
25 pmoles CRX66
560-mM dNTPs (all four)
1 mM MgCl$_2$
1.25 units AmpliTaq polymerase
2–100 ng genomic DNA
in 50 mM KCl, 10 mM Tris, pH 8.3

Cycling Conditions

Standard thermocycling profile for 50–200 μl reaction
1 sec to 94° C.
1 min at 94° C.
1 sec to 65° C.
1 min at 65° C.
1 sec to 72° C.
1 min at 72° C.
Run for 28 cycles

Following the last cycle, the samples are extended for 10 min at 72° C. to extend any partially double-stranded PCR fragments into fully double-stranded DNA.

Analysis of the Amplification Products

Gel Electrophoresis and Blotting

Load 10 μl of PCR reaction + 1 μl gel loading solution [10X Blue Juice: 50% glycerol, 20 mM Tris–Cl (pH 8.0), 2.5 mM EDTA, 0.2% w/v bromophenol blue] onto a 1.5%–2.0% SeaKem GTC agarose gel dissolved into Tris borate EDTA (TBE) (Sambrook et al., 1989) containing 0.2 μg/ml ethidium bromide. Run at 100 V until the dye front is at the bottom of the gel. Gels 15 × 7.5 cm long are suitable for analytical experiments; 15 × 18 cm-long gels are necessary for resolving the alleles for population genetics applications (600 V-hr electrophoresis). It is recommended that the gels be Southern blotted and probed with the HRP-conjugated probe CRX73 (see later discussion and Figs. 2 and 3) to maximize specificity and increase the range of detection. When detecting the alleles on agarose gels, it is also easier to size the alleles on an autoradiogram that is the actual size of the gel than from a Polaroid photograph. Gels are soaked in 0.5 N NaOH, 1.5 M NaCl for 15 min to denature the DNA and tansferred overnight to BioDyne B nylon filter with 20 × SSPE. The transferred DNA is immobilized on the filter by UV irradiation at 55 mJ/cm^2 with a Stratagene Stratalinker.

Sequence of the Probe CRX73

5'—HRP-CAGATAAGCGCTGGCTCAGTC—3'
Hybridization and wash conditions
Hybridization solution: 2X SSPE, 5X Denhardt's, 0.5% SDS
Wash solution: 0.5X SSPE, 0.1% SDS
Use 5 ml of hybridization solution per 40 cm^2 of membrane
Hybridize with 2 pmoles CRX73 per 5 ml hybridization solution for 15 min at 42° C.
Wash 15 min at 42° C. Rinse at RT for 5 min in PBS, develop membrane with Amersham ECL detection reagents (Amersham Part No. RPN2105).

Acrylamide Gel Analysis

Five μl of the PCR reaction was mixed with 1 μl of gel loading buffer provided with the GeneAmp detection gel (Perkin-Elmer, Norwalk, CT). This mixture was applied to 0.8 mm × 32 cm 5% GeneAmp detection gels cast with 1.2X TBE gel buffer, and electrophoresed with 0.6X TBE reservoir buffer (see Fig. 4). Gels were electrophoresed and stained according to instructions in the GeneAmp gel package insert.

Protocol Comparison From the data discussed above and other optimization experiments, we developed a revised protocol for the amplification of the D1S80 VNTR. Figure 5 shows a comparison of this revised protocol with the first amplification protocol published (Kasai et al., 1990). The EtBr-stained gel (panel A) shows that the revised protocol provides improved yield and less preferential amplification.

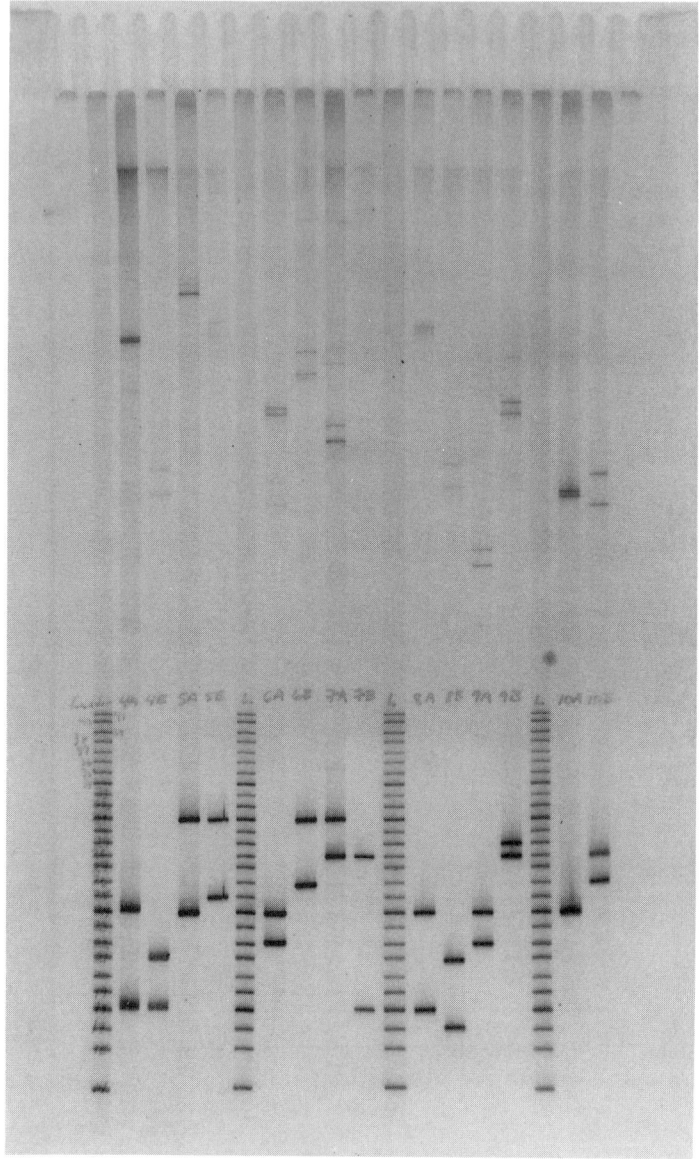

Figure 4 Sizing of D1S80 alleles. 14 DNA samples were genotyped by applying 5 μl of the D1S80 amplification to a 5% acrylamide gel, and using a D1S80 allelic ladder to determine the sizes of the alleles. The ladder ranges from 14 to 41 repeats, with every allele in between being present. The samples are typed by visually matching the samples bands with the alleles comprising the ladder.

The Southern blot of this gel (panel B) shows that the larger allele in lane 9 is quite weakly amplified compared with the same allele in lane 4. There are also some spurious bands above the alleles in lanes 6 and 7 which do not appear in the corresponding samples in lanes 1 and 2 when amplified with the revised protocol. This revised protocol also uses a 1-min polymerase extension at 72° C. instead of an 8-min extension at 70° C., reducing the time required to amplify the sample.

Allelic Sizing Ladders

The D1S80 locus VNTR is an example of how the analysis of a PCR-based VNTR can pose the same problems of an RFLP-defined VNTR. Because the core repeat unit for the D1S80 marker averages 16 base pairs in length, and the alleles range from 400 to 800 base pairs in length, it can be difficult to accurately distinguish alleles that may only differ by one repeat unit, particularly for larger alleles. Conventional molecular weight markers, such as the 123 base pair repeat ladder, do not provide the precision necessary to discriminate D1S80 alleles that differ by one repeat unit. For example, as many as eight D1S80 alleles can fit between two bands of the 123-base pair repeat ladder (data not shown). We have prepared a sizing ladder derived from actual alleles in the populations we have examined to facilitate the typing of amplified samples. The precision this ladder affords provides accurate genotyping of amplified samples for population studies as well as forensic or gene mapping applications. To demonstrate the utility of this ladder for sizing the alleles, a set of samples was coelectrophoresed with the allelic ladder on an acrylamide gel (see Fig. 4). The alleles can be sized by a simple visual match. Also, any allelic variants that may reflect core repeat heterogeneities can also be easily discerned as differences in migration relative to the alleles comprising the ladder.

Summary

The development of amplification procedures for VNTRs has advantages over RFLP analysis of these markers. Because the amplification fragments are relatively small compared with those analyzed by

172 Part Two. Analysis of PCR Products

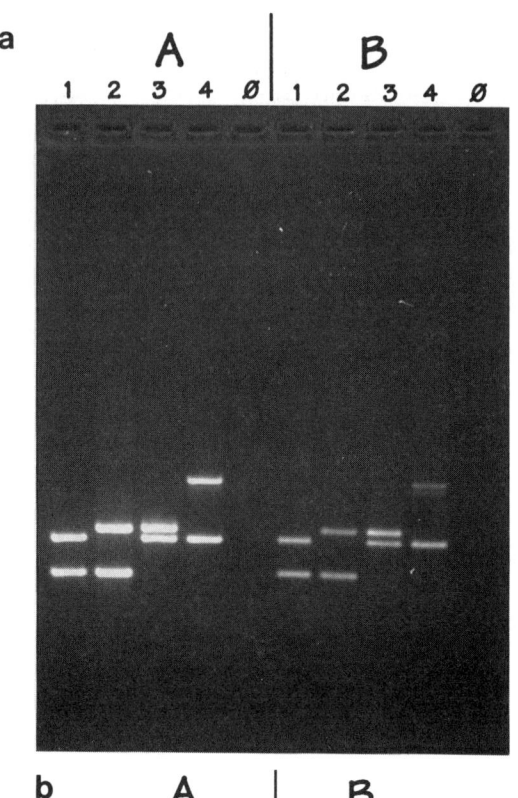

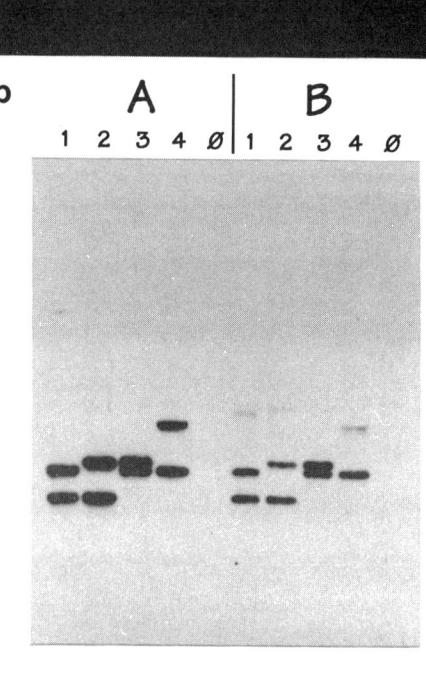

RFLP, alleles which differ by only one repeat unit are potentially better resolved than RFLP fragments. The possibility that a heterozygous sample would be scored as homozygous, owing to "coalescencing" or insufficient resolution of alleles, is reduced; no excess of homozygotes is observed, in contrast to some RFLP markers (Devlin et al., 1990).

When selecting from one or more markers that may be of use for a particular application, it is worthwhile to consider markers with a smaller (e.g. <500 base pairs) rather than a larger range of allele sizes. Smaller alleles amplify more efficiently than large ones, and are less prone to preferential amplification as a result. STRs are advantageous markers in this respect, in that the size range of alleles is typically under 400 base pairs. The dinucleotide STRs are quite frequent in the genome, and the stuttering phenomenon of dinucleotide repeats is well known and affects both the efficiency and specificity of the reaction. The extra "shadow" bands produced by dinucleotide repeat stuttering consume PR constituents that would normally be incorporated into the product of choice, and the shadow bands often make resolving the real allelic bands next to impossible. Litt et al. (1993) describe approaches that can be used to alleviate the stuttering and the resultant shadow bands. Tetranucleotide repeat STRs are less frequent in the genome than dinucleotide repeats, but they are advantageous in that they are much less prone to stuttering. Tetranucleotide repeats do not exhibit these problems nearly to the degree the dinucleotide repeats do and consequently are typically much more efficiently and specifically amplified targets.

When determining the optimum conditions for amplification, some discussion about amplification specificity vs efficiency is appropriate. These two parameters often tend to "conflict" with one another; for example, amplification conditions favoring an increase in overall efficiency are usually accompanied by a concomitant in-

---

Figure 5 Comparison of protocols for amplification of the D1S80 VNTR. A: Two nanograms of DNA from four unrelated individuals (samples 1–4) were amplified for 30 cycles as described in Methods and Materials. ϕ is a negative control reaction with water added instead of DNA. B: Samples were amplified as described by Kasai et al. (1990) except that samples were amplified for 30 instead of 35 cycles. Two nanograms of DNA from the same four unrelated individuals (samples 1–4) as in Panel A were amplified; ϕ is a negative control reaction with water added instead of DNA. Panel A: 10 μl of PCR-amplified DNA was loaded onto a 1.5% agarose gel containing ethidium bromide as described in Methods. Panel B: Southern blot of the same gel probed with CRX73. After immersion in ECL reagent, the filter was exposed to Kodak XAR-5 autoradiography film for 1 min.

crease in nonspecific PCR product. This accumulation of nonspecific product may attenuate the ability to distinguish or resolve the alleles, or may cause the amplification of the alleles themselves to be attenuated (Walsh et al., 1992). Some thought must be given to the desired application of the marker. To be able to clearly distinguish alleles, PCR product specificity is generally the more important parameter, rather than overall efficiency. However, this is not always the case; for example, it is reasonable to optimize for greater efficiency if an additional level of specificity can be provided by an internal hybridization probe. For samples or applications where the DNA is expected to be of very limited quantity, or of poor quality, maximizing amplification efficiency may be of first concern. Ideally, though, one strives for obtaining conditions that provide as much efficiency and specificity as possible, giving greater weight to whichever parameter is the most suitable for the particular application.

Literature Cited

B., Budowle, R. Chakraborty, A. M. Giusti, A. J. Eisenberg, and R. C. Allen. 1991 Analysis of the VNTR locus D1S80 by the PCR followed by high-resolution PAGE. *Am. J. Hum. Genet.* **48**:137.

Devlin, B., N. Risch, and K. Roeder. 1990. No excess of homozygosity at loci used for DNA fingerprinting. *Science* **249**:1416.

Edwards, A., H. Hammond, L. Jin, C. T. Caskey, and R. Chakraborty. 1992. Genetic variation at five trimeric and tetrameric tandem repeat loci in four human population groups. *Genomics* **12**:241.

Kasai, K., Y. Nakamura, and R. White. 1990. Amplification of a variable number of tandem repeats (VNTR) locus (pMCT118) by the polymerase chain reaction (PCR) and its application to forensic science. *J. Forensic Sciences* **35**:1196.

Kimura, A. and T. Sasazuki. Eleventh International Histocompatibility Workshop Reference Protocol for the HLA DNA-Typing Technique, *in* "Proceedings of the Eleventh International Histocompatibility Workshop and Conference" (Kimiyoshi, T., Aizawa, M., and Sasazuki, T., (eds.), Oxford University Press, p. 397. Eleventh International Histocompatibility Workshop, Yokohama, Japan, 1991.

M., Lawler, P. Humphries, and S. R. McCann. 1991. Evaluation of mixed chimerism by *in Vitro* amplification of dinucleotide repeat sequences using the polymerase chain reaction. *Blood* **77**:2504.

Levenson, C., and C.-A. Chang. 1990. Nonisotopically labeled probes and primers, *in PCR Protocols* (Innis, M. A., Gelfand, D. H., Sninsky, J. J., and White, T. J., eds.), pp. 99–112). San Diego, Academic Press.

Litt, M., X. Hauge, and V. Sharma. 1993. Shadow bands seen when typing polymorphic dinucleotide repeats: Some causes and cures. *BioTechniques* **15**(2):280–284.

Nakamura, Y., M. Leppert, P. O'Connell, R. Wolff, T. Holm, M. Culver, C. Martin, E. Fuijmoto, M. Hoff, E. Kumlin and R. White. 1987. Variable number of tandem repeat (VNTR) markers for human gene mapping. *Science* **235**:1616.

Nakamura, Y., M. Carlson, K. Krapcho, and R. White. 1988. Isolation and mapping of polymorphic DNA sequence (pMCT118) on chromosome 1p (D1S80). *Nucleic Acids Res.* **16**:9364.

Negrin, R. S., H.-P. Kiem, I. G. H. Schmidt-Wolf, K. G. Blume, and M. L. Cleary. 1991. Use of the polymerase chain reaction to monitor the effectiveness of *ex Vivo* tumor cell purging. *Blood* **77**:654.

Saiki, R. K., D. H. Gelfand, S. Stoffel, S. J. Scharf, R. Higuchi, G. T. Horn, K. B. Mullis, and H. A. Erlich. 1988 Primer-directed enzymatic amplification of DNA with a thermostable DNA polymerase. *Science* **239**:487.

Sambrook, J., E. F. Fritsch, and T. Maniatis. Preparation of reagents and buffers used in molecular cloning, *in* "Molecular Cloning" (Sambrook, J., ed.), 2nd Ed., Vol. 3, p. B.1. Cold Spring Harbor, NY, 1989.

Sharma, V., and M. Litt. 1992. Tetranucleotide repeat polymorphism at the D21S11 locus. *Hum. Mol. Genet.* **1**:67.

Walsh, P. S., H. A. Erlich, and R. Higuchi. 1992. Preferential PCR amplification of alleles; Mechanisms and solutions. *PCR Meth. Appl.* **1**:241.

Part Three

RESEARCH APPLICATIONS

14

SITE-SPECIFIC MUTAGENESIS USING THE POLYMERASE CHAIN REACTION

Jonathan Silver, Teresa Limjoco, and Stephen Feinstone

The polymerase chain reaction (PCR) has had a major impact on site-specific mutagenesis protocols, as it has in other areas of molecular biology. The major themes of PCR methodology—primer extension, selective amplification, and clever use of freedom to modify the 5' ends of amplifying oligonucleotides—have been applied in mutagenesis schemes to create altered and recombinant DNA molecules with a facility not possible by more classic means. The major limitations of PCR—restriction to molecules of less than several kilobases, and an appreciable cumulative error rate for *Taq* polymerase after multiple PCR cycles—have also affected PCR mutagenesis.

This chapter briefly reviews the major PCR site-specific mutagenesis methods proposed to date; the reader is referred to primary sources for protocol details. We have tried to emphasize the basic ideas which underline the methods described. The order of presentation was chosen for logical clarity and does not necessarily correspond to the order in which the methods were published.

The simplest technique in PCR mutagenesis is to put the desired mutated sequence within one of the oligonucleotides used for amplification. This obviously restricts one to small numbers of base changes from the wild-type template sequence and the altered bases should not be placed at the extreme 3' end of the oligonucleotide because primer extension will be inhibited (Kwok *et al.*, 1990). The addition of restriction enzyme sites to the 5' ends of oligonucleotides (Scharf *et al.*, 1986) can be viewed as an example of this form of mutagenesis.This is frequently done to facilitate cloning of PCR products. Most workers in the field have set the restriction enzyme recognition sequence two or more bases in from the 5' end of an oligonucleotide to promote efficient cutting.

A major limitation of this form of mutagenesis is that the mutated sequence appears near the end of the amplified molecule. Several methods have been proposed to overcome this limitation. An early idea (Kadowaki *et al.*, 1989) was to append to the 5' end of an oligonucleotide the recognition sequence for an enzyme which cuts distal to the recognition sequence (e.g., *Fok*I). The bases added to the 5' end of the oligonucleotide could then be cut away from a PCR product, leaving a "sticky end" overhang. Two PCR products with such ends near the site of mutation could be ligated together, leaving the mutated sequence in the middle of the ligated fragment, with no "extra" bases left behind (Fig. 1). This clever idea was proposed but not actually used for PCR mutagenesis (Kadowaki *et al.*, 1989), so it is not possible to evaluate how well it works in practice.

Another method for "embedding" a mutation inside a PCR product uses "inverse" PCR to amplify an entire plasmid with oligonucleotides whose 5' ends are immediately adjacent to one another on the plasmid DNA (Helmsley *et al.*, 1989) (Fig. 2). One of these oligonucleotides contains the mutated sequence. The amplification product is a linearized version of the plasmid with the mutation at one end. Blund end ligation of this PCR product at low concentration regenerates the circular plasmid. Sequencing of the products of this scheme revealed that *Taq* polymerase often leaves an extra A at the 3' end of DNA strands, necessitating the use of Klenow enzyme to blunt the ends of the PCR product before ligation. This method is probably limited to fairly short plasmids, and represents a tradeoff between ease of cloning after PCR and the risk of introducing *Taq* polymerase errors along the entire length of the plasmid.

The PCR has also been used to expand the variety of mutations which can be introduced via classic, site-specific mutagenesis methods for sequences cloned in single-stranded vectors. In this approach (Wichowski *et al.*, 1990), a fairly arbitrary segment can be substituted for, or inserted into, any portion of a cloned DNA segment by first amplifying the sequence to be inserted using oligonucleotides whose 5' ends contain about 20 bases of the sequence flanking the site(s)

1. Two PCRs: with oligos *1* and *3*; and *2* (with mutation, X) and *4*

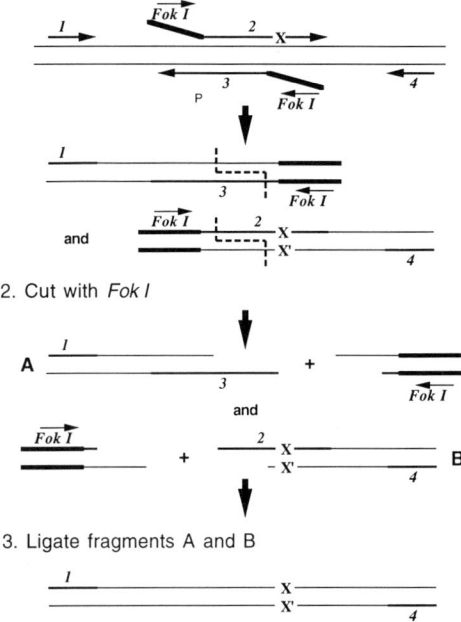

Figure 1 Use of primers containing mutation plus 5' *Fok*I extensions to insert mutation at a nearly arbitrary position in the middle of a DNA segment. X, mutation. Arrows indicate 5'→3' orientation. Primers are bold. 5' extensions are diagonal.

on the parent plasmid where the substitution/insertion is to be made (Fig. 3). A single-stranded version of the PCR product is obtained, for example, by asymmetric PCR (Guyllensten and Erlich, 1988) or use of biotin on one oligonucleotide and an avidin affinity matrix (Mitchell and Merril, 1989). The single-stranded PCR product is then annealed, via the complementary segments derived from the 5' ends of the PCR oligonucleotides, to the complementary strand of wild-type DNA cloned in a single-stranded vector such as M13. The PCR-amplified single strand acts like an oversized mutagenic oligonucleotide in a classic mutagenesis scheme. A limitation of this method is the need for the parent sequence to be cloned in a single-stranded vector.

Other "general" PCR mutagenesis methods rely on "overlap extension" of the ends of PCR-amplified DNA strands (Hiquchi *et al.*, 1988; Ho *et al.*, 1989). Two segments of DNA, one lying "upstream"

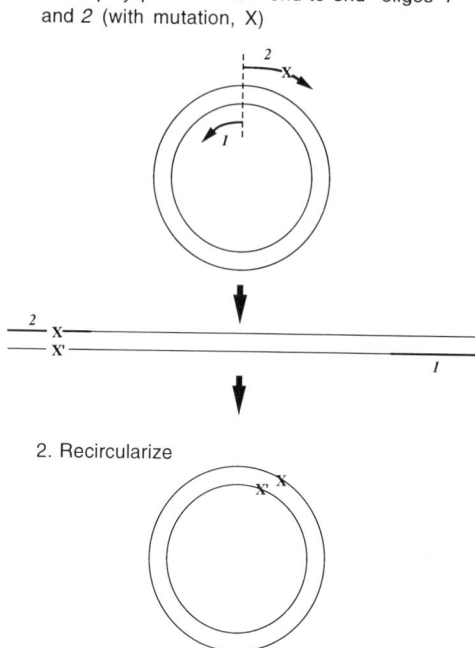

Figure 2 Amplification of whole plasmid and recircularization for arbitrary placement of primer-containing mutation.

and the other "downstream" of a desired mutation, are amplified using oligonucleotides which introduce the desired mutation and generate PCR products which overlap by about 20 or more nucleotides in the region of the mutation (Fig. 4). When these PCR products are melted and annealed to one another, one of the possible products consists of DNA strands with a short duplex overlap at their 3' ends which can be extended by a DNA polymerase to generate a long template with the mutation in the middle. This template can be specifically amplified in a subsequent PCR using the "outer" two oligonucleotides from the first amplification. This method usually requires gel purification of the two primary PCR products in order to eliminate carryover of wild-type template; such carryover would lead to amplification of the unmutated wild-type sequence in the second PCR.

The "overlap" sequence can also be introduced artificially by adding about 20 or more bases to the 5' ends of one (or both) of the

1. PCR template 1 (T1) with oligos 1 and 2 (each with about 20 bases flanking insertion site on template 2 (T2), A and B, respectively)

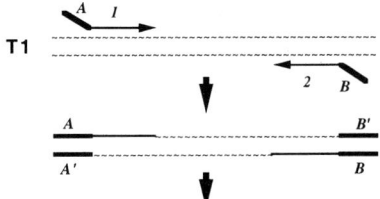

2. Obtain single strand and anneal to single-stranded template 2 (T2)

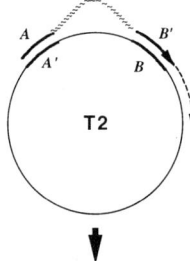

3. Extend, ligate, and transform

Figure 3 Insertion of arbitrary sequence into single stranded vector. Wavy lines denote arbitrary sequence to be inserted.

oligonucleotides used in the first PCR. This allows completely unrelated sequences to be joined precisely and at arbitrary positions by "overlap extension" in a second PCR. More than two segments can be joined in the same fashion, allowing a patchwork of genes to be assembled. If two gene segments to be joined are initially nonoverlapping and exist on separate pieces of DNA, their fusion can be achieved in a single PCR containing both templates, the "outer" oligonucleotides, and a single chimeric oligonucleotide to generate the requisite overlapping intermediate product (Yon and Fried, 1989). This method has the advantage of reducing the number of amplification cycles and thereby the risk of unintended mutations introduced by DNA polymerase during the PCR.

Several variants of the basic overlap extension methods have been proposed to avoid amplifying unmodified, wild-type template. In one method (Sarkar and Sommer, 1990), a primary PCR is used to generate a DNA fragment having the desired mutation at one end; one

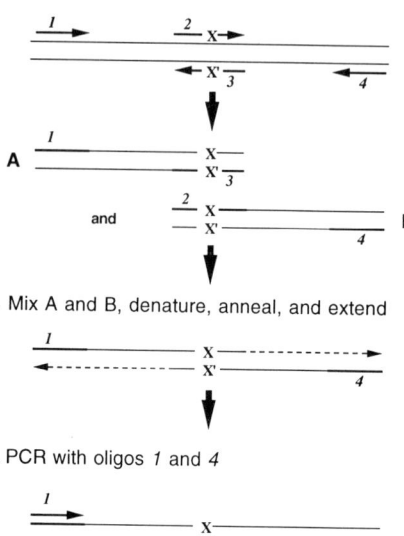

Figure 4 Standard "overlap extension" mutagenesis.

strand of this PCR product is then used as a "megaprimer" in a subsequent PCR containing the wild-type template and a distal primer oriented so that the mutation becomes trapped in the middle of a larger segment of amplified DNA (Fig. 5). The megaprimer strand of the first PCR reaction does not have to be purified for the method to work. Since the megaprimer contains the mutation, all PCR products derived from it contain the mutation as well. This method requires efficient removal of the primary oligonucleotides to prevent amplification of the wild-type sequence.

Oligonucleotides can be removed by gel purification of the first PCR product, or with filtration devices such as Centricon 100 (Amicon, Beverly, MA), or by gel exclusion chromatography, for example, using Sephacryl S300 spun columns (Pharmacia Labs, Piscataway, NJ). The yield from the "megaprimer" PCR would not be expected to be as great as the yield from the first PCR unless all of the "megaprimer" were converted to product in the second PCR. Nevertheless, the idea of using (single strands of) PCR products as primers in subsequent PCRs is important and has implications for gene walking

1. PCR with oligos 1 and 2

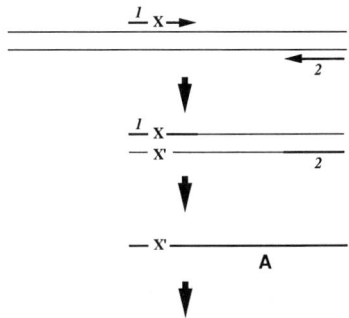

2. Second PCR uses strand A of first PCR product as a "megaprimer" with oligo 3

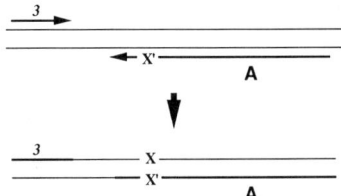

Figure 5 "Megaprimer" method for inserting mutation.

and jumping experiments where such techniques could possibly eliminate the need for synthesis of new oligonucleotide primers.

Another variant of the overlap extension method for mutagenesis uses a primer with an arbitrary 5' extension of about 20 nucleotides in the primary PCR, in combination with an oligonucleotide containing the desired mutation (Nelson and Long, 1989) (Fig. 6). The primary PCR product is used to prime the wild-type template; its primer extension product (which contains the mutation) can be specifically amplified in a second PCR using a distal primer and a primer consisting of only the 5' 20 base extension from the nonmutagenic primer in the first PCR. Use of the extension primer ensures that the mutation-containing primer extension product, and not the wild-type template, is amplified in the second PCR.

Several caveats apply to all of the PCR mutagenesis schemes described here. The first is that inadvertent polymerase errors may be introduced during amplification. The literature contains varying estimates of the error rate of *Taq* polymerase, from about 1 per 10^4 bases incorporated (Tindall and Kunkel, 1988) to < 1 per 10^5 bases

1. PCR with oligos *1* and *2* (with about 20 arbitrary bases at 5' end, A)

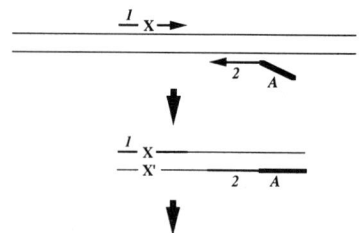

2. Prime template with PCR product and extend

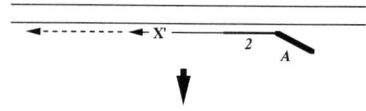

3. PCR with oligos *3* and *A*

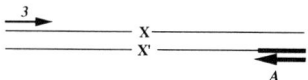

Figure 6 "Megaprimer" with 5' extension to avoid amplification of unmodified template.

(Gelfand and White, 1990). The error rate may depend on template sequence as well as the precise conditions of the PCR (nucleotide concentration in particular). Applications in which the total PCR product can be used without cloning are, in general, insensitive to *Taq* errors, since such errors affect only a small fraction of the bases at any one position. However, in applications in which the PCR product must be cloned prior to use, the proportion of *molecules* which are free of error is the important variable, not the proportion of correct bases.

Since errors are cumulative, after 30 cycles of polymerization of a 300-base segment, at the 1 per 10^4 error rate, only about one-third of the product molecules would be free of inadvertent errors; and after 60 cycles this number would drop to about 1 in 10. These calculated error rates are higher than have generally been observed in mutagenesis experiments; nevertheless, they stress the importance of minimizing the number of PCR cycles and the length of the segment to be amplified.

When it is important to identify clones of PCR products with the correct sequence, a simple screening procedure to eliminate clones which obviously have an incorrect sequence is useful to reduce the number of clones which have to be sequenced. One such screening method which appears promising in the authors' laboratory is analysis of single-stranded conformational polymorphisms (Orita et al., 1989) in PCR products from individual clones (Fujita et al., 1992).

Because of the cumulative nature of errors in PCR, thermostable polymerases with reliably lower error rates than Taq would be of considerable utility for mutagenesis experiments. Several heat-stable polymerases with $3' \rightarrow 5'$ exonuclease proofreading activity have recently become commercially available. Another approach to reducing polymerase errors is to use a nonthermostable polymerase with a considerably lower error rate, such as T4 DNA polymerase, adding fresh enzyme after each denaturation step (Keohavong and Thilly, 1989).

A related problem is that unexpected base changes have sometimes been observed in cloned PCR products *within the sequence determined by the oligonucleotides* (J. Silver and O. Nahor, unpublished observations). Some of these changes have been internal deletions; these are unlikely to have arisen from oligonucleotides with internal deletions because such oligonucleotides would not be expected to be made during the chemical synthesis. One possible explanation is that incomplete removal of protecting groups from the phosphoramidite bases leads to excision of the altered bases and repair in bacteria. (This process could also explain the occasional difficulty in cloning PCR products in bacteria.)

Unfortunately, there is at present no convenient way to determine if these blocking groups are completely removed from a synthetic oligonucleotide. Use of phosphoramidites with shorter deblocking times (FOD phosphoramidites, Applied Biosystems) might be advantageous. Another strategy is to use, whenever possible for cloning PCR products, restriction sites in the amplified DNA outside of the sequence determined by the oligonucleotide primers; this will remove possibly chemically aberrant sequences derived from at least the outer PCR primers.

As in all experimental systems, the mutagenesis methods described here do not always work, and there may be some problematic sequences which for obscure reasons are refractory to PCR mutagenesis. However, the more common experience is that PCR mutagenesis works well and rapidly. The methods described here demonstrate the remarkably increased power and flexibility that PCR has brought to the creation of recombinant and mutant DNA molecules.

Literature Cited

Fujita, K., J. Silver, and K. Peden. 1992. Changes in both gp120 and gp41 can account for increased growth potential and expanded host range of Human Immunodeficiency Virus Type I. *J. Virology* **66**:4445–4451.

Gelfand D. H., and T. J. White. 1990. Thermostable DNA polymerases. In: *PCR protocols* (eds. M. A. Innis, D. H. Gelfand, J. J. Sninsky, and T. J. White), pp. 129–141. Academic Press, San Diego.

Guyllensten, U. B., and H. A. Erlich. 1988. Generation of single-stranded DNA by the polymerase chain reaction and its application to direct sequencing of the HLA-DQA locus. *Proc. Natl. Acad. Sci. U. S. A.* **85**:7652–7656.

Hemsley, A., N. Arnheim, M. D. Toney, G. Cortopassi, and D. Galas. 1989. A simple method for site-directed mutagenesis using the polymerase chain reaction. *Nucleic Acids Res.* **17**:6545–6551.

Higuchi, R., B. Kummel, and R. K. Saiki. 1988. A general method of *in vitro* preparation and specific mutagenesis of DNA fragments: study of protein and DNA interactions. *Nucleic Acids Res.* **16**:7351–7367.

Ho, S. N., H. D. Hunt, R. M. Morton, J. K. Pullen, and L. R. Pease. 1989. Site directed mutagenesis by overlap extension using the polymerase chain reaction. *Gene* **77**:51–59.

Kadowaki, H., T. Kadowaki, E. Wondisford, and S. L. Taylor. 1989. Use of polymerase chain reaction catalyzed by Taq DNA polymerase for site-specific mutagenesis. *Gene* **76**:161–166.

Keohavong, P., and W. G. Thilly. 1989. Fidelity of DNA polymerases in DNA amplification. *Proc. Natl. Acad. Sci. U. S. A.* **86**:9253–9257.

Kwok, S., D. E. Kellogg, N. McKinney, D. Spasic, L. Goda, C. Levenson, and J. J. Sninsky. 1990. Effects of primer-template mismatches on the polymerase chain reaction: Human immunodeficiency virus type 1 model studies. *Nucleic Acids Res.* **18**:999–1005.

Mitchell, L. G., and C. R. Merril. 1989. Affinity generation of single-stranded DNA for dideoxy sequencing following the polymerase chain reaction. *Anal. Biochem.* **178**:239–242.

Nelson, R. M., and G. L. Long. 1989. A general method of site-specific mutagenesis using a modification of the *Thermus aquaticus* polymerase chain reaction. *Anal. Biochem.* **180**:147–151.

Orita, M., Y. Suzuki, T. Sekiya, and K. Hayashi. 1989. Rapid and sensitive detection of point mutations and DNA polymorphisms using the polymerase chain reaction. *Genomics* **5**:874–879.

Sarkar, G., and S. S. Sommer. 1990. The "megaprimer" method of site-directed mutagenesis. *BioTechniques* **8**:404–407.

Scharf, S. J., G. T. Horn, and H. A. Erlich. 1986. Direct cloning and sequence analysis of enzymatically amplified genomic sequences. *Science* **233**:1076–1078.

Tindall, K. R., and T. A. Kunkel. 1988. Fidelity and synthesis by the *Thermus aquaticus* DNA polymerase. *Biochemistry* **27**:6008–6013.

Wichowski, C., S. Emerson, J. Silver, and S. Feinstone. 1990. Construction of recombinant DNA generated by the polymerase chain reaction: Its application to chimeric hepatitis A virus/Polio virus subgenomic cDNA. *Nucleic Acids Res.* **18**:913–918.

Yon, J., and M. Fried. 1989. Precise gene fusion by PCR. *Nucleic Acids Res.* **17**:4895.

15

EXACT QUANTIFICATION OF DNA–RNA COPY NUMBERS BY PCR–TGGE

Jie Kang, Peter Schäfer, Joachim E. Kühn, Andreas Immelmann, and Karsten Henco

The ability to precisely quantify low copy-numbers of a specific nucleic acid template in very limited amounts of sample is important in both basic and applied medical research. The combination of polymerase chain reaction (PCR) and temperature gradient gel electrophoresis (TGGE) provides a rapid and reliable system for quantification (Henco and Heibey, 1990).

The strategy is based on coamplification of the template and an internal standard of known copy number (Fig. 1). This standard is identical with the template to be quantified, except for a single-base substitution. Because of identical priming sites and almost identical sequences, template and standard are coamplified homogenously, with the relative amounts of each product type at any point accurately refecting the original ratio prior to amplification. Following extensive cycling and prior to TGGE analysis, a limited amount of labeled standard DNA is added to the amplicon. After a thermal denaturation and renaturation step, the labeled molecules form homoduplexes with the amplified standard and heteroduplexes with the amplified template. Subsequently, TGGE separates the two species of duplexes on the basis of their different thermal stability.

The accuracy of this quantification strategy is very high, with a

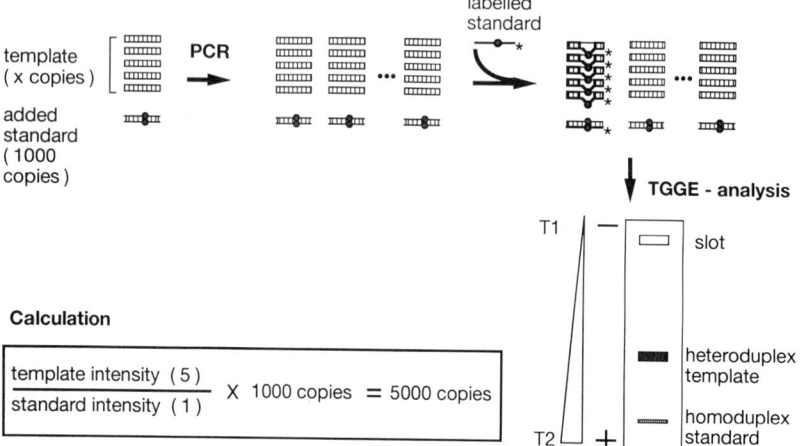

Figure 1 Schematic strategy of PCR–TGGE quantification.

variability of <15%. In addition to quantification, PCR–TGGE detects PCR artifacts and template mutants. Thus, PCR–TGGE is ideal for the detection and characterization of viral infections: it allows the quantification of virus titers, transcriptional activities, viremic or silent states, and the control of therapy by detection of mutated resistant populations. "False negative" results are directly visible; "false positive" results can be excluded by the "cutoff" approach.

Using this method, we determined the plasmid copy numbers in *Escherichia coli* during a fermentation. As an example for virus quantification, we determined quantitatively the copy number of human cytomegalovirus (HCMV) in urine of neonates with connated HCMV infection. We also demonstrated the quantification of interleukin-2 (IL-2) mRNA in human lymphocytes.

PCR–TGGE Quantification System

Selection of Primers and Synthesis of the Standard

A typical target for PCR–TGGE quantification is shown schematically in Fig. 2. Target fragments containing 100–300 bp are most suitable for TGGE analysis. The primers P1 and P2 are used for

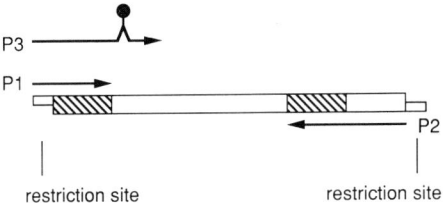

Figure 2 Target segment for PCR–TGGE quantification. The primer pair P1 and P2 are used for quantification analysis. The primer P3 is used to introduce the base substitution in the standard sequence.

quantification analysis. The primer P3 is used exclusively to construct the standard sequence. For easy cloning and labeling of the standard, all primers contain a 5′ overhang restriction site. In order to increase the difference in the electrophoretic mobility of homo- and heteroduplexs, a G:C clamp (a stretch of G:C base pairs that are very stable) should be added to the primer P2 (Myers et al., 1989).

The point mutation within the standard sequence can be conveniently introduced by a primer P3 carrying a mismatched base. To exclude any potential difference in the amplification efficiency between standard and template, the base exchange in the standard sequence should be checked so that it will not create or destroy an inverted repeat. By forming hairpin structures, repeats in the amplified sequence may influence amplification kinetics (unpublished data). Ideally, an A:T pair should be turned into a T:A, or a G:C into a C:G. The standard is cloned into a plasmid vector after PCR amplification using primers P3 and P2. For quantification of DNA template, the standard DNA is prepared as plasmid DNA, purified by high-performance liquid chromatography (HPLC), and calibrated by UV spectrophotometry.

For quantification of RNA templates, standard RNA is used as a control both for cDNA synthesis and PCR amplification. Primer sequences located on separate exons are chosen in order to distinguish between amplicons derived from cDNA and from contaminating genomic DNA. Standard RNA can be synthesized in a "runoff" reaction from transcription vectors.

The standard DNA is labeled on one strand only. The radioactive labeling is performed by a "filling-in" reaction with [^{32}P]deoxyribonucleoside triphosphate (dNTP) (Sambrook et al., 1989). The specific activity should be 0.5–1 μCi/pmole standard. Nonradioactive labeling procedures such as fluorescent labeling are also suitable.

PCR–TGGE Quantification Protocol

To include all manipulations in the standardization strategy, the biological sample can be mixed directly with the calibrated standard (Fig. 3). After preparation of the template, in DNA quantification the mixture is incubated with a restriction enzyme that does not cleave the fragment internally, in order to optimize first-round PCR synthesis. Following extensive PCR (30–40 cycles) (Saiki et al., 1988), an aliquot of the amplicon is transfered into another tube containing approximately 1 fmole ^{32}P-labeled standard DNA (0.2–1 × 10^4 cpm) and loading buffer [1 × : 4 M urea, 200 mM Morpholino Propane Sulfonic Acid (MOPS) (pH 8.0), 5 mM EDTA, dye marker]. After denaturation at 98°C for 5 min and renaturation at 50°C for 10–15 min, the probes are put into TGGE, which is per-

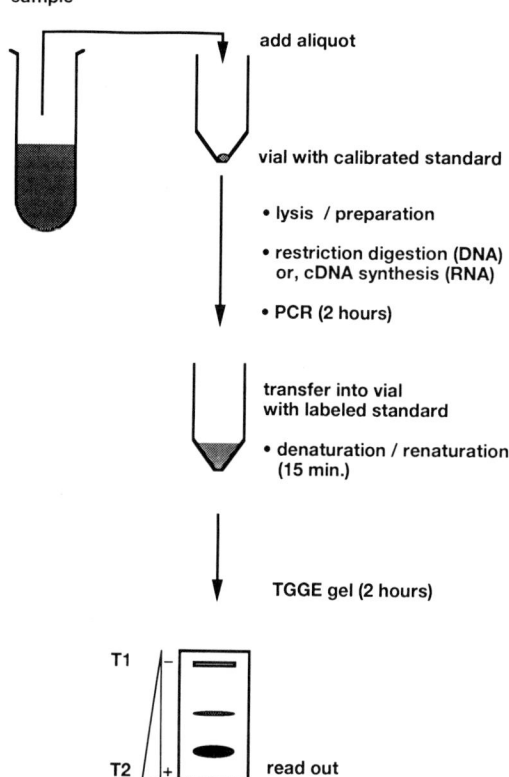

Figure 3 Schematic protocol of PCR–TGGE quantification.

formed according to the manufacturer's instruction (QIAGEN GmbH, Hilden, Germany). The gel contains 8% polyacrylamide (stock 30/0.5), 8 M urea, 20 mM MOPS (pH8.0), 1 mM EDTA, 2% glycerol, 0.03% (w/v) ammonium persulfate (APS), and 0.17% (v/v) n,n,n',n'-tetramethyl ethylene diamine (TEMED). The electrophoresis buffer consists of 20 mM MOPS (pH 8.0) and 1 mM EDTA. The gel is usually run at 300 V for about 2 hr using a gradient between 20° and 60°C. Following TGGE, the gel is fixed by incubation in 10% ethanol /0.5% acetic acid for 15 min, then dried and exposed to X-ray film at -70°C for 1–5 hr. The intensities of the homo- and heteroduplexes on the autoradiogram are measured by scanning and subsequent computer processing (Kang et al., 1991). The ratio between these intensities and the copy number of the standard present in the mixture is used to calculate the copy number of the DNA template in the biological sample (Fig. 1).

To quantify RNA templates such as mRNA or viral RNA, the standard is incorporated as RNA. Both template and standard RNAs are first transcribed into cDNA, and then analyzed using the same procedure as in DNA quantification.

Experimental Results

Determination of Plasmid Copy Numbers in *Escherichia coli*

The copy number of pUC19 in *E. coli* strain HB101 during a benchscale fermentation was quantified (Kang et al., 1991). Defined subgenic segments of the ampicillin resistance gene (amp fragment) of pUC19 and the pyruvate kinase I gene of HB101 (PK fragment) were selected as target fragments. Plasmid pBR322 containing a transition from A to G within the amp fragment was used as standard for plasmid quantification. The PK standard was introduced by a primer carrying an A instead of a T.

Ten microliters of a diluted cell culture were added to a microfuge vial containing 6×10^8 copies of amp standard and 2×10^6 copies of PK standard. After lysis (5 min at 95°C in 0.5% NP40 / 0.5% Tween 20) and Rsa I digestion, the mixture was divided into two aliquots and amplified with PK and amp primers, respectively. Four microliters of the PCR product were transferred into another tube containing ^{32}P-labeled template and analyzed by TGGE. The calculation of the plasmid copy number per *E. coli* genome during the fermentation was as described in Kang *et al.* (1991).

To determine the accuracy of this quantification system, we performed 25 independent analyses (Fig. 4). The quantitative evaluation of the overall accuracy results in a variability of $\leq 15\%$.

Quantitative Detection of Human Cytomegalovirus in Clinical Specimens

A 121-bp fragment of the HCMV *gp58* gene (Cranage *et al.*, 1986) was used as target sequence. One hundered microliters of urine samples from neonates were boiled for 10 min. A range of volumes were added to 10^4 copies of linearized standard DNA, and then assayed directly by PCR–TGGE. In a typical case of connatal HCMV infection shown in Fig. 5, 1.3×10^7 copies of HCMV DNA per milliliter urine were determined. The method was described in detail in Schäfer *et al.* (1993).

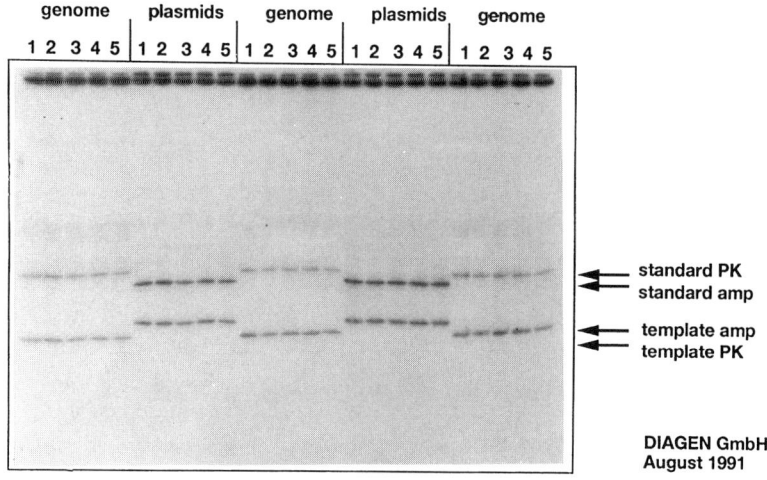

Figure 4 Interassay reproducibility of PCR–TGGE quantification. At 2 and 4 hr of fermentation of *E. coli* HB101 containing pUC19, the copy number of genome (PK) and plasmid (amp) in the culture was quantified in five independent analyses (lanes 1–5). At 6 hr of fermentation, the copy number of genome in the culture was quantified five times. Each assay was internally calibrated with 6×10^8 (amp), and 2×10^6 (PK) standard sequences. The interassay variation was determined as $\pm 10\%$. The temperature gradient of TGGE was 40°–60°C. The gel was run at 300 V for 105 min.

15. Exact Quantification of DNA–RNA Copy Numbers   195

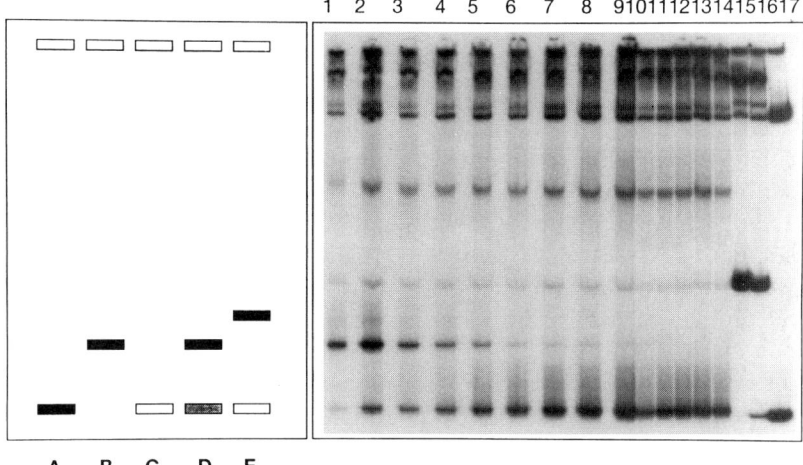

Figure 5 PCR–TGGE quantification in virus diagnosis. On the left, the possible results of PCR–TGGE analysis using a distinct copy number for a standard (cutoff) are drawn: (A) No infection (template copies ≪ cutoff, no false positive). (B) Viremia (template copies ≫ cutoff). (C) False negative (no PCR reaction). (D) Quantitative result, copy number determination (both template and standard sequences with comparable signals). (E) Viremia with new mutation. On the right, the copy number of CMV in urine of a CMV-positive neonate was determined. Lanes 1–14: 10^4 copies of standard and 5 μl, 2.5 μl, 1 μl, 0.5 μl, 0.25 μl, 0.1 μl, 0.05 μl, 0.025 μl, 1×10^{-2} μl, 5×10^{-3} μl, 1×10^{-3} μl, 5×10^{-4} μl, 1×10^{-4} μl, 5×10^{-5} μl, respectively. Lane 15: negative control. Lane 16: labeled standard after denaturation/renaturation (single strand mainly). Lane 17: labeled standard native (double strand). The high-molecular-weight bands refer to labeled fragments from the vector part of the cloned standard. The TGGE was run for 3.5 hr using a temperature gradient between 20 and 60°C.

If PCR is used for diagnosis, the possibility of false positives has to be carefully considered. Several methods have been proposed to prevent false positive results caused by PCR contamination (Heinrich, 1991). In our approach, a series of template dilutions were analyzed with a fixed amount of the standard. The real positive result was confirmed by a shift in intensity from the hetero- to the homoduplex band.

If DNA sequences of the amplicon differ between individual viral isolates and the known prototype laboratory virus strain, the heteroduplex band does not migrate with the expected electrophoretic mobility. This is a special feature of TGGE, which separates DNA fragments on the basis of their size *and* sequence.

Quantification of IL-2 mRNA

We quantified IL-2 mRNA in human lymphocytes before and after induction with phytohemagglutinin (PHA). Two sets of primers were used in a nested amplification strategy. The primary reaction mixture of 20 μl volume contained 1 pmole of each primer of the primer set I, 1 U of avian myeloblastosis virus (AMV) reverse transcriptase, 1 unit *Taq* polymerase, 200 ng total RNA, and 10^4 copies of IL-2 standard RNA. After 30 min of cDNA synthesis at 42°C, 10 cycles were run. The nested reaction mixture of 80 μl volume contained 50 pmoles of each primer of the primer set II and 1.5 U of *Taq* polymerase. This was added to the primary mixture, then amplified for a further 35 cycles. Figure 6 shows that after 26 hr of PHA induction, the IL-2 mRNA was expressed to a level 18-fold higher than uninduced mRNA.

Discussion

The PCR–TGGE quantification system provides a rapid and exact method to quantify DNA or RNA copy numbers in biological sam-

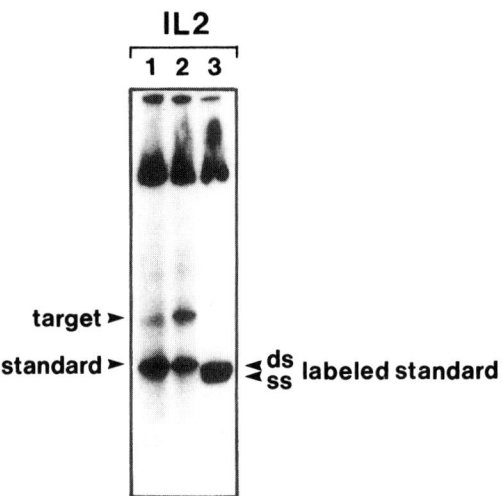

Figure 6 Quantification of IL-2 mRNA in human lymphocytes. Lane 1: total RNA from uninduced cells. Lane 2: total RNA from cells 26 hr after phytohemagglutinin induction. Lane 3: negative control. The TGGE was run for 2.5 hr with a 30/60°C gradient.

ples. Selected subgenic fragments are analyzed with the aid of an internal standard that is identical to the template to be quantified, except for a single-base substitution. This calibrated standard is present during the entire assay procedure. Therefore, all limitations inherent in the PCR reaction will affect both the template and the standard to the same extent. This allows a precise quantification of the template initially present in the biological specimen. The standard serves also as an internal positive control. A failure in the experimental assay, that is, in the PCR reaction, is directly detected, thus avoiding false negative results. After PCR, an aliquot is mixed with labeled standard. The label forms homoduplexes with the internal standard and heteroduplexes with the template in a ratio identical to that of their initial copy numbers. TGGE is then used to separate the homoduplexes from the heteroduplexes by means of their different thermal stabilities.

Several methods using the strategy of an internal standard have been reported (Wang et al., 1990; Gilliland et al., 1990; Becker-Andre and Hahlbrock, 1989). One approach is to use an internal standard with a restriction site but a sequence otherwise identical to the target template. The PCR products of the standard and the template can be distinguished by restriction enzyme digestion (Gilliland et al., 1990; Becker-Andre and Hahlbrock, 1989). However, the formation of heteroduplexes between the mutant standard and the wild-type template is unavoidable in practice. Such heteroduplexes would not be cleaved by the restriction enzyme. This phenomenon can cause an error exceeding a factor of ten. In our system, heteroduplex formation does not influence the quantification results at all.

On TGGE, the homoduplex of the standard sequence always migrates faster than all heteroduplexes potentially formed between the template and the labeled standard. Mutations in the target sequence are indicated by one or more shifted heteroduplex bands on TGGE. Neither natural mutants nor PCR misincorporations remain undetected. The detection of viral strain selection induced, for example, by chemotherapy (a resistant mutant) is possible. PCR–TGGE can thus quantify the copy number and detect mutations in one procedure. This method is ideal for detecting and characterizing viral infections.

Acknowledgments

We thank Jutta Harders and Thorsten Klahn for computer calculations on the thermodynamic DNA stabilities, Susanne Welters for excellent technical assistance, and Joanne Crowe for useful discussions.

Literature Cited

Becker-Andrea, M. and K. Hahlbrock. 1989. *Nucleic Acids Res.* **17**:9437–9446.

Cranage, M. P., T. Kouzarides, A. D. Bankier, S. Satchwell, K. Weston, P. Tomlinson, B. Barrel, H. Hart, S. E. Bell, A. C. Minson, and G. L. Smith. 1986. Identification of the human cytomegalovirus glycoprotein B gene and induction of neutralizing antibodies via expression in recombinant vaccinia virus. *EMBO J.* **5**:3057–3063.

Gilliland, G., S. Perrin, K. Blanchard, and F. Bunn. 1990. *Proc. Natl Acad. Sci. U.S.A.* **87**, 2725–2729.

Heinrich, M. 1991. PCR carry-over, *Biotech Forum Europe* **8**:594–597.

Henco, K. and M. Heibey. 1990. Quantitative PCR: the determination of template copy numbers by temperature gradient gel electrophoresis (TGGE). *Nucleic Acids Res.* **19**:6733–6734.

Kang, J., A. Immelmann, S. Welters, and K. Henco. 1991. Quality control in the fermentation of Recombinant cells. *Biotech Forum Europe* **8**:590–593.

Myers, R. M., V. C. Sheffield, and D. R. Cox. 1989. Mutation detection by PCR, GC-clamps, and denaturing gradient gel electrophoresis. In: *PCR Technology: Principles and Applications for DNA Amplification,* (ed. H. S. Erlich), pp. 71–78. Stockton Press, New York.

Saiki, R. K., D. H. Gelfand, S. Stoffel, S. J. Scharf, R. Higuchi, G. T. Horn, K. B. Mullis, and H. A. Erlich. 1988. Primer-directed enzymatic amplification of DNA with a thermostable DNA polymerase. *Science* **239**:487–491.

Sambrook, J., E. F. Fritsch, and T. Maniatis. 1989. *Molecular Cloning: A Laboratory Manual.* Cold Spring Harbor Laboratory, Cold Spring Harbor, NY.

Schäfer, P., R. W. Braun, K. Henco, J. Kang, T. Wendland, and J. E. Kühn. 1993. Quantitative determination of human cytomegalovirus target sequences in peripheral blood leukocytes by nested polymerase chain reaction and temperature gradient gel electrophoresis. *J. General Virol.* **74**:2699–2707.

Wang, A. M., M. V. Doyle, and D. F. Mark. 1990. Quantification of mRNA by the polymerase chain reaction. In: *PCR Topics* (ed. A. Rolfe), pp. 3–8. Springer Verlag.

16

THE *IN SITU* PCR: AMPLIFICATION AND DETECTION OF DNA IN A CELLULAR CONTEXT

Ernest F. Retzel, Katherine A. Staskus, Janet E. Embretson, and Ashley T. Haase

Introduction

The researcher faces many conceptual and technical challenges in understanding the pathogenesis of infection caused by taxonomically diverse microorganisms, the latter because of the variety of histological, immunological, and molecular techniques that are required for an analysis. Two central issues in viral pathogenesis, namely viral persistence and the slow progression of diseases resulting from persistent infections, are particularly good examples of the problems the investigator faces. In slow and persistent viral infections, the macromolecules involved in the initiation and progression of disease are often present in vanishingly small quantities and in only a minor population of cells or tissues. In addition, the intracellular levels and composition of these macromolecules may vary from infected cell to infected cell and with the stage of disease. Consider the slow infections caused by lentiviruses, members of the subfamily of "complex" retroviruses (Cullen, 1991) which includes the human immunodeficiency virus (HIV), the causative agent of AIDS in humans (Fauci, 1988), and visna-maedi virus, the etiological agent of neurological and pulmonary diseases in sheep (Haase, 1986).

Upon initial infection of the host, the virus replicates permissively, incites a humoral and cellular immune response, and establishes virus-host cell relationships in which the viral genome (the provirus) frequently integrates into the genome of its host and, for the most part, is harbored in a transcriptionally quiescent state (Haase, 1975; Haase et al., 1977). As a result of this downregulation of viral gene expression, the infected cells are able to escape immune surveillance by the host. The virus is then able to persist and spread throughout the organism via these infected cells (a Trojan horse mechanism of dissemination) (Peluso et al., 1985). At any given point, a small percentage of these covertly infected cells is induced to productive levels of virus gene expression, with two pathogenic consequences. First, increased expression of viral genes directly (cytopathic effects of virus production) or indirectly (elimination by host defenses) results in dysfunction and elimination of infected cells, and second, the virus produced perpetuates new infection and maintains the viral burden which, over an extended period of time, leads to damage of organ systems and death.

To gain insight into the pathogenesis and epidemiology of these slow infections and to develop better means of diagnosing and monitoring progressive infections, one clearly must be able to detect and enumerate latently infected cells. Nucleic acid hybridization and solution-phase PCR techniques are powerful in their own right and demonstrably useful in analyzing viral infections, but they are limited to studies of nucleic acids isolated from a population of cells and thus produce a single result averaged over infected and uninfected members. Single-cell techniques, such as in situ hybridization (Haase et al., 1984; Haase, 1987) and in situ transcription (Tecott et al., 1988), used in combination with cytochemistry and immunohistochemistry (Brahic et al., 1984), provide additional information about the cellular and tissue context of infection; however, these methods are of limited value in addressing the latency issue because a single copy of provirus cannot be convincingly and routinely detected.

Since the viral nucleic acid within the infected cell in latent infections is below the level of routine detection (Haase, 1987), attempts to develop nucleic acid amplification technologies applicable to single cells in tissues and in circulation are both logical and essential. This is a difficult and complex problem, however, because of the many variables involved in generating, retaining, and detecting amplified sequences within a morphologically identifiable cell (Haase *et al.*, 1990; Staskus *et al.*, 1991a). For example, a fixation and pretreatment protocol that preserves cellular morphology and antigens yet allows amplification reactants to reach the nucleic acid substrate must be developed. The cellular milieu, maintained as a consequence of the fixation procedure, dramatically affects the efficiency of amplification because it introduces potential interference from protein contaminants and from the cross-linking of proteins to nucleic acid substrates. These problems are compounded by the general reduction in efficiency characteristic of solid-phase systems. In this case, the diffusion of reaction components into the fixed and cross-linked cell environment most likely proceeds at a slower rate than in solution, and may require pretreatments to permeabilize the fixed structure. Finally, the amplified product DNA (amplimer) must be retained within the source cell for detection with isotopic or nonisotopically labelled probes. Because of these antithetical requirements of substrate accessibility and product retention, strategies must be developed which will satisfy both.

In this chapter, we describe the continuing evolution of the *in situ* application of the polymerase chain reaction (ISPCR), from its origin in cell culture (Haase *et al.*, 1990) and *in vivo* model systems (Staskus *et al.*, 1991a) to its present uses in pathogenesis. *In situ* amplification can be used to analyze both adherent and nonadherent cells fixed in solution (Haase *et al.*, 1990), or with cells or tissues that have been fixed and embedded in paraffin (Staskus *et al.*, 1991a). This technique has proven successful in documenting the reservoir of latently infected cells in natural infections of both visna-maedi virus (Staskus *et al.*, 1991a) and HIV (Embretson *et al.*, 1993).

Primer Strategy

In general, the design of a primer set is critical for the PCR, and for the *in situ* PCR, the special considerations alluded to in the introduction make this step even more crucial. The relatively small size of the amplification product frequently chosen for the PCR (200–500 bp) is within the size class of molecules frequently chosen as probes for *in situ* hybridization (Haase *et al.*, 1984; Haase *et al.*, 1987); consequently, this size of nucleic acid product easily diffuses out of the permeabilized fixed cell at a rate very likely enhanced by the high temperatures of the PCR. On the other hand, fixation of

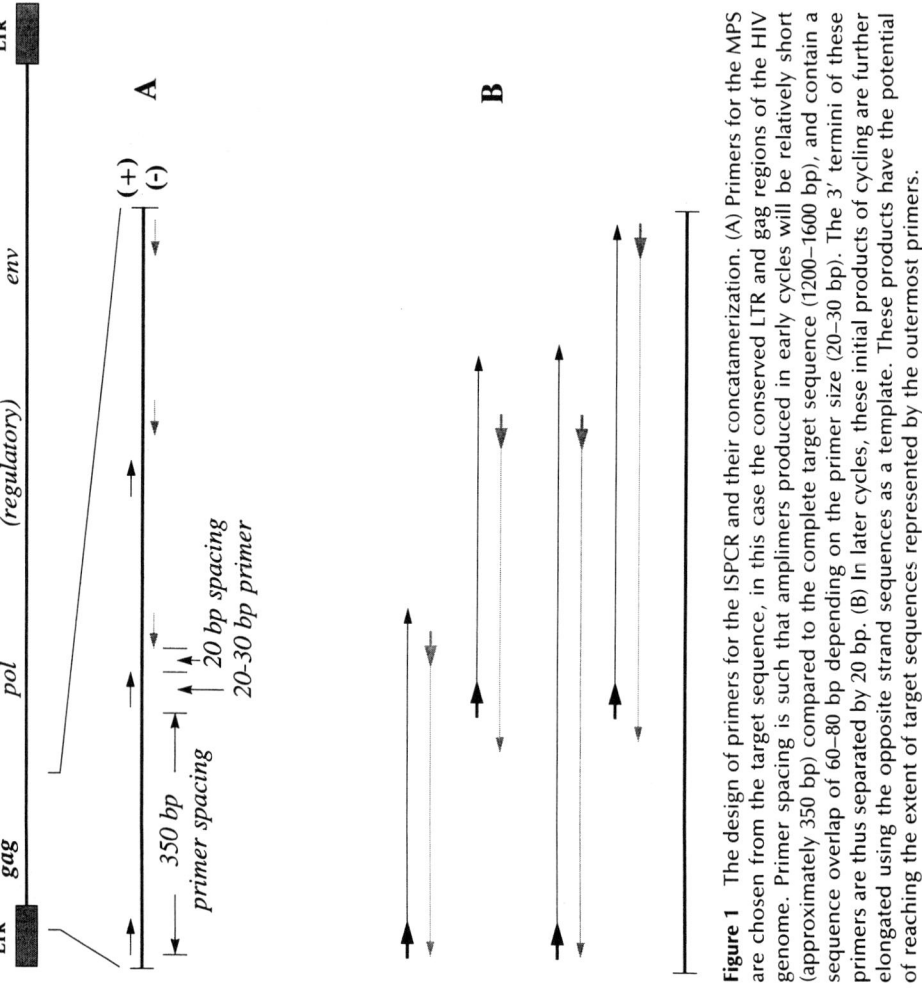

Figure 1 The design of primers for the ISPCR and their concatamerization. (A) Primers for the MPS are chosen from the target sequence, in this case the conserved LTR and gag regions of the HIV genome. Primer spacing is such that amplimers produced in early cycles will be relatively short (approximately 350 bp) compared to the complete target sequence (1200–1600 bp), and contain a sequence overlap of 60–80 bp depending on the primer size (20–30 bp). The 3' termini of these primers are thus separated by 20 bp. (B) In later cycles, these initial products of cycling are further elongated using the opposite strand sequences as a template. These products have the potential of reaching the extent of target sequences represented by the outermost primers.

the cell greatly reduces the already inefficient amplification of larger targets such that the retention problem cannot be simply remedied by choosing a pair of primers that will give rise to a product DNA that will be better retained within the cell (900–1600 bp). To resolve these dilemmas, we designed a set of multiple primers that generate individual reaction products which (1) are small, and therefore amplified with a greater efficiency in fixed cells (approximately 350 bp); and (2) at the same time can be combined to form products of sufficient size to be retained within the cell.

The multiple primer set (MPS) is composed of three to four primer pairs (6 to 8 oligonucleotides of appropriate sense), each 20 to 30 nucleotides in length, representing approximately 1200–1600 bp of contiguous sequence information from the relatively conserved 5'-terminal of the viral genome (see Fig. 1). Extension of the primers gives rise to products with overlapping cohesive terminals that can form, through base-pairing interactions, larger DNA fragments (up to 1600 bp) that will be retained inside the fixed cell. Until later cycles, the stoichiometry of the reactions is such that the products are primarily extensions of the MPS. However, as the oligonucleotide primers are utilized, it is likely that PCR products themselves will function as primers for further elongations and amplification, so that large covalently linked fragments will accumulate in the fixed cell as targets for detection by *in situ* hybridization. Other strategies for product retention utilize the incorporation of biotinylated (Komminoth *et al.*, 1992) or digoxigenin-conjugated (Patterson *et al.*, 1993) nucleotides that are thought to make the DNA more bulky and less able to diffuse from the cell.

In the MPS, no two primers share identical or complementary sequence information, and are slightly offset (approximately 20 bases), so that the primers must elongate in order to pair with adjacent primer sets in successive cycles. As with the design of sequencing primers (Staskus *et al.*, 1991b), these primers are chosen from highly conserved sequences (when the information is available), and tested for similarity to both strands of sequence information represented by the complete GenBank nucleotide sequence database, using pattern matching and similarity programs (Lawrence *et al.*, 1989); any primer having significant similarity in the 3' terminal 6–10 nucleotides to any gene that may potentially be present in the sample is rejected. Inverted repeats, and the consequent potential to form hairpins, are also tested for, using stem-and-loop designations which would have an extremely low probability of existing under the conditions of the amplification reaction (stems, four base pairs; loops, two nucleotides); again, sequences that fail this test are generally rejected. In addition, the primers are so designed as to incorporate, to the extent feasible, comparable GC content and the criteria of Kwok (Kwok et al., 1990) and Rychlik (Rychlik et al., 1990). In the overall reaction scheme, primer pairs must be capable of producing appropriate products in standard liquid-phase PCR with purified template. Optimization of the ISPCR, however, must be done in the context of the MPS and the target cell or tissue and not be based on solution-phase primer pair reactions.

Protocols

Preparation of Samples

Fixation

We and others have shown that *in situ* amplification of nucleic acids is possible in cells and tissues that have been fixed under a variety of conditions. This is an important issue, since it is not always possible to control fixation, especially in the case of archival material. The type and concentration of fixative, time and temperature of fixation, type of tissue, size of tissue fragment, etc., all contribute to the extent of fixation of a particular sample. Therefore, prior to thermal cycling, it is often necessary to balance an appropriate pretreatment against the extent of fixation. Generally, we find that, with highly fixed material, it is necessary to be more aggressive in pretreatments to permeabilize cells and tissues so that the reaction components will have access to target sequences. When fixation is under our control, we fix with freshly prepared 4% paraformaldehyde for a relatively brief period of time and, as a result, in many instances we are able to omit pretreatment.

Cells fixed in suspension

Individual cells, from culture or isolated from an organism, may be subjected to ISPCR in solution phase or while attached to a glass slide. For amplification in solution, fix the cells in suspension as follows (Haase et al., 1990). The cells are trypsinized (if necessary), pelleted in a clinical centrifuge (600 xg, 5 min), and resuspended and washed with calcium- and magnesium-free phosphate-buffered saline (PBS–CMF). After the cells are pelleted and washed in PBS–CMF a second time, they are pelleted and resuspended in freshly prepared 4% (w/v) paraformaldehyde (in PBS–CMF) or Streck tissue fixative (STF) (Streck Laboratories, Inc., Omaha, NE) and fixed for 20–30 min at room temperature. [Note: STF is incompatible with phosphate-containing buffers; therefore, when using STF, substitute Tris-buffered saline (TBS) for PBS–CMF.] The cells are then pelleted (1600 xg, 5 min), resuspended in PBS-CMF, and transferred to PCR reaction tubes. Finally, the cells are pelleted (1600 xg, 5 min) and resuspended in PCR reaction mixture for thermal cycling. If the cells are not cycled immediately, they are resuspended in a small amount of PBS–CMF; 25 volumes of 70% (v/v) ethanol are added and then are stored at 4°C. Before thermal cycling, the cells are pelleted, resuspended, and rehydrated in PBS–CMF for 20 min at room temperature, transferred to PCR reaction tubes, pelleted, and then resuspended in PCR reaction mixture.

Cells fixed on slides

The cells are trypsinized, pelleted, and washed as described in the previous section. The final cell pellet is resuspended in PBS–CMF and deposited onto silane-coated glass slides by cytocentrifugation (450 rpm, 5 min). Alternatively, an appropriate number of cells, in a small volume of PBS–CMF (e.g., 3–5 × 10^4 cells/10 μl), can be spotted onto the slides and air dried in a laminar-flow hood. The latter allows many samples to be processed on one slide. The dried slides are immersed and fixed in freshly prepared 4% (w/v) paraformaldehyde or STF for 20–30 min at room temperature. The slides are then washed in PBS–CMF for 5 min, dehydrated through a series of graded ethanol solutions (70%, 80%, absolute; 5 min each), and air dried.

Tissues or cells embeded in paraffin

Tissue fragments (1 cm × 1 cm × 3–4 mm, or smaller) are fixed in freshly prepared 4% paraformaldehyde or STF for 4–6 hr at room temperature. The tissue is transferred to 80% ethanol for 3 days and then placed in embedding cassettes and processed for embedding in paraffin. We use an automated tissue processor with a protocol that does not include an initial incubation period in fixative: 80% ethanol, 95% ethanol (three times), absolute ethanol (three times), toluene (three times), paraffin (four times); 30–45 min at each step. Following the embedding process, 6–10 μm serial sections are cut and deposited onto silane-coated glass slides. The slides are placed in a 60°C oven for 1–2 hr, deparaffinized with two 10-min washes (with agitation) in xylenes followed by two 10-min washes in absolute ethanol, and air dried.

As an alternative to the protocol for individual cells described in the preceding sections, cells from culture or body fluids may be aggregated by clotting with fresh serum and fibrinogen or solidified in an agarose matrix and then fixed, processed, embedded, and sectioned as described for tissues. Cells prepared in this fashion would amplify under conditions which approximate those of sectioned tissue.

Pretreatment

In our experience, samples that have been fixed in 10% buffered formalin (Staskus et al., 1991a), 4% paraformaldehyde, or STF for brief periods do not require pretreatment before thermal cycling. However, if in situ amplification is unsuccessful or if we know that the samples have been in fixative for extended periods of time, we experiment with pretreatment using a series of proteinase K concentrations. Make a set of solutions containing 20 mM Tris–HCl (pH 7.4), 2 mM CaCl$_2$, and proteinase K in increasing concentrations (e.g., 1, 5, 20, 50 μg/ml), equilibrated to 37°C. Incubate a series of slides in these solutions for a selected period of time (5–30 min), wash the slides twice in 20 mM Tris–HCl (pH 7.4) for 5 min, dehydrate the slides through graded alcohol solutions, and air dry.

In Situ PCR

PCR Reaction Mixture and Thermal Cycling Parameters

As is the case for solution-phase PCR, for a given target and primer set, the optimal concentrations of reaction components and optimal thermal cycling parameters must be empirically determined. It is wise to begin with conditions that are optimal for solution-phase PCR using purified target DNA. Our reaction mixture contains 10 mM Tris–HCl (pH 8.3), 50 mM KCl, 0.01% (w/v) gelatin, 1.5–2.5 mM MgCl$_2$, 200 μM each deoxyribonucleoside triphosphate (dNTP), 1 μM each external primer, 0.01 μM each internal primer, and 0.2 U/μl *Thermus aquaticus* (*Taq*) DNA polymerase. When incorporating biotinylated nucleotide to ballast the product, we reduce the concentration of deoxyadenosine triphosphate (dATP) in the

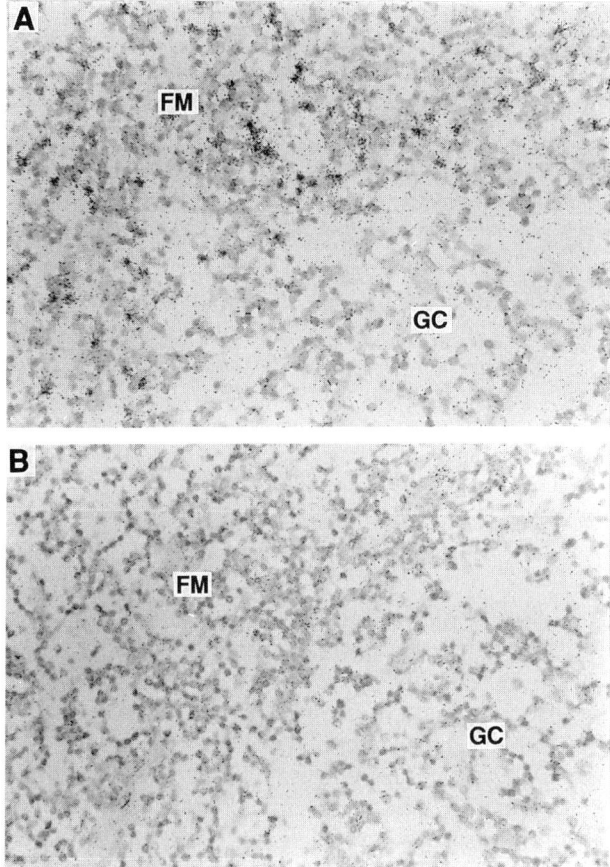

Figure 2 *In situ* amplification and detection of HIV DNA in infected lymph node. A lymph node biopsy from an HIV-infected individual was fixed in 4% paraformaldehyde, embedded in paraffin, sectioned, and analyzed by ISPCR for the distribution of HIV DNA-containing cells. A multiple primer set (sequences derived from HIV HXB2 complete genome) containing six primers (3 sense and 3 antisense) representing approximately 1270 nucleotides of the *gag* gene was used in the amplification [5'-GGGTGCGAGAGCGTCAGTATTAAGCGGGGG-3' (bases 338–367), 5'-GTCAGCCAA-AATTACCCTATAGTGCAGAAC-3' (bases 717–746), 5'-GTGATATGGCCTGATGTACC-ATTTGCCCCT-3' (bases 780–751), 5'-TAGTAAGAATGTATAGCCCTACCAGCATTC-3' (bases 1153–1182), 5'-ACATAGTCTCTAAAGGGTTCCTTTGGTCCT-3' (bases 1225–1196), 5'-AATCTTTCATTTGGTGTCCTTCCTTTCCAC-3' (bases 1611–1582)]. The two external primers were present at a final concentration 1 μM and the four internal primers at 0.01 μM. PCR reaction mixture was applied and the slides were cycled [40c: 90°C for 30 sec, 60°C for 30 sec, 72°C for 3 min]. The slides were hybridized with ^{35}S-labeled double-stranded DNA probe, coated with nuclear track emulsion and exposed for 14 days. (A) 20–30% of the CD4$^+$ cells (demonstrated by double label experiments; Embretson, 1993) located predominantly in the follicular mantle zone (FM) surrounding the germinal center (GC) are positive for HIV DNA (silver grains over nuclei). Occasional CD4$^+$ cells containing HIV DNA are seen within the GC. Published results show that fewer than 1% of these cells also contain viral RNA (Embretson, 1993). (B) Controls lacking *Taq* polymerase in the reaction mixture background levels of grains.

reaction to 180 µM and add biotin-14-dATP to a final concentration of 20 µM.

Much of the developmental work (Haase et al., 1990; Staskus et al., 1991a; Embretson et al., 1993) was done with a multiple primer set in equimolar concentration. More recently, we have begun to bias the concentrations of external versus internal primers. This approach was influenced by the finding (Lewis et al., 1994) that primers proximal to each other inhibit the production of amplimers from the distal primers when amplified simultaneously in solution. We are at present routinely using a bias of 50 : 1 or 100 : 1 for the external primers relative to internal primers (see Fig. 2). It may, in fact, be that equimolar primers contribute to the perceived inefficiency of ISPCR (that is, small product is efficiently synthesized but diffuses away from the site of amplification more readily); however, it should also be noted that using only the external primers dramatically reduces efficiency of synthesis in our hands. While biasing may be of use, the MPS is still essential for consistent results.

We have been successful using a variety of thermal cycling parameters. Earlier work from our group (Haase et al., 1990; Staskus et al., 1991a; Embretson et al., 1993) involved extended elongation times. While the complex architecture of the cell may decrease the efficiency of the polymerase, other factors may be involved as well. Much of this work was done in oil heated by air convection; these long incubations may have been essential, and optimization of those reaction conditions indicated precisely that. However, contemporary cyclers designed specifically for slides allow more efficient heat transfer, and consequently, we have been able to successfully reduce elongation times to those more in keeping with solution reactions. Our original conditions [total of 50 cycles: 94°C for 2 min, 44°C (20-mer primers) for 2 min, 72°C for 15 min] used long incubations (Haase et al., 1990; Staskus et al., 1991a); more recently, however, we are finding that shorter incubations and lower denaturation temperatures [40 cycles: 90°C for 30 sec, 60°C (30-mer primers) for 30 sec, 72°C for 3 min] not only produce similar apparent efficiencies of amplification, but improve the localization of product within the target cells.

Cells in Solution

Cells that have been fixed in suspension are resuspended in PCR reaction mixture in standard PCR reaction tubes. The suspension is overlaid with mineral oil and subjected to thermal cycling employing hot start. Following the PCR, the cells are pelleted (16,500 ×g, 5 min), the oil and reaction supernatant are removed, and the pellet is resuspended in PBS–CMF and transferred to a 1.5-ml microcentrifuge tube. The cells are then pelleted (16,500 ×g, 5 min), resuspended in PBS–CMF, and deposited by cytocentrifugation or spotting onto silane-coated slides. More recently, we have included a 10-min postfixation in 4% paraformaldehyde following the PBS–CMF washes.

Cells or Tissues on Slides

The PCR reaction mixture (minus polymerase) is assembled and heated to 94°C for 5 min. The tube is spun briefly in a microcentrifuge to collect condensation from the cap and walls, *Taq* polymerase is added, and the tube is returned to 94°C. The reaction mixture is then pipetted onto the dehydrated cells or tissue on the slide and the environment is sealed in such a way as to prevent evaporation and concentration of reaction components. The volume of reaction mixture required is dependant on the sealing method and thermal cycling instrument chosen and may vary from 10 to 75 μl. The reaction chambers vary from glass coverslips, sealed to the slide with a layer of mineral oil or nail polish, or plastic chambers which are sealed to the slide by means of a high-temperature adhesive (Gene Tec Corp., Durham, NC) [for use with the BioTherm (Bio-Therm Corp., Fairfax, VA), Coy (Coy Corp., Grass Lake, MI), MJ Research (MJ Research, Inc., Watertown, MA) and other slide-dedicated instruments or adapted tube thermal cyclers], to a stainless steel–silicon cover assembly that is pressure-fitted onto thick (1.2 mm) glass slides (Perkin–Elmer Corp., Norwalk, CT). Advantages and disadvantages exist for each system and much depends on personal preference and experience. At this point in the development of the technology, none of them is clearly superior to the others, and the advantages of these dedicated machines range from convenience to capacity. Under appropriate conditions, all are capable of producing satisfactory results.

Following thermal cycling, the slides are rinsed briefly with PBS–CMF, fixed for 10 min in 4% paraformaldehyde, washed twice in PBS–CMF (5 min each time), dehydrated through graded alcohol solutions, and air dried.

Controls should include normal or uninfected cells and tissues, omission of the polymerase from the reaction, omission of the primers from the reaction, amplification with heterologous primers, hybridization with a heterologous probe, DNase treatment prior to detection, and hybridization for DNA without prior amplification.

Detection

DNA that results from amplification is detected by *in situ* hybridization. Direct incorporation of modified nucleotides for detection purposes is not recommended because of the propensity of the PCR for generating nonspecific products. Probes may be composed of RNA (generated by *in vitro* transcription), double-stranded DNA (generated by random priming, nick translation, or PCR) or single-stranded DNA (generated by asymmetric PCR or oligonucleotide synthesis), and may be radiolabeled or nonisotopic (Schowalter *et al.*, 1989; Sambrook *et al.*, 1989; Mertz and Rashtchian 1994; Peng *et al.*, 1994; Kashio *et al.*, 1994). In general, the probes should not contain primer sequences to avoid the detection of nonspecific product and reduce background.

Prior to hybridization, the slides are acetylated by placing them in 0.1 M triethanolamine (pH 8.0) and adding acetic anhydride, dropwise with agitation, to a final concentration of 0.25% (v/v). After a 10-min incubation, the slides are rinsed briefly in deionized water and dehydrated. To denature the product DNA, the slides are then placed in 95% formamide, $0.1 \times$ standard saline citrate (SSC) at 65°C for 15 min, quenched in ice-cold $0.1 \times$ SSC for 2 min, rinsed briefly in deionized water, and dehydrated. The hybridization mixture [10% (v/v) dextran sulfate, 50% (v/v] deionized formamide, 20 mM HEPES (pH 7.4), 1 mM EDTA, $1 \times$ Denhardt's medium, 0.1 mg/ml polyA, 0.6 M NaCl, 100 μM aurintricarboxylic acid, 200–250 μg/ml yeast RNA, 100 mM dithiothreitol (DTT), 1×10^5 dpm/μl ^{35}S-labeled double-stranded DNA probe ($0.8 - 1 \times 10^9$ dpm/μg specific activity)] is pipetted onto the samples that are then sealed with siliconized glass coverslips and rubber cement. Following hybridization for 1–3 days at 37°C and removal of the coverslips, the slides are washed with $2 \times$ SSC at 55°C for 1 hr, followed by a formamide-containing wash buffer [50% (v/v)] formamide, 20 mM HEPES (pH 7.4), 0.6 M NaCl, 1 mM EDTA, 5 mM DDT) at room temperature for 1–3 days. The slides are washed in $2 \times$ SSC for 5 min and dehydrated through graded alcohol solutions (70%, 80%, absolute; 5 min each) containing 0.3 M NH$_4$ acetate. The slides are dipped in photographic emulsion at 43°C (Kodak NTB-2 which has been diluted 1 : 1 with 0.6 M NH$_4$

acetate) air dried, sealed in lightproof boxes with desiccant, and placed at 4°C for an appropriate period of exposure (Haase et al., 1984; Haase, 1987).

Literature Cited

Brahic, M., A. T. Haase, and E. Cash. 1984. Simultaneous *in situ* detection of viral RNA and antigens. *Proc. Natl. Acad. Sci. U.S.A.* **81:** 5445–5448.

Cullen, B. R. 1991. Human immunodeficiency virus as a prototypic complex retrovirus. *J. Virol.* **65:** 1053–1056.

Embretson, J., M. Zupancic, J. L. Ribas, A. Burke, P. Rácz, K. Tenner-Rácz, and A. T. Haase. 1993. Massive covert infection of helper T lymphocytes and macrophages by human immunodeficiency virus during the incubation period of AIDS. *Nature* **362:** 359–362.

Fauci, A. S. 1988. The human immunodeficiency virus: infectivity and mechanisms of pathogenesis. *Science* **239:** 617–622.

Haase, A. T. 1975. The slow infection caused by visna virus. *Curr. Top Microbiol. Immunol.* **72:** 101–156.

Haase, A. T. 1986. Pathogenesis of lentivirus infections. *Nature* **322:** 130–136.

Haase, A. T. 1987. Analysis of viral infections by *in situ* hybridization. In *In situ hybridization—applications to neurobiology* (Symposium Monograph). (K. L. Valentino, J. H. Eberwine, and J. D. Barchas, eds.), pp. 197–219, Oxford University Press, New York.

Haase, A. T., L. Stowring, O. Narayan, D. Griffin, and D. Price. 1977. The slow persistent infection caused by visna virus: role of host restriction. *Science* **195:** 175–177.

Haase, A. T., M. Brahic, L. Stowring, and H. Blum. 1984. Detection of viral nucleic acids by *in situ* hybridization. In *Methods in virology*, Vol. VII (K. Maramorosch and H. Koprowski, eds.), pp. 189–226, Academic Press, New York.

Haase, A. T., E. F. Retzel, and K. A. Staskus. 1990. Amplification and detection of lentiviral DNA inside cells. *Proc. Natl. Acad. Sci. U.S.A.* **87:** 4971–4975.

Kashio, N., S. Izumo, S. Ijichi, K. Hashimoto, F. Umehara, I. Higuchi, H. Nakagawa, M. C. Yoshida, E. Sato, T. Moritoyo, A. T. Haase, and M. Osame. 1995. Detection of HTLV-I provirus in infiltrated mononuclear cells of spinal cord lesions from HAM/TSP patients by *in situ* polymerase chain reaction. Submitted for publication.

Komminoth, P., A. A. Long, R. Ray, and H. J. Wolfe. 1992. *In situ* polymerase chain reaction detection of viral DNA, single-copy genes, and gene rearrangements in cell suspensions and cytospins. *Diag. Molec. Path.* **1:** 85–97.

Kwok, S., D. E. Kellogg, N. McKinney, D. Spasic, L. Goda, C. Levenson, and J. J. Sninsky. 1990. Effects of primer-template mismatches on the polymerase chain reaction: human immunodeficiency virus type 1 model studies. *Nucleic Acids Res.* **18:** 999–1005.

Lawrence, C. B., T. Shalom, and S. Honda. 1989. A comprehensive software package for nucleotide and protein sequence analysis for UNIX systems. Molecular Biology Information Resource, Technical Report, Baylor College of Medicine, Houston, TX.

Lewis, A. P., M. J. Sims, D. R. Gewert, T. C. Peakman, H. Spence, and J. S. Crowe. 1994. Taq DNA polymerase extension of internal primers blocks polymerase chain reactions allowing differential amplification of molecules with identical 5' and 3' ends. *Nucleic Acids Res.* **22:** 2859–2861.

Mertz, L. M., and A. Rashtchian. 1994. Nucleotide imbalance and polymerase chain reaction: effects on DNA amplification and synthesis of high specific activity radiolabeled DNA probes. *Anal. Biochem.* **221:** 160–165.

Patterson, B. K., M. Till, P. Otto, C. Goolsby, M. R. Furtado, L. J. McBride, and S. M. Wolinsky. 1993. Detection of HIV-1 DNA and messenger RNA in individual cells by PCR-driven *in situ* hybridization and flow cytometry. *Science* **260:** 976–979.

Peluso, R., A. T. Haase, L. Stowring, M. Edwards, and P. Ventura. 1985. A Trojan horse mechanism for the spread of visna virus in monocytes. *Virology* **147:** 231–236.

Peng, H., T. A. Reinhart, E. F. Retzel, K. A. Staskus, M. Zupancic, and A. T. Haase. 1995. Single cell transcript analysis of human immunodeficiency virus gene expression in the transition from latent to productive infection. *Virology.* In press.

Rychlik, W., W. J. Spencer, and R. E. Rhoads. 1990. Optimization of the annealing temperature for DNA amplification in vitro. *Nucleic Acids Res.* **18:** 6409–6412.

Sambrook, J., E. F. Fritsch, and T. Maniatis. 1989. Molecular cloning: A laboratory manual. Cold Spring Harbor Laboratory Press, Cold Spring Harbor, NY.

Schowalter, D. B., and S. S. Sommer. 1989. The generation of radiolabeled DNA and RNA probes with polymerase chain reaction. *Anal Biochem.* **177:** 90–94.

Staskus, K. A., L. Couch, P. Bitterman, E. F. Retzel, M. Zupancic, J. List, and A. T. Haase. 1991a. *In situ* amplification of visna virus DNA in tissue sections reveals a reservoir of latently infected cells. *Microbial Pathogenesis* **11:** 67–76.

Staskus, K. A., E. F. Retzel, E. D. Lewis, J. L. Silsby, S. St. Cyr, J. M. Rank, S. W. Wietgrefe, A. T. Haase, R. Cook, D. Fast, P. T. Geiser, J. T. Harty, S. H. Kong, C. J. Lahti, T. P. Neufeld, T. E. Porter, E. Shoop, and K. Zachow. 1991b. Isolation of replication-competent molecular clones of visna virus. *Virology* **181:** 228–240.

Tecott, L. H., J. D. Barchas, and J. H. Eberwine. 1988. *In situ* transcription: specific synthesis of complementary DNA in fixed tissue sections. *Science* **240:** 1661–1664.

17

Y CHROMOSOME-SPECIFIC PCR: MATERNAL BLOOD

Diana W. Bianchi

The ability to isolate fetal cells from maternal blood allows prenatal genetic diagnosis to be performed by maternal venipuncture, at no risk to the fetus. Although male lymphocytes were reported in maternal metaphase spreads obtained from the blood of pregnant women more than 20 years ago (Walknowska et al., 1969), their rarity when compared with the number of maternal cells present precluded further work. The ratio of fetal to maternal cells in maternal peripheral blood is on the order of $1 : 10^5$ to $1 : 10^7$ (Ganshirt-Ahlert et al., 1990; Chueh and Golbus, 1990; Price et al., 1991). Prior to the development of PCR, research in this field was limited by the technical challenges of collecting enough fetal cells for analysis. With the advent of PCR, however, the extremely small number of fetal cells in a given maternal sample is no longer an obstacle if the target sequence being amplified is absent in the mother. We, as well as other groups (Wachtel et al., 1991), have performed PCR amplification of sequences unique to the Y chromosome to verify the presence of fetal cells in blood samples from women carrying male fetuses. Theoretically, any paternally inherited DNA polymorphism can be amplified and detected in fetal cells. Examples of fetal autosomal genes demonstrated in maternal blood include *HLA-DR4* (Yeoh et al., 1991), *HLA-*

DQα (Geifman-Holtzman et al., 1995), Hemoglobin Lepore-Boston (Camaschella et al., 1990), and Rhesus D (Lo et al., 1993a).

Although the early diagnosis of male fetal sex by PCR amplification of maternal blood for Y chromosomal sequences is of potential clinical use in families at risk for X-linked disorders (Lo et al., 1989,1990,1993b), we are concerned about the possibility of fetal cells persisting from prior pregnancies (Ciaranfi et al., 1977; Schroder et al., 1974; Bianchi et al., 1993a) and the consequent problem of false positive diagnosis. In our laboratory, we perform Y chromosomal PCR amplification to correlate the presence of an amplified band of DNA with a male fetus as part of a continuing assessment of the efficacy and purity of a variety of fetal cell separation techniques (Bianchi et al., 1993b,1994). Our fetal cell diagnostic work is based on a combination of fetal cell enrichment by multiparameter flow sorting, followed by PCR for sequences present on the Y chromosome. This will become especially important in the future, when fetal cells isolated from maternal blood are used to diagnose conditions inherited in an autosomal recessive manner, where the presence of the maternal genome will complicate interpretation of results. Improvement in fetal cell purity (on the order of 90%) will be necessary.

The potential passage of various cells from the fetus to the mother is being studied by several laboratories. Specific examples of cell types include the trophoblast (Mueller et al., 1990; Adinolfi, 1991), "trophoblast-like" cells (Bruch et al., 1991), the lymphocyte (Iverson et al., 1981; Yeoh et al., 1991), and the granulocyte (Wessman et al., 1992). We are focusing our efforts on the erythroblast, as this cell type is unlikely to be maternal in origin (Bianchi et al., 1990).

Protocol

Flow Cytometry

Venous blood (20 ml) is collected in citrate dextrose (ACDA) from a pregnant woman. The blood is diluted 1:3 with Hanks' balanced salt solution (HBSS), layered over a Ficoll-Hypaque column (Pharmacia Labs, Piscataway, NJ), and centrifuged at 2000 rpm for 20 min at room temperature in a Sorvall RT 6000 centrifuge. The mononuclear cell layer is removed, washed with phosphate-buffered saline

(PBS), and centrifuged at 1400 rpm for 10 min at 4°C. The supernatant is discarded and the cell pellet is incubated with a 1:10 dilution of fluorescence-conjugated antierythroblast antibody in PBS on ice for 30 min. The cells are washed once more in PBS prior to flow cytometry.

Fetal cell analysis and sorting are performed on a Becton-Dickinson FACS Star Plus with a Lysis II software program. The gain is standardized manually using fluorescent beads and a fluorescence-conjugated antibody to an antigen not expressed on human cells, keyhole limpet hemocyanin (Becton-Dickinson Monoclonals, San Jose, CA). A small aliquot of the patient's mononuclear cells is incubated with the control antibody to determine background fluorescence. In the actual sort, fluorescent (antierythroblast antibody positive) and nonfluorescent (antierythroblast antibody negative) cells are determined by physical separation on a logarithmic scale. Cells and sheath fluid are flow sorted into 1.5-ml centrifuge tubes. The tubes are then centrifuged, the sheath fluid is removed, the cells are resuspended in 16 μl dH$_2$O, and refrigerated at 4°C.

PCR

Because the presence of extraneous male DNA will result in inaccurate interpretation of experimental results, extreme caution must be observed in all phases of the protocol. We prepare all reagents for PCR in a hood with separate pipettors. We change gloves frequently. We do not allow known sources of male DNA inside the hood. Amplified products are analyzed in a separate area.

We have had good results amplifying a 291-bp target sequence recognized by the probe p 49a. Probe p 49a is a low-copy sequence that maps to Y q 11.2 and is absolutely specific for the Y chromosome (Lucotte et al., 1991). The primers we use are: 49a - 06 :5' CTT-TCT-TTT-CAG-GCA-TTT-CCT-GCT-TAT and 49a -07 :5' GTT-CTA-CAG-AAA-AGT-TAT-TGC-CAA-GTAT.

The PCR is set up as follows: a master mix is made up for ($n + 1$) reactions, where n equals the number of samples to be amplified. The master mix is based on a 25-μl reaction volume. For each reaction, the master mix contains 2.5 μl of 10X buffer (Cetus), 4 μl of nucleotides (0.2 μM final concentration), 1 μl of each primer (0.5 μM final concentration), and 2.5 units of Taq DNA polymerase.

The sorted cells are boiled for 5 min. The 16 μl of denatured cell lysate are added to the master mix and capped with 50 μl of mineral oil. Each experiment has a negative (reagent) control with no added DNA, and positive controls consisting of 500 ng genomic female DNA with a range of added male DNA (0.01 to 100 ng).

Amplification consists of 30 cycles at 94° for 1 min, 62° for 2 min, and 72° for 3 min. The extension at 72° proceeds for 10 min in the final cycle. After amplification, the samples are loaded in a 20-ml 2% agarose minigel with 250 ng/ml of ethidium bromide and separated by electrophoresis at 70 mV for 60 min (model Horizon 58, Bethesda Research Laboratories, Bethesda, MD). The gels are inspected under ultraviolet light, photographed, denatured in 0.5 M NaOH–0.5 M NaCl for 5 min, and neutralized in three 5-min changes of 0.4 M Tris HCl–0.08 M Tris base–0.6M NaCl. The DNA is then transferred to Hybond N$^+$ filters in 20X SSC [0.10 M NaCl–0.01 M Na citrate (pH 7)] for 4 hr to overnight. The filters are rinsed briefly in 2X SSC, soaked in 0.4 M NaOH for 20 min, baked in a vacuum oven at 80°C for 20 min, layered with mesh, and placed in the following buffer at 42°: 5% sodium dodecyl sulfate (SDS), 50% formamide, 20 μg/ml sheared salmon sperm DNA, 5% dextran sulfate, 6X SSC, 5X Denhardt's solution. Probe p 49a is oligonucleotide labeled with ^{32}P. The filters are incubated overnight with 5×10^5 counts of probe per milliliter, washed in three changes of 0.2X SSC–0.4% SDS at 42°C and three changes of the same solution at 55°C. After drying, the filters are exposed to Kodak X-Omat AR film at -70°C with Cronex (DuPont Co., Wilmington, DE) intensifying screens. Autoradiographs are developed 1–3 days after exposure. An example of PCR amplification of the Y chromosome sequence in cells obtained from maternal blood is shown in Fig. 1.

Discussion

We have described one strategy for detecting rare (male) fetal DNA sequences in maternal blood. Infrequently, a large enough fetomaternal hemorrhage occurs that permits the direct detection of amplified fetal DNA on an agarose gel under ultraviolet light. In general, however, most fetal DNA is detectable only on autoradiographs. We use this approach because it is sensitive enough to detect one fetal cell and the signal-to-noise ratio can be controlled by the duration of film exposure. More important, it allows us to determine the extent of background false positive amplification. Other investigators use

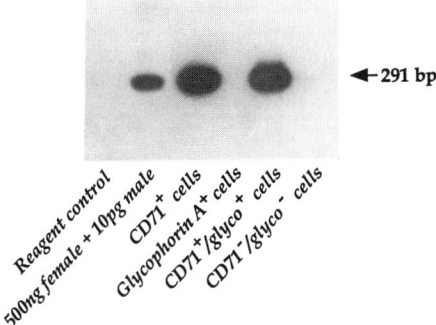

Figure 1 Autoradiograph of amplified DNA in fetal cells sorted from the blood of a pregnant woman. The fetus was male. CD71 is an antibody to the transferrin receptor and is present in all cells that actively incorporate iron. Glycophorin A is an antibody specific for red cells. Male fetal DNA is detected in CD71$^+$ and CD71$^+$/Glyco$^+$ cells, which are likely to be erythroblasts. Male fetal DNA is not detected in Glyco$^+$ cells because the majority of them are not nucleated. The CD71$^-$/Glyco$^-$ cells are predominantly maternal lymphocytes.

nested primers (Lo et al., 1989; Kao et al., 1992; Suzumori et al., 1992), but this technique involves the transfer of only a small part of the original reaction mixture to a second test tube. We are currently optimizing quantitative PCR techniques to determine the numbers of maternal and fetal cells remaining after a variety of cell separation methods (Bianchi et al., 1994).

Literature Cited

Adinolfi, M. 1991. On a non-invasive approach to prenatal diagnosis based on the detection of fetal nucleated cells in maternal blood samples. *Prenatal Diagn.* **11**:799–804.

Bianchi, D. W., A. F. Flint, M. F. Pizzimenti, J. H. M. Knoll, and S. A. Latt. 1990. Isolation of fetal DNA from nucleated erythrocytes in maternal blood. *Proc. Natl. Acad. Sci. U.S.A.* **87**:3279–3283.

Bianchi, D. W., A. P. Shuber, M. A. DeMaria, A. C. Fougner, and K. W. Klinger. 1994. Fetal cells in maternal blood: Determination of purity and yield by quantitative polymerase chain reaction. *Am. J. Obstet. Gynecol.* **171**:922–926.

Bianchi, D. W., S. Sylvester, G. K. Zickwolf, M. A. DeMaria, G. J. Weil, O. H. Geifman. 1993a. Fetal stem cells persist in maternal blood for decades post-partum. *Am. J. Human Genet.* **53**(suppl): A 251.

Bianchi, D. W., G. K. Zickwolf, M. C. Yih, A. F. Flint, O. H. Geifman, M. S. Erikson, and J. M. Williams. 1993b. Erythroid-specific antibodies enhance detection of fetal nucleated erythrocytes in maternal blood. *Prenatal Diagn.* **13**:293–300.

Bruch, J. F., P. Metezeau, N. Garcia-Fonknechten, Y. Richard, V. Tricottet, B-L. Hsi, A. Kitzis, C. Julien, and E. Papiernik. 1991. Trophoblast-like cells sorted from peripheral maternal blood using flow cytometry: A multiparametric study involving transmission electron microscopy and fetal DNA amplification. *Prenatal Diagn.* **11**:787–798.

Camaschella, C., A. Alfarano, E. Gottardi, M. Travi, P. Primignani, F. Caligaris Cappio, G. Saglio. 1990. Prenatal diagnosis of fetal Hemoglobin Lepore-Boston disease on maternal peripheral blood. *Blood* **75**:2102–2106.

Chueh, J., and M. S. Golbus. 1990. Prenatal diagnosis using fetal cells in the maternal circulation. *Semin. Perinatol.* **14**:471–482.

Ciaranfi, A., A. Curchod, and N. Odartchenko. 1977. Survie de lymphocytes foetaux dans le sang maternel post-partum. *Schweiz. Med. Wschr.* **107**:134–138.

Ganshirt-Ahlert, D., M. Pohlschmidt, A. Gal, P. Miny, J. Horst, and W. Holzgreve. 1990. Ratio of fetal to maternal DNA is less than 1 in 5000 at different gestational ages in maternal blood. *Clin. Genet.* **38**:38–43.

Geifman-Holtzman, O., E. J. Holtzman, T. J. Vadnais, V. E. Phillips, E. L. Capeless, and D. W. Bianchi. 1995. Detection of fetal HLA-DQα sequences in maternal blood: A gender-independent technique of fetal cell identification. *Prenatal Diagn.* in press.

Iverson, G. M., D. W. Bianchi, H. M. Cann, and L. A. Herzenberg. 1981. Detection and isolation of fetal cells from maternal blood using the fluorescence-activated cell sorter (FACS). *Prenatal Diagn.* **1**:61–73.

Kao, S.-M., G.-C. Tang, T.-T. Hsieh, K.-C. Young, H.-C. Wang, and C.-C. Pao. 1992. Analysis of peripheral blood of pregnant women for the presence of fetal Y chromosome-specific ZFY gene deoxyribonucleic acid sequences. *Am. J. Obstet. Gynecol.* **166**:1013–1019.

Lo, Y-M. D., J. S. Wainscoat, M. D. G. Gillmer, P. Patel, M. Sampietro, and K. A. Fleming. 1989. Prenatal sex determination by DNA amplification from maternal peripheral blood. *Lancet* **ii**:1363–1365.

Lo, Y.-M. D., P. Patel, M. Sampietro, M. D. G. Gillmer, K. A. Fleming, and J. S. Wainscoat. 1990. Detection of single copy fetal DNA sequence from maternal blood. *Lancet* **335**:1463–1464.

Lo, Y.-M. D., P. J. Bowell, M. Selinger, I. Z. Mackenzie, P. Chamberlain, M. D. G. Gillmer, T. J. Littlewood, K. A. Fleming, and J. S. Wainscoat. 1993a. Prenatal determination of fetal RhD status by analysis of peripheral blood of rhesus negative mothers. *Lancet* **341**:1147–1148.

Lo, Y.-M. D., P. Patel, C. N. Baigent, M. D. G. Gillmer, P. Chamberlain, M. Travi, M. Sampietro, J. S. Wainscoat, K. A. Fleming. 1993b. Prenatal sex determination from maternal peripheral blood using the polymerase chain reaction. *Hum. Genet.* **90**:483–488.

Lucotte, G., F. David, and M. Mariotti. 1991. Nucleotide sequence of p 49a, a genomic Y-specific probe with potential utilization in sex determination. *Mol. Cell. Probes* **5**:359–363.

Mueller, U. W., C. S. Hawes, A. S. Wright, A. Petropoulos, E. DeBoni, F. A. Firgaira, A. A. Morley, D. R. Turner, and W. R. Jones. 1990. Isolation of fetal trophoblast cells from peripheral blood of pregnant women. *Lancet* **336**:197–200.

Price, J. O., S. Elias, S. S. Wachtel, K. Klinger, M. Dockter, A. Tharapel, L. P. Shulman, O. P. Phillips, C. M. Meyers, D. Shook, and J. L. Simpson. 1991. Prenatal diagnosis using fetal cells isolated from maternal blood by multiparameter flow cytometry. *Am. J. Obstet. Gynecol.* **165**:1731–1737.

Schroder, J., A. Tiilikainen, and A. de la Chapelle. 1974. Fetal leukocytes in the maternal circulation after delivery. *Transplantation* **17**:346–354.

Suzumori, K., R. Adachi, S. Okada, T. Narukawa, Y. Yagami, and S. Sonta. 1992. Fetal cells in the maternal circulation: Detection of Y-sequence by gene amplification. *Obstet. Gynecol.* **80**:150–154.

Wachtel, S., S. Elias, J. Price, G. Wachtel, O. Phillips, L. Shulman, C. Meyers, J. L. Simpson, and M. Dockter. 1991. Fetal cells in the maternal circulation: isolation by multiparameter flow cytometry and confirmation by polymerase chain reaction. *Hum. Reprod.* **10**:1466–1469.

Walknowska, J., F. A. Conte, and M. M. Grumbach. 1969. Practical and theoretical implications of fetal/maternal lymphocyte transfer. *Lancet* **i**:1119–1122.

Wessman, M., K. Ylinen, and S. Knuutila. 1992. Fetal granulocytes in maternal venous blood detected by *in situ* hybridization. *Prenatal Diagn.* **12**:993–1000.

Yeoh, S. C., I. L. Sargent, C. W. G. Redman, B. P. Wordsworth, and S. L. Thein. 1991. Detection of fetal cells in maternal blood. *Prenatal Diagn.* **11**:117–123.

18

GENOMIC SUBTRACTION

Don Straus

Studies of genes defined by mutant phenotypes have led to rapid progress in fields that include developmental biology, cell cycle control, oncology, and hereditary disease. However, it is very difficult to isolate phenotypically defined genes from all but a few experimental organisms. The most generally applicable method for cloning genes corresponding to mutations—chromosome walking—is time-consuming, labor intensive, and can be impeded by unclonable sequences or by blocks of repetitive DNA. Genomic subtraction provides an efficient alternative to chromosome walking that can be used if a deletion mutation is available.

Genomic subtraction purifies DNA fragments that are present in one sample but absent in a related one. When applied to wild-type DNA and DNA from a homozygous deletion mutant, genomic subtraction purifies the wild-type DNA that is missing in the deletion mutant. The method is easy to use and, compared with related methods (Barr and Emanuel, 1990; Bautz and Reilly, 1966; Bjourson and Cooper, 1988; Kunkel et al., 1985; Lamar and Palmer, 1984; Nussbaum et al., 1987; Welcher et al., 1986; Wieland et al., 1990), yields impressive enrichments (Straus and Ausubel, 1990; Sun *et al.*, 1992a,b).

Overview of Genomic Subtraction

Genomic subtraction purifies DNA that is missing in a deletion mutant by removing from wild-type DNA the sequences that are present in both the wild-type and the deletion mutant genomes. The DNA that corresponds to the deleted region remains. Enrichment for the deleted sequences is achieved by allowing a mixture of denatured wild-type and biotinylated deletion mutant DNA to reassociate (Fig. 1,A). After reassociation, the biotinylated sequences are removed by binding to avidin-coated beads (Fig. 1,B). This subtraction process is then repeated several times (Fig. 1,C). In each cycle, the unbound wild-type DNA from the previous round is hybridized with fresh biotinylated deletion mutant DNA. The unbound DNA from the final cycle is ligated to adaptors (Fig. 1,D) and amplified by using one strand of the adaptor as a primer in the polymerase chain reaction (PCR; Fig. 1,E). The amplified sequences can then be used to probe a genomic library (Fig. 1,F) or they can be cloned directly (Fig. 1,G).

Genetic Prerequisites

Wild-type restriction fragments that cannot reassociate with deletion mutant DNA will be isolated by genomic subtraction. Thus, for the method to work, the deletion mutation must be homozygous. There must exist a restriction fragment in the wild-type sample that is composed of unique sequences and that is entirely deleted in the mutant. The use of genomic subtraction is therefore appropriate for deletions that are several times larger than the average restriction fragment (~250 bp) and that occur at nonessential sites.

It is demanding to apply the method to complex genomes because genomic subtraction entails several rounds of reassociation during which single-copy DNA strands must reanneal almost completely. Using the protocol described here, we have efficiently isolated DNA

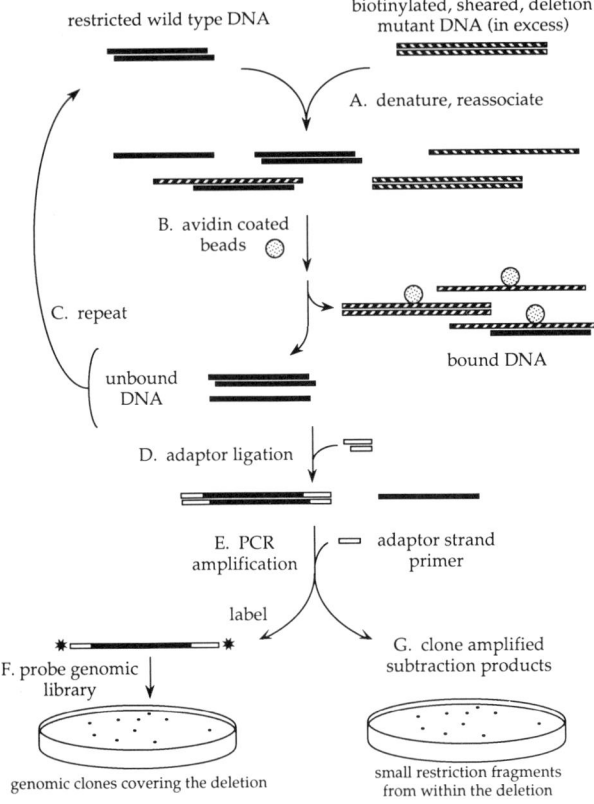

Figure 1 A schematic representation of genomic subtraction.

corresponding to 5-kb deletions from organisms with genome sizes up to 10^8 bp using 1-day long reassociation reactions. Achieving similar enrichments in a comparable time frame using organisms with complex genomes, such as humans or maize, requires achieving higher $C_o t$ values. We have developed a related method, RFLP subtraction, that efficiently isolated RFLPs [including some deletions from mouse [Rosenberg et al., 1994].

Applications

Genomic subtraction has been used to efficiently isolate DNA corresponding to chromosomal deletions in organisms from several kingdoms. We originally tested the method using a strain of yeast containing a characterized 5-kb deletion mutation representing about 1/4000 of the genome (Straus and Ausubel, 1990). Purification of the sequences corresponding to the deletion required three rounds of subtraction, 3 days, and 30 μg of genomic DNA. Figure 2A shows that the products of genomic subtraction correspond in size to the fragments from the cloned 5-kb sequence that is missing in the deletion mutant. Figure 2B demonstrates that the subtraction prod-

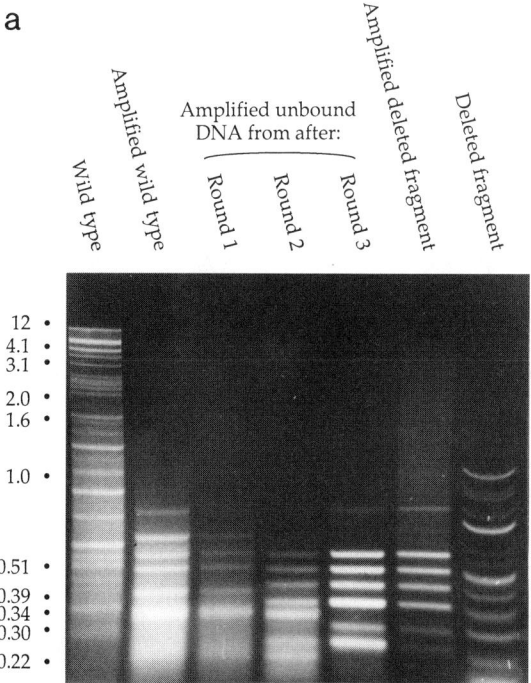

Figure 2A Gel electrophoresis of the DNA products of genomic subtraction in 2% agarose. From left to right: wild-type yeast DNA cut with Sau3A (0.75 μg); amplified wild-type Sau3A-digested yeast DNA; amplified unbound DNA from after the first, second, and third cycles of genomic subtraction; amplified Sau3A digest of the 5-kb Bgl II fragment that is missing in the deletion mutant; the deleted 5-kb Bgl II fragment digested with Sau3A. Comparing the products obtained after three rounds of subtraction with the amplified fragments derived from the cloned deleted fragment shows that genomic subtraction purified the desired DNA. The first and last lanes contain fragments that were not capped with adaptors and were thus 48 bp shorter and migrated faster than comparable fragments in the other lanes.

b

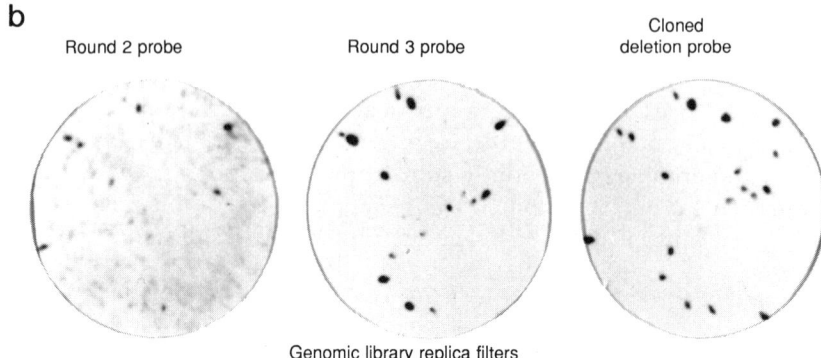

Figure 2B An autoradiogram of three replica filters made from a plate containing about 1X 10^5 yeast genomic plasmid clones. The filters were hybridized with the labeled products of the amplification of unbound DNA from after either the second (round 2 probe) or third (round 3 probe) subtraction cycles, or with the labeled, cloned 5-kb Bgl II fragment that corresponds to the fragment that is missing in the deletion mutant. The products obtained after three rounds of subtraction identify clones that hybridize to the cloned deleted fragment. Note that some of the same clones are identified by a probe made after two rounds of subtraction before purification was achieved (see lane "round 2" in Fig. 2A).

ucts can accurately identify genomic clones corresponding to a deletion mutation even if only partial purification is achieved.

Genomic subtraction has been used to clone the region of DNA that contains a previously uncharacterized virulence gene that is deleted in a mutant strain of *Yersinia pestis*, the bacterium that causes bubonic plague (J. Shao, D. Straus, J.Michel, F. Ausubel, unpublished results). The method has also been used to isolate the *ga-1* gene, which encodes an enzyme involved in gibberellin biosynthesis, from the plant *Arabidopsis* (Sun et al., 1992a,b). The deletion at *ga-1* spans 5 kb, the equivalent of about 1/20,000 of the *Arabidopsis* genome.

Although the protocol is presented in the context of isolating DNA corresponding to deletion mutations, genomic subtraction has other applications. For example, DNA from a pathogen could be purified from an infected sample by subtraction with an uninfected sample. The methodology could be used for differential cloning from cDNA libraries or for comparing the genomes of two related organisms. For example, we isolated fragments from a pathogenic *Escherichia coli* K1 strain that do not hybridize to a nonpathogenic *E. coli* K12 strain (G. Juang, D. Straus, F. Ausubel, G. Church, unpublished results).

Protocol

Reagents

EE: 10 mM (Na)EPPS, 1 mM EDTA (pH 8.0)
1 M (Na)EPPS (pH 8.25): used because the pk$_a$ (8.0 at 20°C) changes only slightly with temperature (as opposed to Tris buffer).

<u>126.15 g [N-[2-hydroxyethyl]-piperazine-'-[3-propane-sulfonic acid] (EPPS) acid (Sigma chemical Co., St. Louis MO)
395 ml H$_2$O
13.2 ml concentrated NaOH (51% by acidimetry)</u>
500 ml 1 M (Na)EPPS (pH 8.25; pH of a 10 mM solution = 8.0)

EEN: 0.5 M NaCl, 10 mM (Na)EPPS, 1 mM EDTA (pH 8.0)
Photobiotin acetate (2 µg/µl in pure H$_2$O)
Fluoricon avidin polystyrene assay particles: IDEXX Corp. (Westbrook, ME); 0.7–0.9 µ diameter
Ultrafree microcentrifuge filter unit: Millipore UFC3 OGV 00, SpinX: Costar RTY 8160 or equivalent.
Sunlamp for photobiotinylation: GE model RSM/H (275-W RSM bulb) or equivalent.

Some Notes on the Procedure

The time required for each round of reassociation depends on the complexity of the genome and the concentration of genomic DNA. We use reassociation times of forty to fifty times the time required for half of the single-copy sequences to reanneal. A rough estimate of the recommended time, based on the reassociation kinetics of *E. coli* DNA and the effect of 1 M NaCl on the rate of annealing (Britten et al., 1972), is:

$$\text{recommended reassoc time (hr)} = 50 \times t_{1/2(\text{unique})} \approx \frac{5 \times 10^{-7} \times X}{C_\Delta}$$

Where X is the complexity of the unique sequence component of the genome in base pairs and C_Δ is the concentration of deletion mutant DNA in micrograms per microliter.

We include a trace of end-labeled, single-stranded, exogenous DNA (just enough to follow the sample with a hand-held monitor) in the subtraction sample. This DNA simulates the desired unbound DNA and can help prevent losses and allow calculation of yields. We use a synthetic 84-mer that has been end labeled with polynucleotide kinase and [α – ^{32}P]ATP. The labeled DNA should (1) be long enough to resemble a wild-type DNA fragment, (2) be single stranded (to mimic the fate of the sticky single-stranded molecules in the sample, and (3) must not hybridize to the genome at 1M NaCl, 65°C.

Ethanol precipitations can be done entirely at room temperature providing the samples contain at least 20 μg of yeast tRNA. After being incubated for 5 min, the precipitate is collected by spinning for 5 min at 12,000 g. The pellets (which should always be large enough to be visible) are washed with 100% ethanol before drying. Carefully observe the pellets after the precipitations and lyophilizations. Is there extra material? (If so, there may be contaminating macromolecules in the genomic DNA.) Are the pellets too small? (Add some tRNA carrier; or, if the pellet is spread over the side of the tube, be careful to resuspend it all.) Be sure that the pellets are truly resuspended. Follow the sample by frequently monitoring counts. Check the tubes and filters for adsorbed radioactivity.

Controls

Ideally, DNA from a characterized deletion mutant should be used to test the method. If no such mutant is available, a deletion mutation can be mimicked by adding an equimolar amount of double-stranded, exogenous, nonhomologous vector or viral DNA to the genomic DNA before restriction digestion. Genomic DNA lacking the exogenous DNA then mimics a deletion mutant that is missing some unique sequence DNA. We have used ϕx174 DNA or adenovirus DNA in such control experiments. Subtracting the "spiked" genomic DNA from biotinylated genomic DNA should yield products enriched in the exogenous DNA fragments. The enrichment can be followed by analyzing the amplified products obtained after each round of subtraction. The cloned deletion DNA or the exogenous DNA is labeled and hybridized to dot blots containing the subtraction samples and standards composed of serial dilutions of unlabeled probe DNA. Passing the amplified samples over Sephadex G-50 spin columns (or the equivalent) is required to obtain full signals on the dot blots. A negative control subtraction in which wild-type DNA is used in both subtraction samples (i.e., both the biotinylated and nonbiotinylated samples) can identify fragments that do not correspond to a deletion, but that nevertheless are incompletely removed by subtraction.

It is important to make sure that the biotinylated DNA binds efficiently to the avidin coated beads. It is easy and efficient to check the biotinylated DNA biotinylated DNA in the following way. Add 10 μg of heat denatured biotinylated DNA and 20 μg of carrier yeast tRNA (in a total volume of 100 μl of EE) to 100 μl of 5% AP beads and process the sample as in steps 12–17 in the section "First Round of Subtraction." After ethanol-precipitating the filtrate, resuspend the unbound DNA and load on an agarose gel alongside 100 ng, 200 ng, 500 ng, and 1 μg of heat denatured biotinylated DNA. If, for example, the amount of unbound DNA is comparable to the 100 ng standard, then 99% of the biotinylated DNA was bound to the beads. Binding end-labeled biotinylated DNA to the avidin-coated beads is another good way to check the biotinylation and affinity chromatography steps. Be sure to remove all of the unincorporated label from the DNA before this test.

Controls relevant to the adaptor ligation and amplification steps are described in the protocol itself.

Preparation

Genomic DNA Isolation

DNA for genomic subtraction should be pure (i.e., free of contaminating exogenous DNA, RNA, ribonucleotides, polysaccharides and proteins). Extra care taken at this stage is worthwhile. After RNase treatment we repeatedly spool the DNA out of 50% EtOH to eliminate (biotinatable) nucleotides. We calculate genomic DNA concentrations by comparing band intensities after gel electrophoresis of the genomic DNA samples and several dilutions of bacteriophage λ DNA of known concentration. (Do not rely on absorbance measurement.) For organisms that are rich in polysaccharides and/or glycoproteins, we have found cetyltrimeth ammonium bromide (CTAB) extractions most useful (Murray and Thompson, 1980; Rogers and Bendich, 1985).

It is desirable to start with several hundred micrograms of deletion mutant DNA. If this DNA is scarce, however, 100 μg will suffice to conduct two genomic subtraction experiments. Several micrograms of wild-type DNA are also necessary.

Shearing and Biotinylating Deletion Mutant DNA

If possible, biotinylate 200 μg of sheared genomic DNA. This will provide extra DNA to work with if you need to repeat a subtraction or trouble shoot. A typical experiment might require 50 μg of biotinylated deletion mutant DNA. Nucleic acids are precipitated with 2 vol of ethanol. Precipitates are washed with 100% ethanol.

1. Sonicate 200 μg of DNA in 500 μl of EE buffer. We immerse a sealed microcentrifuge tube containing the sample in the water-filled cup of a Branson sonifier (Cell Disruptor 200) and sonicate for 10 sec at an output setting of 7. A sonicating tip can also be used, if care is taken not to contaminate samples with exogenous DNA.
2. Check the sheared DNA on a 1% agarose gel. The center of the smear should be about 3000 bp. Avoid generating fragments that are smaller than 500 bp. Since only about 1 in 100 nucleotides react with photobiotin (Forster et al., 1985), small fragments may escape biotinylation, leading to lowered subtraction efficiency.
3. Denature the sheared DNA by boiling for 2 min.
4. Add 50 μl of 3 M NaOAc. Ethanol precipitate, wash, dry.
5. Resuspend the pellet in 200 μl of *pure water* so that the DNA concentration is 1 μg/μl. Avoid Tris buffers.
6. Mix 200 μl of 2 μg/μl photobiotin (in H_2O) with 200 μl of 1 μg/μl sheared DNA (in H_2O) in a microcentrifuge tube.
7. Pipette four 100-μl aliquots of the DNA–photobiotin mixture into the bottom of four 2-ml microcentrifuge tubes (with wide-angle conical bottoms). It is important to keep the depth of the sample shallow. If the sample is deep, only the top layer will react fully. Any polypropylene vessel can be used as long as the depth of the DNA sample is $\leq$ 2 or 3 mm.
8. Place the samples (with open caps) in a rack 10 cm below the sun lamp in an ice bath in a polypropylene tray for 15 min. The lamp gets quite hot and can melt Styrofoam racks and ice buckets so be careful with placement of these items.
9. Combine the aliquots of biotinylated DNA. Add 40 μl (0.1 vol) 1 M Tris (pH 9.0).
10. Extract four times with *n*-BuOH (saturated with H_2O). Pull off the top *n*-BuOH layer (which is reddish, at first) saving the bottom, aqueous layer.
11. Bring the sample volume to 0.45 ml with H_2O. Add 50 μl 3 M NaOAc. Ethanol precipitate, wash, dry. The pellet obtained after ethanol precipitation should be a dark reddish-brown color.
12. Resuspend in the biotinylated sheared deletion mutant DNA at 2.5 μg/μl in 80 μl 2.5X EE.

Preparing the Wild-Type DNA

A few micrograms of the wild-type DNA are cut with a restriction enzyme that recognizes a 4-bp restriction site. The enzyme chosen must generate ends that are compatible with the oligonucleotide adaptor that will be ligated to the DNA products of genomic subtraction.

1. Cut a few micrograms of the wild-type DNA with a restriction enzyme that recognizes a 4-bp site (10-fold overdigestion).
2. Make sample 0.3 M NaOAc, 10 mM EDTA, and extract with a 1:1 mixture of phenol and chloroform.
3. Add 20 μg of ytRNA. Ethanol precipitate, wash, dry.
4. Resuspend the DNA at 0.1 μg/μl in EE.

Adaptor Preparation

1. Using T4 polynucleotide kinase, phosphorylate 25 μg of the oligonucleotide whose 5' end will be covalently linked to the genomic DNA (Maxam and Gilbert, 1980).
2. Anneal 25 μg of the kinased strand with 25 μg of the complementary oligonucleotide strand (which is not phosphorylated at the 5' end) in 50 μl of EE. Heat the mixture to 100°C for 2 min in a glass beaker containing 200 ml of water. Allow the bath to cool to room temperature by placing it on a laboratory bench.
3. If you intend to clone the PCR-amplified products, phosphorylate the 5' end of some of the oligonucleotide that corresponds to the adaptor strand whose 3' end will be ligated to genomic DNA. This will later be used as a primer in the PCR reaction.

Preparing the Affinity Matrix

We wash the affinity matrix to remove the azide and any free avidin. Each round of affinity chromatography consumes 100 μl of 5% suspension of AP beads. If five rounds of subtraction are applied to four samples, 2 ml of beads will be needed. DNA sticks nonspecifically to the beads when the salt concentration is low, unless SDS is present. Always use EEN (0.5 M NaCl) when the beads are in contact with DNA. The beads tend to float in salt solutions, therefore the preparative washes are done in EE. Then the washed beads are resuspended in EEN.

1. Put 1.25 ml of a 5% suspension of AP beads into each of four microcentrifuge tubes.
2. Spin for 1 min at 12,000 rpm in a microcentrifuge.
3. Pull off the supernatant. Try to minimize the amount of beads removed with the supernatant (some loss is inevitable).
4. Resuspend the pellet in 1.5 ml EE by repipetting or vortexing.
5. Wash the pellet twice more with 1.5-ml EE washes.
6. Resuspend the beads representing 5 ml of the original 5% suspension in 4 ml of EEN (assuming a 20% loss of beads during washing, this results in a 5% suspension).

Genomic Subtraction

First Round of Subtraction

1. Mix in a 500-μl microcentrifuge tube:
   2.5 μl of 0.1 μg/μl wild-type DNA restriction fragments (ends defined by a 4-bp recognition site)
   4 μl of 2.5 μ/μl sheared photobiotinylated deletion mutant DNA
   1 μl labeled exogenous single-stranded DNA ($\approx 10^5$ cpm; see earlier notes)
   2 μl of 20 μg/μl yeast tRNA
2. Boil 1 min.
3. Lyophilize.
4. Resuspend in 4 μl H$_2$O. (Inspect to make sure DNA is truly resuspended.)
5. Add 1 μl 5 M NaCl. Mix. Spin.
6. Count sample (Cerenkov method).
7. Carefully transfer sample to a fresh tube.
8. Cover sample with several drops of paraffin oil.
9. Incubate reassociation reaction at 65°C (see earlier notes for determining incubation times).
10. Bring reassociation reaction to 100 μl with EEN. Mix by vortexing.
11. Pull off the paraffin oil carefully. It is alright to leave the last bit—it disappears during the subsequent manipulations.
12. Add 100 μl of washed 5% AP beads in EEN to 100 μl hybridization.

13. Incubate for 10 min at room temperature.
14. Pipette sample/bead mixture into the cup of a microcentrifuge filter unit.
15. Spin the microcentrifuge filter unit for 30 sec in a microcentrifuge.
16. Wash the tube used for the hybridization and the pipette tip with 200 μl EEN and add this to the cup of the filter unit. Spin 30 sec. (The volume of the filtrate in the filter unit tube is now ≈ 400 μl.)
17. Ethanol precipitate, wash, dry.

Subsequent Rounds of Subtraction

1. Resuspend the pellet in 4 μl of 2.5 μg/μl sheared photobiotinylated deletion mutant DNA in 2.5X EE. Rinse the sides of the tube with the biotinylated DNA. Centrifuge the sample for 10 sec. Check to make sure that the pellet is completely resuspended.
2. Transfer the sample to a new 500-μl tube. The transfer ensures that all of the wild-type DNA is mixed with photobiotinylated DNA. Otherwise, if some of the wild-type DNA were stuck high on the wall of the old tube, it might not participate in the hybridization reaction. This DNA would not stick to the beads in the next step and would lower the enrichment factor. Follow the counts to make sure that losses are minimal during the transfer.
3. Boil 1 min.
4. Add 1 μl 5 M NaCl. Mix. Spin.
5. Count sample (Cerenkov). Save an aliquot of the sample for subsequent analysis by diluting 0.5 μl of the sample in 10 μl EE.
6. Carefully transfer sample to a fresh tube.
7. Incubate the sample at 65°C under paraffin oil as earlier.
8. Collect unbound DNA as before.

Capping the Unbound DNA with Adaptors

Capping the subtraction product with adaptors performs a dual role. It renders the subtraction products amplifiable and it eliminates a major class of contaminants from the amplified subtraction products. Wild type DNA fragments that correspond to fragments in the deletion mutant but that inefficiently hybridize to the biotinylated deletion mutant DNA contaminate the products of subtraction. Snapback DNA, for example, will form intramolecular hairpins to the exclusion of intermolecular duplexes. There may also be wild type DNA fragments with low T_ms that fail to efficiently hybridize to the biotinylated DNA. DNA that fails to hybridize to the biotinylated DNA for either of these reasons will not have sticky ends which can be ligated to the adaptor. If the adaptor is not ligated to the DNA it will not be amplifiable. The desired products, in contrast, will have two sticky ends, will ligate to the adaptor, and will be amplifiable. Before ligating adaptors, we ensure that fragments with low T_ms are completely denatured by including a high temperature incubation.

Several control ligations should be included. The adaptor ligation is tested by self-ligating 1 μg of adaptor DNA. Electrophoretic comparison [5% (3:1) NuSieve agarose] of the adaptor self-ligation products and the unligated adaptor determines the ligation efficiency. Ligating adaptors to vector DNA digested with the same enzyme used to cut the wild-type DNA is useful for testing the ability of the adaptors to make DNA amplifiable.

1. Resuspend the lyophilized unbound DNA from the final round of subtraction in 10 μl EE.
2. Mix in a 500-μl tube:
   5 μl of unbound DNA (1/2 of the sample)
   29 μl H$_2$O
   10 μl 5 M NaCl
   5 μl 10X EE
   1 μl of 20 μg/μl yeast tRNA
3. Incubate at 80°C for 30 min.
4. Add 100 μl EE. Ethanol precipitate, wash, dry.
5. Resuspend in 5 μl EE.
6. Mix in a 500-μl tube:
   2.5 μl DNA from previous step
   2 μl of 25 ng/μl adaptor (phosphorylated only at the 5' end that will be joined to the genomic DNA)
   4μl TE
7. Incubate at 37°C for 5 min (this step denatures sticky ends).
8. Add:
   1 μl 10X ligation buffer
   0.5 μl T4 ligase (1 Weiss unit/μl)
9. Incubate at room temperature for 30 min.

10. Heat the ligation reaction for 10 min at 70° to eliminate enzymatic activity.

Amplification

We amplify the unbound DNA by performing 35–50 PCR cycles composed of three segments: 30 sec at 94°C, 30 sec at 55°C, and 3 min at 72°C, with a final incubation of 10 min at 72°C. We include 0.5 μg of one strand of the adaptor (the strand whose 3' end was ligated to the wild-type genomic DNA) as a primer in a 50-μl reaction. Be sure to use the correct strand of the adaptor as a primer. If the amplified DNA will be directly cloned, the primer should be phosphorylated. Alternatively, the amplified products can be phosphorylated.

Amplify the following control samples also: (1) 10 fg of digested vector DNA, capped with adaptors, (2) 10 fg of wild-type DNA, capped with adaptors, (3) a positive control for amplification that uses different primers, and (4) a sample that has the primer corresponding to the adaptor strand but no template. We often see "no template" artifact bands which vary in size from sample to sample but which do not appear if the sample contains amplifiable template.

Analyzing the Products of Genomic Subtraction

Several methods for analyzing the products of genomic subtraction have been used. The labeled products of subtraction can be used to probe a genomic library (Fig. 2B). This approach is superior to probing Southern blots because it greatly reduces background caused by traces of repetitive sequences in the probe and the presence of random sequences from the genome in the probe. Furthermore, after screening a genomic library with the amplified subtraction products, one obtains a genomic clone that corresponds to the deleted sequences.

It is often useful to clone the products of subtraction. Probing of Southern or dot blots with the resulting individual cloned fragments indicates whether such fragments do indeed correspond to deletion mutations. Probing DNA from the F_2 progeny of crosses between mutant and wild-type parents with a cloned deletion fragment can show whether these genetic linkage of the deletion to the mutant gene.

Discussion

Genomic subtraction resembles cDNA subtraction methods in that it finds DNA fragments that occur in one sample but that are absent in another. The methods pose different technical problems, however, in part because genomic DNA is double stranded while cDNA and mRNA are single stranded. Single-stranded cDNA cannot self-anneal in traditional single-stranded cDNA subtraction, but wild-type DNA strands *can* self-anneal in genomic subtraction. Self-annealed wild-type strands cannot associate with biotinylated deletion mutant strands and thus lower the enrichment achieved per round of subtraction. Genomic subtraction, then, requires multiple rounds of subtraction to generate high enrichments for deleted sequences, while cDNA subtraction requires only one or two rounds.

Our method is designed to optimize rapid processing of several samples in parallel. Our method for affinity chromatography is especially efficient in this regard. Furthermore, in our hands, this technique removes biotinylated fragments more efficiently than the phenol–chloroform–streptavidin extraction method (Sive and St. John, 1988; Barr and Emanuel, 1990). There are several alternatives to photobiotinylation. We have labeled our DNA for subtractions by amplifying with biotinylated primers (Rosenberg *et al.*, 1994). This method is convenient but leads to only 97% removal of the biotinylated DNA compared to the near quantitative removal of the biotinylated DNA that we achieve using photobiotin. We have tested many affinity matrices and found the beads used here to be superior in that they efficiently bind biotinylated DNA (many matrices bind less than 90% of the input DNA) and have a high binding capacity.

We have achieved the best results when we have ligated adaptors onto the wild-type DNA after genomic subtraction. Attempts at making the wild-type DNA amplifiable by ligating adaptors onto the digested DNA before subtraction have, in general, not led to high enrichments unless steps are added to remove nonhybridizing subtraction products (Rosenberg *et al.*, 1994). A number of adaptors have been used successfully in genomic subtraction experiments. We currently use adaptors that are compatible with blunt-ended restriction fragments at one end and that have a nonpalindromic 3' overhang at the other end. New adaptors and primers should be well tested. Adaptors should dimerize efficiently without generating higher order multimers [ligate 1 μg of the adaptor to itself and analyze on a 5% NuSieve (3:1) agarose gel]. Primers should not efficiently amplify fragments from genomic DNA that is not capped with adaptors.

We have developed two methods that extend the range of the genomic subtraction technique. Large numbers of unique sequence RFLPs (restriction fragment length polymorphisms) can be isolated from complex genomes using RFLP subtraction (Rosenberg et al., 1994). We have also developed a powerful new method called Targeted RFLP Subtraction that isolates dense RFLPs tightly linked to any type of mutation—even point mutations—as long as the mutation has a scorable phenotype (manuscript in preparation).

Acknowledgments

I thank Fred Ausubel for his generous, cheerful, and constant support during the development of genomic subtraction. I also thank members of my laboratory and Fred Ausubel's laboratory for many useful discussions. I am grateful to Michael Rosenberg for helpful comments on the manuscript. The work has been supported by grants from Hoechst, AG to Massachusetts General Hospital and by USDA-CRGO Grant 91-37300-6501 to DS.

Literature Cited

Barr, F., and B. Emanuel. 1990. Application of a subtraction hybridization technique involving photoactivatable biotin and organic extraction to solution hybridization analysis of genomic DNA. *Anal. Biochem.* **186**:369–373.

Bautz, E. K. F., and E. Reilly. 1966. Gene-specific messenger RNA: Isolation by the deletion method. *Science* **151**:328–330.

Bjourson, A. J., and J. E. Cooper. 1988. Isolation of Rhizobium-loti strain-specific DNA sequences by subtraction hybridization. *Appl. Environ. Microbiol.* **54**:2852–2855.

Britten, R., D. Graham, and B. Neufeld. 1972. Analysis of repeating DNA sequences by reassociation. *Methods Enzymol.* **29**:363–418.

Forster, A. C., J. L. McInnes, D. C. Skingle, and R. H. Symons. 1985. Non-radioactive hybridization probes prepared by the chemical labelling of DNA and RNA with a novel reagent, photobiotin. *Nucl. Acids Res.* **13**:745–761.

Kunkel, L. M., A. P. Monaco, W. Middlesworth, H. D. Ochs, and S. A. Latt, 1985. Specific cloning of DNA fragments absent from the DNA of a male patient with an X chromosome deletion. *Proc. Natl. Acad. Sci. U.S.A.* **82**:4778–4782.

Lamar, E. E., and E. Palmer. 1984. Y-encoded, species-specific DNA in mice: evidence that the Y chromosome exists in two polymorphic forms in inbred strains. *Cell.* **37**:171–177.

Maxam, A. M., and W. Gilbert. 1980. Sequencing end-labeled DNA with base-specific chemical cleavages. *Methods Enzymol.* **65**:499–560.

Murray, M. G., and W. F. Thompson. 1980. Rapid isolation of high molecular weight plat DNA. *Nucleic Acids Res.* **8**:4321–4325.

Nussbaum, R. L., J. G. Lesko, R. A. Lewis, S. A. Ledbetter, and D. H. Ledbetter 1987. Isolation of anonymous DNA sequences from within a submicroscopic X chromoscomal deletion in a patient with choroideremia, deafness, and mental retardation. *Proc. Natl. Acad. Sci. U.S.A.* **84**:6521–6525.

Rogers, S. O., and A. J. Bendich. 1985. Extraction of DNA from milligram amounts of fresh, herbarium and mummified plant tissues. *Plant Mol. Biol.* **5**:69–76.

Rosenberg, M., M. Przybylska, and D. Straus. 1994. RFLP subtraction: a method for making libraries of polymorphic markers. *Proc. Natl. Acad. Sci. U.S.A.* **91**:6113–6117.

Sive, H., and T. St. John. 1988. A simple subtractive hybridization technique employing photoactivatable biotin and phenol extraction. *Nucleic Acids Res.* **16**:10937.

Straus, D., and F. M. Ausubel. 1990. Genomic subtraction for cloning DNA corresponding to deletion mutations. *Proc. Natl. Acad. Sci. U.S.A.* **87**:1889–1893.

Sun, T.-P., H. M. Goodman, and F. M. Ausubel. 1992a. Cloning the *Arabidopsis thaliana ga-1* locus by genomic subtraction. *Plant Cell.* **4**:119–128.

Sun, T.-P., D. Straus, and F. Ausubel. 1992b. Cloning *Arabidopsis* genes by genomic subtraction. In: *Methods in arabidopsis research* (ed. C. Koncz, J. Schell, and N. H. Chua), pp. 331–341. World Scientific Publishing Co., Singapore.

Welcher, A. A., A. R. Torres, and D. C. Ward. 1986. Selective enrichment of specific DNA, cDNA and RNA sequences using biotinylated probes, avidin and copper-chelate agarose. *Nucleic Acids Res.* **14**:10027–10044.

Wieland, I., G. Bolger, G. Asouline, and M. Wigler. 1990. A method for difference cloning: gene amplification following subtractive hybridization. *Proc. Natl. Acad. Sci. U.S.A.* **87**:2720–2724.

19

DNA AMPLIFICATION-RESTRICTED TRANSCRIPTION TRANSLATION (DARTT): ANALYSIS OF *IN VITRO* AND *IN SITU* PROTEIN FUNCTIONS AND INTERMOLECULAR ASSEMBLY

Erich R. Mackow

DNA amplification-restricted transcription translation, or DARTT, has been used to describe the PCR-based amplification of protein encoding genes and truncations coupled to the rapid analysis of protein functions following transcription and translation of the encoded protein. Transcriptional initiation signals are added to genes during amplification, and runoff RNA transcripts are synthesized (Saiki *et al.*, 1985; Stoflet *et al.*, 1988; Holland and Innis, 1990). PCR-generated RNAs are used to prime *in vitro* translation reactions with or without PCR-/added, ATG translational initiation signals (Sarkar and Sommer, 1989). DARTT has been used to define functionally active portions of polypeptides truncated at the amino and carboxy termini (Mackow *et al.*, 1990). Initially DARTT was used to define the minimum polypeptides required to recognize neutralizing monoclonal antibodies (Fig. 1). Subsequently, it has been used to define the amino acids necessary for oligomerization of protein subunits and for the assembly of viral particles (Clapp and Patton, 1991). More recently, DARTT has been extended to generating truncated polypeptides in transfected tissue culture cells by providing the Studier (Studier and Moffatt, 1986) cloned T7 RNA polymerase in

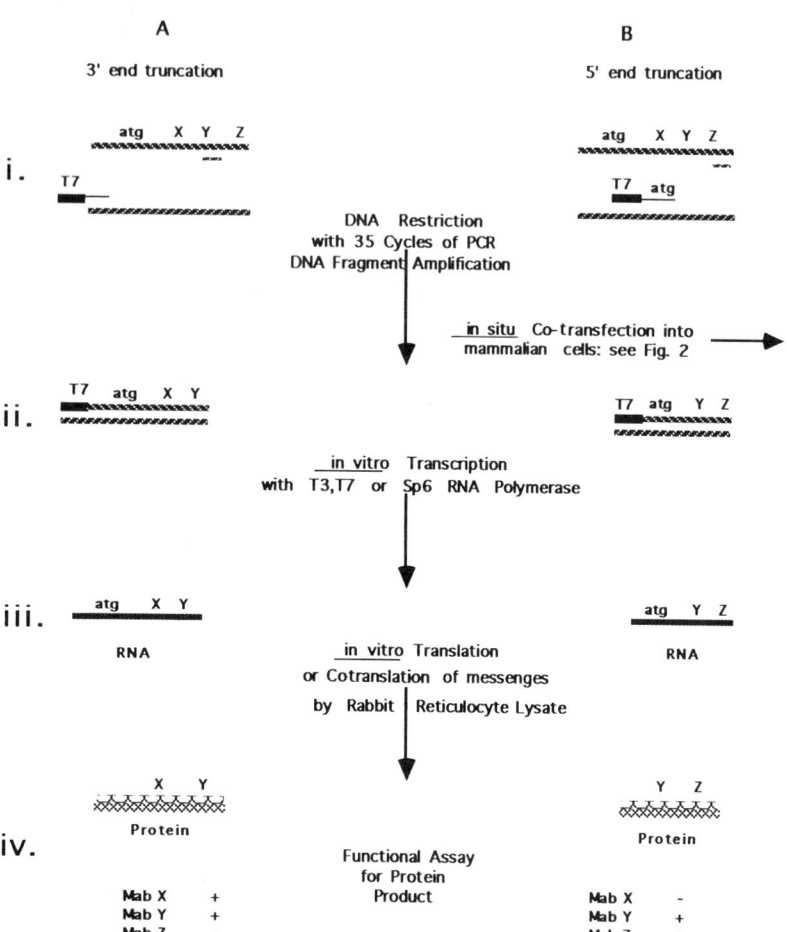

Figure 1 DARTT strategies for protein-function analysis. (i) DNA encoding a single protein is depicted by two bars with 5'–3' orientation specified by diagonals. X, Y, and Z represent functional regions of the encoded polypeptide and are preceded by a translational initiation codon (atg). Primers that delimit the region to be amplified are shown to anneal to their complementary DNA strands. In (A), the 3'-end limiting primer anneals at an internal position in the gene, and the 5'-end primer contains a noncomplementary T3 polymerase transcription initiation site (solid bar) upstream from a complementary sequence (thin line) and utilizes the native translational start site of the protein. In (B), the primer pair includes the 3' end of the gene but the upstream primer contains a noncomplementary T3 polymerase transcription initiation site and an ATG translational initiation site, the latter being in frame with the adjacent sequence. (ii) PCR amplification between primer pairs results in DNA that is restricted at the 3' end (A) or the 5' end (B) which includes both transcriptional and translational initiation elements. (iii) T3 RNA polymerase is used to synthesize plus-stranded RNA *in vitro* that is specific to the amplified DNA fragment. The transcribed DNA contains native (A) or new (B) translational initiation signals. (iv) RNA is translated into the

19. Analysis of Protein Functions and Intermolecular Assembly

trans (Fig. 2). This extended use of DARTT should allow the rapid definition of protein processing and assembly signals within the cell as well as the identification of cell regulatory or enzymatically active polypeptides. A brief synopsis of the tissue culture adaptation of DARTT follows the *in vitro* protocol. The ability to rapidly generate truncated polypeptide products via PCR in either an *in situ* or *in vitro* system is the advantage of the DARTT system for defining functionally active peptide fragments. In this chapter we present both an *in vitro* approach for using DARTT to identify functionally active polypeptides and one version of an *in situ* DARTT application.

Rotaviruses have a single outer capsid spike protein, VP4, which is required for cellular attachment and viral entry. The VP4 protein is cleaved by trypsin into two new peptides, VP8 and VP5, in order to make the virus infectious. VP4 is also a target of the immune system, and antibodies which bind VP4 neutralize the virus. It is important to determine the location of peptides involved in viral neutralization since identified peptides define functional portions of VP4 which could be further used to enhance the neutralizing immune response to rotavirus and to study viral entry and uncoating. We had previously defined the location of mutations which abolished binding of neutralizing monoclonal antibodies (N-MAbs) by a mutational analysis of the VP4 neutralization sites (Mackow et al., 1988). By cultivating a rotavirus in the presence of VP4-specific N-MAbs, isolating viral mutants which grew in the presence of the antibodies, and sequencing the gene 4 segments, we defined single-base, single amino acid substitutions in VP4 which allowed the virus to escape neutralization (Mackow et al., 1988). However, we synthesized peptides to the regions comprising the neutralization-defined mutation, and these peptides were not recognized by the selecting antibodies. In other words, the mutational analysis did not delineate a short peptide which could be recognized by the N-MAb, and suggested that functionally a larger, possibly conformationally determined epitope was required for N-MAb binding. As a result, we were prompted to define functional N-MAb binding regions in VP4 by an alternative method which would allow us to assay binding to large VP4 peptides. Our solution was to synthesize and assay VP4 polypeptide fragments via DNA amplification-restricted transcription and translation (Mackow et al., 1990). A generalized protocol for applying DARTT

---

encoded polypeptide in rabbit reitculocyte lysates *in vitro*. The polypeptide product is then assayed for the presence of retained functions (antigenicity, binding, catalysis, etc.) that exist in the full-length protein.

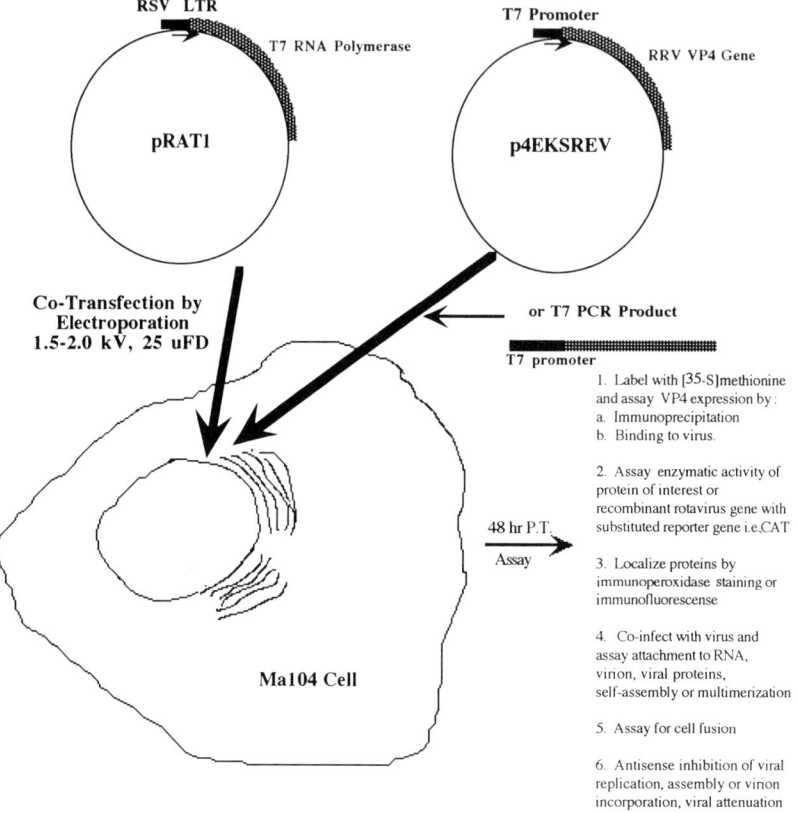

Figure 2 *In Situ* application of DARTT. PCR amplified products for the expression of proteins in *in vitro* systems (described in Figure 1) can be applied to *in situ* expression studies by introducing gene truncations into cells either directly or in plasmids. DARTT amplified genes or plasmids can be introduced into cells by a variety of transfection procedures including electroporation (depicted) and liposome mediated techniques. *In situ* expression of T3 or T7 driven genes occurs by the coexpression of T3 or T7 polymerases in the same transfected cells. RNA polymerases can be expressed constitutively from stably selected cell lines, inducibly from IPTG regulated plasmids and cell lines or by coinfection with T7 vaccinia viruses (Fuerst 1987a,b). One plasmid for expressing T7 RNA polymerase intracellularly which drives transcription of contransfected T7 driven genes is the pRAT plasmid constructed by Kevin Ryan (St. Judes Childrens Hospital, Memphis, TN). This plasmid places the T7 RNA polymerase under control of the Rous Sarcoma Virus (RSV) promoter. Transcription from the RSV LTR drives T7 RNA polymerase synthesis and in turn T7 RNA polymerase transcribes RNA from T7 promoters upstream of genes introduced into the cell. Low levels of protein synthesis are produced by this system, but intracellular functions and binding studies can be performed by introducing DARTT derived truncations into the T7 expressing cells. A list of possible uses for *in situ* DARTT are listed.

to any gene segment is described in the following paragraphs. The outcome of the VP4 binding analysis was the definition of the minimum polypeptides recognized by N-MAbs. In this case, the polypeptides recognized were large, approximately 200-amino acid, VP4 fragments. We have since used this information to baculovirus express the minimum VP4 polypeptides and found that the expressed proteins are antigenic and immunogenic, and that the immune response to the expressed protein includes antibodies that neutralize the virus.

The DARTT assay described here is one way of generating polypeptide fragments for a variety of functional analyses. DARTT polypeptides are generally large (too large to synthesize) and can vary in size, depending upon the ability to analyze the *in vitro* product. DARTT lacks the problems associated with limited proteolytic digestions or expensive, potentially futile, mimitope or peptide synthesis, and could be used before the synthesis of short peptides.

DARTT Protocol

Rationale

DARTT is essentially the rapid synthesis and functional analysis of protein products encoded by PCR fragments (Mackow *et al.*, 1990). DNA is amplified and restricted by PCR while at the same time adjacent transcriptional and/or translational initiation sites are added (Fig. 1). Amplified fragments are then subjected to two *in vitro* steps: (1) Transcription of the segment into RNA and (2) translation of the RNA into a radiolabeled polypeptide. Proteins can then be analyzed for various enzymatic and binding functions by reacting the expressed proteins with other ligands and substrates, or by coexpressing associated ligands with the truncations.

Technique

PCR Reaction

DNA or RNA templates are prepared for PCR or RT–PCR as described by previous studies (Saiki *et al.*, 1985; Kawasaki, 1990). Initially, one oligonucleotide to the 3' terminus of a cloned gene and one oligo upstream from the plasmid transcriptional initiation site (at the clone's translational initiation site) should be synthesized and used to amplify the full-length coding sequences. Alternatively, a full-length RT–PCR product can be obtained by including a transcriptional element (see later discussion) in the 5' oligonucleotide, along with 17 or more bases homologous to the RNA segment. Of course, if one is starting from RNA, a reverse transcriptase step must be included prior to PCR (Kawasaki, 1990). The idea is to test out the functional properties of the full-length protein before continuing on to protein subfragments. Once DARTT has synthesized a full-length protein (see later discussion) and whatever function to be assayed has tested positively, oligonucleotides for 5' or 3' end truncations can be synthesized. Usually 3'-end truncations are made first since they are short relative to the 5'-end oligos, which contain transcription initiation sites (see later discussion) (Fig. 1A). Standard PCR conditions and annealing temperatures will have to be worked out, but typically 10 ng of each oligo along with at least 10 ng of template DNA are subjected to denaturation at 94°C for 1 min, annealing at 42°C for 1 min, and polymerization at 72°C for 3 min for 35 cycles as standard starting conditions. *Taq* DNA polymerase (Perkin–Elmer) concentrations and buffer conditons for PCR and RT–PCR have been previously published or are provided by the manufacturer and should be followed initially (Saiki *et al.*, 1985; Kawasaki, 1990). Following PCR, DNA is phenol extracted and spun in a Microcon 100 (Amicon) in order to remove unincorporated oligos (especially those containing transcriptional initiation sites), deoxyxnucleoside triphosphates and buffer. Alternatively, gel filtration (spun columns), selective ethanol precipitation (ammonium acetate), or silica binding can be used to clean amplified products.

Oligonucleotide Primers

Transcriptional and translational signals can be added to any RNA or DNA gene segment by including them in the oligonucleotides used for PCR or RT–PCR. An additional method for adding transcriptional signals to coding segments is oligonucleotide ligation to existing PCR products (Landegren *et al.*, 1990). This technique works provided the oligos used for PCR are selectively phosphorylated at the translation initiation terminus and both the downstream oligo and the oligo containing the transcriptional element are dephosphorylated or the 5'-termini are blocked. Fragments can be reamplified by priming with the added transcriptional signal and the downstream oligonucleotide. A 5' terminal block can also serve as a purification scheme for PCR products if the oligo is biotinylated (Levenson and

Chang, 1990). The residual PCR oligos can be removed as above, or streptavidin agarose beads can be used to isolate ligated, 5′-biotinylated PCR products.

T7, T3, and SP6 transcriptional signals have all been used successfully in DARTT (Stoflet et al., 1988; Sarkar et al., 1989; Mackow et al., 1990). Preference for one or another should be based on the vector and orientation of the initial clone. Since T7 polymerase can be provided in trans intracellularly (see later discussion) (Fig. 2), if subsequent tissue culture assays will be performed, the addition of T7 transcriptional sequences to PCR products would be an advantage. We have used the published minimal binding sequences for T7 and T3 polymerases (ATTAACCCTCACTAAAG, T3 and AATACGACTCACTATAG, T7) as well as extended sequences (GAATTAACCCTCACTAAAGGG, T3 and GTAATACGACTCACTATAGGG, T7) as transcription initiation signals added to PCR oligos. Each works with indistinguishable differences in translated products.

We have also tried to optimize translational initiation signals by adding Kozak's (1987) consensus sequences to various internal translational initiation sites. Unfortunately, we have not been able to enhance translational efficiency by changing initiation codon contexts to ATGG or the GCCATGG sequences (Kozak, 1987) as we first thought. Instead some ATGs which are not at the native site of translational initiation work quite well and some do not, suggesting that mRNA secondary structures may be affecting translational initiation (Kozak, 1986). In our hands, VP6 which is synthesized to high levels during a rotavirus infection is similarly efficiently translated *in vitro* whereas VP4 which is synthesized at low levels intracellularly is comparatively inefficiently translated *in vitro*. Translational enhancement can usually be achieved by changing the magnesium or potassium acetate concentrations in the translation reaction, and can be optimized for each RNA following a poor initial translation. Canine microsomes also enhance the translation of many messenges, including those without signal sequences. Denaturation (68°C, 5 min) of long RNAs mightalso enhance their ability to be translated. Although we have not tried pretreating RNAs with methyl mercury hydroxide, dimethyl sulfoxide, formamide, or formaldehyde, any of these techniques might be used to denature RNAs and provide for more efficient translations.

Transcription and Translation

Amplified products (0.5 to 1μg) are added to *in vitro* transcription reactions following the manufacturer's recommendations (Stratagene T7 and T3 polymerases, or Ambion Inc.'s Megascript) for buffer and rXTP concentrations Diethylpyrocarbonate (DEPC)-treated water and reagents are used for everything. Reactions are generally extended to 3 to 6 hr and 0.5 U/μl RNAsin is included in the reactions. At times we have also increased rXTP concentrations to 0.5 to 1 mM in order to increase the amount of RNA product. Transcription can be monitored by radiolabeling products and separating them on formaldehyde gels, by trichloroacetic acid (TCA)-precipitable counts, or by directly assaying the protein products following transcription. At the end of transcription, reactions are diluted to 200 μl and RNA is extracted with phenol. The aqueous solution is removed to a new tube and single-stranded RNA is precipitated by adding an equal volume of 4 M LiCl (DEPC treated). The solution is left on ice 1–2 hr and the RNA is pelleted for 5 min at 14,000 rpm. Pellets are resuspended in water and reprecipitated with 0.1 volumes 3 M sodium acetate (pH 5.2) and 3 volumes of ethanol. Alternative isopropanol, ammonium acetate, sodium acetate, or sodium chloride precipitations may also used at the first or second precipitation steps. Lithium chloride precipitation removes most of the free rXTPs, does not precipitate DNA, and as a result has been our method of choice when time was not an issue.

The amount of RNA to use in the translation reaction will depend on the template, the protein to be synthesized, and the salt and magnesium conditions of the translation. In general, we start out by using 0.2 to 0.5 of our transcription reactions (0.5–5 μg) in initial 25 μl translation reactions in the presence of 1.8 mCi/ml [^{35}S] l-methionine. We evaluate translations by mini-PAGE analysis. A successful translation generally yields a radiolabeled protein band, with the appropriate molecular weight by fluorography, from 1–2 μl of the translation. Be aware that some batches of reticulocyte lysate contain endogenous mRNA. The manufacturer usually discusses this and the molecular weight of the protein product. However, when successful transcriptions are used to program translations, the endogenous band is usually not seen. We have successfully used reticulocyte lysates from several sources (Ambion, Austin, Tx., Promega, Madison, Wisc., Stratagene, La Jolla, CA) for *in vitro* translation reactions, initially following the prescribed conditions. Some of the lysates lend themselves to varying the conditions of salt and magnesium in the translation, but the concentrations of the others can be varied just as easily from stocks of magnesium or potassium acetate.

Discussion

Functional Assays for Protein Products

Translated proteins can be tested for functional activity. We initially characterized the minimal binding domains of neutralizing monoclonal antibodies to the VP4 protein of *Rhesus* rotavirus (RRV). Monoclonal antibodies and hyperimmune serum were reacted with radiolabeled VP4 polypeptides made by DARTT and, following immunoprecipitation with protein A Sepharose, binding was assessed by PAGE analysis of radiolabeled products (Mackow et al., 1990). The failure of smaller DARTT products to be recognized by N-MAbs defined the minimum peptides required for recognition. In the case of VP4, two groups of monoclonal antibodies were found. Each of the VP8-specific MAbsbound the amino acid 55 to 222 fragment of VP4 (Mackow et al., 1990). VP5-specific N-MAbs all required amino acids 247 to 474 for recognition. In some way, these large polypeptides maintain a conformation which presents each of the N-MAb epitopes of either the variable VP8 region or the conserved and cross-reactive VP5 region. These results have allowed us to focus on the conserved VP5 polypeptide as an immunogen and to evaluate this polypeptide in eliciting an N-MAb immune response and enhancing the immune response to VP5 following a live rotavirus vaccination.

DARTT protocols have since been used by Clapp and Patton (1991) to define the amino acids of the rotavirus VP6 protein (the major inner capsid structural protein) involved in trimerization and a separate amino acid region required for assembly of VP6 into single-shelled particles (Clapp et al., 1991). We have recently coexpressed and are in the process of self-assembling the four major structural proteins of the group B rotavirus, Adult Diarrheal Rotavirus (ADRV), into particles *in vitro*. We are in the process of dissecting binding domains on these proteins via DARTT and studying the binding of distantly related (20–30 identical) group A and B rotavirus structural proteins. A myriad of possibilities exist for DARTT applications and their adaptation to various systems and questions of protein function. One interesting use of DARTT is the evaluation of antisense RNA fragments as blockers of protein expression. It should be possible to optimize and define the minimum effective antisense segments needed for translational inhibition *in vitro* and apply these antisense fragments to an *in situ* assay (Fig. 2).

Intracellular Adaptation of DARTT

An alternative use of DARTT is to introduce the PCR-restricted fragments or plasmids into cells and allow the cells to transcribe and translate the amplified product into protein (Fig. 2). This can now be done easily through the ability to express T7 RNA polymerase in mammalian cells and through efficient transfection procedures. Studier et al., (1986) have presented the cloning of T7 RNA polymerase and the use of this protein to drive efficient transcription from T7 promoters in trans (Studier et al., 1986). T7 polymerase containing plasmids has been introduced into bacteria for high-level expression of proteins under the control of the T7 promoter (Studier et al., 1986). Similarly, T7 RNA polymerase has been recombined into vaccinia virus by Fuerst et al., (1987a,b) and this technological advance has permitted the use of bacterial T7 promoters as transcription initiators in mammalian cells (Fuerst et al., 1987a,b). Cells transfected with a plasmid containing a T7 promoter-driven gene are infected with the T7 vaccinia virus, and proteins from the plasmid are synthesized (Fuerst et al., 1987a,b). We have used an ideologically similar construct called pRAT1, constructed and provided to us by Kevin Ryan (St. Jude Children's Research Hospital, Memphis, TN), in which the T7 RNA polymerase has been cloned into plasmid pSV2-neo in place of the neomycin resistance gene and behind the rous sarcoma virus promote (LTR). When this plasmid and a plasmid containing a T7-driven gene are cotransfected into Ma104 cells, the encoded proteins under control of the T7 promoter are expressed. In addition, T7-driven PCR products can be introduced into cells along with pRAT1 and expressed intracellularly. This allows for the rapid expression of truncated or mutated proteins in mammalian cells and hence the analysis of intracellular interactions and binding functions. We have expressed the VP4 and VP6 proteins of group A and B rotaviruses by this technique and we are in the process of coexpressing and studying the assembly of the four major rotavirus structural proteins (VP2, VP4, VP6, and VP7).

When plasmids or PCR products are used for transfection, it is important to efficiently introduce the DNAs into cells. In our hands, electroporation has proven to be much more efficient than liposomes, calcium phosphate, or diethylaminoethyl transfection protocols. We have electroporated Ma104 cells essentially as described by Liljestrom et al., (1991) following trypsinization and subsequently replated them on six-well plates (2 pulses, 25 uFD, 1.5 kV, 0.2 cm path in 1 ml PBS, calcium and magnesium free, 5–20 μg of each

DNA and $0.5-1 \times 10^6$ cells) (Liljestrom *et al.*, 1991). Electroporation results in the cotransfection and expression of proteins from approximately 50% of monolayer cells as assayed by immunoperoxidase staining of rotavirus proteins.

In addition to protein staining, we have assayed the function of a single group B rotavirus polypeptide, AD6F1. The AD6F1 protein shares identity with the Newcastle disease virus and respiratory syncytial virus fusion proteins. Since group B rotaviruses also form syncytia, we tested the ability of the AD6F1 protein to cause cell fusion by introducing it into cells as described in Fig. 2. Interestingly, AD6F1 and pRAT1-transfected cells contain a large number of syncytia-like multinucleated cells when compared with cells electroporated with pRAT1 alone. We are in the process of confirming these results, optimizing the fusion assay, making antibodies to AD6F1 to determine its association with fused cells and its intracellular localization, as well as truncating the protein to delineate active portions of the protein via *in situ* DARTT.

Intracellular applications of DARTT make this technique an extremely powerful method by which to study protein functions. In the near future, intracellular protein processing and targeting, T-cell epitope mapping, the functions of proteins in transcriptional enhancement and gene regulation, translational modifications, leukotriene binding, cell fusion, and many other properties might be addressed by the *in situ* adaptation of DARTT.

Acknowledgments

We thank Kevin Ryan for invaluable advice and discussions, as well as for constructing and providing pRAT1. Sincere thanks to Rick Chou, Robin Werner-Eckert, Mary Ellen Fay, and Mark Yamanaka for their input into *in vitro* and *in situ* DARTT. Thanks also to the Molecular Biology Core Facility (Northport, VA) for the use of equipment and the synthesis of oligonucleotides. This work was supported by a Merit Award from the Veterans Administration and by a National Institutes of Health Basic Research Science Grant and R01AI31016.

Literature Cited

Clapp, L. L., and J. T. Patton 1991. Rotavirus morphogenesis: domains in the major inner capsid protein essential for binding to single-shelled particles and for trimerization. *Virology* **180(2)**:697–708.

Fuerst, T. R., P. L. Earl, and B. Moss. 1987a. Use of a hybrid vaccinia virus-T7 polymerase system for expression of target genes. *Mol. Cell. Biol.* **7**:2538–2544.

Fuerst, T. R., E. G. Niles, F. W. Studier, and B. Moss. 1987b. "Eukaryotic transient-expression system based on recombinant vaccinia virus that synthesizes bacteriophage T7 RNA polymerase." *Proc. Natl. Acad. Sci., U.S.A.* **83**:8122–8126.

Holland, M. J., and M. A. Innis 1990. *In vitro* transcription of PCR templates In: *PCR protocols: A guide to methods and applications* (eds. M. A. Innis, D. H. Gelfand, J. J. Sninsky, and T. J. White) pp. 169–176 Academic Press, San Diego, CA.

Kawasaki, E. S. 1990. Amplification of RNA. In: *PCR protocols: A guide to methods and applications* (eds. M. A. Innis, D. H. Gelfand, J. J. Sninsky, and T. J. White) pp. 21–27. Academic Press, San Diego, CA.

Kozak, M. 1986. Influences of mRNA secondary structure on initiation by eukaryotic ribosomes. *Proc. Natl. Acad. Sci. U.S.A.* **83**:2850–2854.

Kozak, M. 1987. At least 6 nucleotides preceding the AUG initiator codon enhance translation in mammalian cells. *J. Mol. Biol.* **196**:947–950.

Landegren, U., R. Kaiser, and L. Hood. 1990. Oligonucleotide ligation assay. In: *PCR protocols: A guide to methods and applications* (eds. M. A. Innis, D. H. Gelfand, J. J. Sninsky, and T. J. White), pp. 92–98. Academic Press, San Diego, CA.

Levenson, C., and C. Chang. 1990. Nonisotopically labeled probes and primers. In: *PCR protocols: A guide to methods and applications* (eds. M. A. Innis, D. H. Gelfand, J. J. Sninsky, and T. J. White), pp. 99–112. Academic Press, San Diego.

Liljestrom, P., S. Lusa, D. Huylebroeck, and H. Garoff. 1991. In vitro mutagenesis of a full length cDNA clone of semliki forest virus: the small 6,000-molecular-weight membrane protein modulates virus release. *J. Virol.* **65**:4107–4113.

Mackow, E. R., R. D. Shaw, S. M. Matsui, P. T. Vo, M. N. Dang, and H. B. Greenberg. 1988. The rhesus rotavirus gene encoding protein VP3: location of amino acids involved in homologous and heterologous rotavirus neutralization and identification of a putative fusion region. *Proc Nat. Acad. Sci. U.S.A.* **85(3)**:645–649.

Mackow, E. R., M. Y. Ymamnaka, M. N. Dang, and H. B. Greenberg. 1990. DNA amplification-restricted transcription-translation: rapid analysis of rhesus rotavirus neutralization sites [published erratum appears in Proc. Natl. Acad. Sci. U.S.A. 1990 June 87(11):4411]. *Proc. Nat. Acad. Sci. U.S.A.* **87(2)**:518–522.

Saiki, R., S. Scharf, F. Faloona, K. B. Mullis, G. T. Horn, H. A. Erlich, and N. Arnheim. 1985. Enzymatic amplification of β-globin genomic sequence and restriction site analysis for diagnosis of sickle cell anemia. *Science* **230**:1350–1354.

Sarkar, G., and S. S. Sommer 1989. Access to a messenger RNA sequence or its protein product is not limited by tissue or species specificity. *Science* **244**:331–334.

Stoflet, E. S., D. D. Koeberl, G. Sarkar, and S. S. Sommer. 1988. Genomic amplification with transcript sequencing. *Science* **239**:491–494.

Studier, F. W., and B. A. Moffatt. 1986. Use of Bacteriophage T7 polymerase to direct selective high-level expression of cloned genes. *J. Mol. Biol.* **189**:113–130.

20

DNA AND RNA FINGERPRINTING USING ARBITRARILY PRIMED PCR

John Welsh, David Ralph, and Michael McClelland

Arbitrarily primed PCR (AP–PCR) provides an information-rich fingerprint of either DNA or RNA (Welsh and McClelland, 1990; Williams et al., 1990; Liang and Pardee, 1992; Welsh et al., 1992a). The method is based on the selective amplification of sequences that, by chance, are flanked by adequate matches to an arbitrarily chosen primer. Arbitrary primers have a specific but arbitrarily chosen sequence, such as the M13 universal sequencing primer. The term "random" primer should be avoided because random primers have 4-fold degeneracy at each position; for example, the random hexamer 5'-NNNNNN-3'.

In arbitrarily primed PCR, DNA synthesis is initiated at low stringency. Complementarity with the template in the first seven or eight bases at the 3' end of the primer must be nearly perfect for productive priming. The numbers of matches at the 3' end needed for productive priming are similar for longer primers (e.g., 20-mers) and shorter primers (e.g., 10-mers). A few of these sites exist, stochastically, on opposite strands of the template, within several hundred nucleotides of each other. The products of this reaction are then resolved by gel electrophoresis, resulting in a reproducible pattern, or fingerprint. Because the priming events during the initial low-stringency cycles depend on the nucleotide sequence of the primer, and this sequence has been chosen arbitrarily, the amplified sequences represent an arbitrarily chosen small sample of the template DNA or RNA. Fingerprinting using other primers allows an essentially limitless array of anonymous sequences to be compared. Such fingerprinting has been referred to as random amplified polymorphic DNA (RAPD) (Williams *et al.*, 1990). When RNA is used, the fingerprints have also been referred to as "differential display" (Liang and Pardee, 1992). In practice, fingerprinting of RNA or DNA differs only in the initial use of reverse transcription for first-strand synthesis during RNA fingerprinting. First-strand cDNA may be made by using oligo(dT) priming from the poly(A) tail of mRNAs, or by arbitrary priming, followed by an arbitrarily primed second-strand synthesis. We discuss the advantages and disadvantages of these approaches in the section dealing with RNA fingerprinting.

If two or more templates differ, their arbitrarily primed PCR reaction products also differ. In RNA fingerprinting, differences reflect differential gene expression (when properly controlled for polymorphisms among individuals). Differences in DNA fingerprints reflect either polymorphisms that have accumulated over evolutionary time or mutations that have developed during the life of the organism. (Other recent sources of information on arbitrarily primed PCR of DNA include Bowditch *et al.*, 1993; Williams *et al.*, 1993; Clark and Lanigan, 1993; and Yu *et al.*, 1993.) We have recently reviewed RNA fingerprinting (McClelland and Welsh, 1994).

Arbitrarily Primed PCR Fingerprinting of DNA

Arbitrarily primed PCR fingerprinting of DNA is useful for genetic mapping, taxonomy, phylogenetics, and for the detection of mutations. Arbitrarily primed PCR has been used to detect dominant polymorphic markers in genetic mapping experiments (Williams *et al.*, 1990). Several hundred anonymous markers derived from arbitrarily primed PCR have been placed on the mouse genetic map (Nadeau *et al.*, 1992; Serikawa *et al.*, 1992; Woodward *et al.*, 1993). Sobral and colleagues have placed more than 200 markers on the sugarcane genetic map (Al-Janabi *et al.*, 1993), and similar maps have been constructed for pine (Neale and Sederoff, 1991), *Arabidopsis* (Reiter *et al.*, 1992), and lettuce (Kesseli *et al.*, 1994). Genetic markers linked to measurable phenotypes can be detected using arbitrarily primed PCR fingerprinting combined with bulked segregant analysis (Michelmore et al., 1991). Kubota et al. (1992) used arbitrarily primed PCR to detect gamma ray-induced damage in fish embryos. Recently, Perucho and colleagues have used this method as a molecular cytology tool to reveal somatic genetic alterations, including allelic losses or gains (Peinado et al., 1992), and deletion or insertion mutations of one or a few nucleotides (Ionov et al., 1993) which occur during tumor development and progression. Polymorphisms detected by arbitrarily primed PCR can also be used as taxonomic markers in population studies, epidemiology, and historical ecology for both prokaryotic and eukaryotic systems (e.g., Welsh and McClelland, 1990, 1991a,b; Welsh et al., 1991; Chalmers et al., 1992; Welsh et al., 1992b; Welsh and McClelland, 1993; Bachmann, 1994; Haymer and McInnis, 1994; Lynch and Milligan, 1994; Mendel et al., 1994; Thomas et al., 1994; Williams et al., 1994; Woods et al., 1994).

What is the nature of the sequences amplified by arbitrarily primed PCR? The majority of amplified bands usually originate from unique sequences rather than from repetitive elements, and while the intensities of different bands within the same fingerprint vary independently of each other, the intensity of a particular band between fingerprints is *proportional* to the concentration of its corresponding template sequence. This can be seen in the F_1 progeny of diploids at heterozygous loci. The semiquantitative nature of arbitrarily primed PCR allows the degree of aneuploidy of a tumor cell genome to be determined by comparing the intensities of arbitrarily primed PCR bands with those from the normal diploid genome from the same individual (Peinado et al., 1992). These experiments require careful adjustment of the template DNA concentration, because the relative intensities of bands within a single arbitrarily primed PCR fingerprint are determined, to some extent, by the initial template concentration. It is strongly recommended that each genomic fingerprint be generated at several DNA concentrations, to reveal any products that show a significant dependence on template concentration.

DNA Preparation and Primer Choice

In this section we present a generic protocol for fingerprinting DNA. The DNA may be prepared by virtually any conventional protocol. For some work, where the same DNA samples will be used extensively, such as for genetic mapping or molecular cytology, ultrapure DNA is absolutely necessary. Also, the DNA of some species must be purified away from inhibitors of *Taq* polymerase. For other experiments, for example, environmental or nosocomial surveys, where there are many samples and each sample will be fingerprinted only once or a few times, less labor-intensive DNA preparation methods may be desirable. In general, when difficulties in generating robust and reproducible fingerprints are encountered, variation in DNA quality among samples is often the culprit. Therefore one should always use the most rigorous purification procedures that are practical. For crude preparations, it is usually wise to check the DNA on agarose gels to see that other methods of measuring concentration, for example, UV absorption, are accurate and to determine the DNA quality. Care should be taken to avoid overloading the agarose gels.

Primer design is not difficult for arbitrarily primed PCR and similar methods. In general, it is necessary to avoid primers with extensive secondary structure and those that have complimentary 3' ends. Most people use primers from 10 to 20 nucleotides in length. Short primers require lower temperature cycling and have a significant advantage in that they are cheaper than longer primers. Longer primers can be designed with special applications in mind and may work more readily for reamplification of products from gels, but otherwise have no apparent advantages over 10-mers for fingerprinting. Primers can be used in pairwise combinations so, for example, 55 fingerprints can be generated using 10 primers in all single and pairwise combinations (Welsh *et al.*, 1991). Kits of 10-mer primers are available from many companies, including Genosys, Woodlands, TX. However, the cheapest primers available are those sitting in someone's freezer down the hall, which they do not intend to use again.

We have frequently found that primer concentration affects the quality of the fingerprint. The optimum primer concentration depends on the sequence of the primer and often on the particular primer preparation. It is therefore necessary to either (1) empirically determine the optimum for each primer or (2) screen primers for those that work best at a single concentration. For genetic mapping, where a primer is to be used only once, screening many primers at a standard concentration seems to be the most efficient approach. If fingerprinting in many different systems is desired, a good strategy is to optimize the primer concentration for a large-scale primer preparation. Optimum primer concentrations generally fall between 0.3 and 10 μM.

Longer primers generally generate a larger number of visible fragments than do 10-mers. However, truncated forms of *Taq* polymerase (the equivalent of *E. coli Pol*I Klenow fragment) such as Stoffel fragment (Perkin–Elmer) and KlenTaq (AB Peptides, St. Louis, MO) allow 10-mers to generate almost as many products as 20-mers produce using AmpliTaq. Also, the *Taq* Stoffel fragment gives better results with a greater fraction of arbitrarily selected 10-mers than does AmpliTaq (Sobral and Honeycutt, 1993).

The issue of the number of products generated is important. First, the more products visible, the more information can be gleaned from a single experiment. However, perhaps even more important, if only a few products are efficiently amplified by a particular primer, then the presence or absence of a polymorphic product can affect the amplification of other products in the same reaction. This "context effect" will result in the presence or absence of some bands being dependent on the presence or absence of other bands. This effect should be reduced or eliminated if no particular band that varies in the fingerprint contributes more than a small amount to the mass. That is, the more bands that are visible in the fingerprint, the less likely it is that the fingerprint will be influenced by the presence or absence of a particular product. Thus 20-mers may be preferred unless 10-mers are used in conjunction with the Stoffel fragment of *Taq* polymerase. If 10-mers are used with AmpliTaq, then only the more complex fingerprints should be scored.

Figure 1 presents an example of a fingerprint of *Staphylococcus aureus* genomes. Note that some products are concentration dependent and thus small differences in the quality or concentration of two templates can lead to spurious differences in the arbitrarily primed PCR pattern. Such problems are eliminated by the use of two DNA concentrations, as shown in this picture.

The fact that fingerprints should be performed at two concentrations differing by 2-fold or more is the single most important point that is routinely neglected by users of this technology. One cannot know if the difference between two genomic fingerprints is real if the experiment is not controlled for DNA quality and quantity. Thus, every experiment must include fingerprinting for *at least two* concentrations of genomic DNA for each individual. Any differences between individuals that do not occur at both genomic DNA concentrations are rejected. Too many concentration-dependent differences should lead to concern about the DNA or reagents.

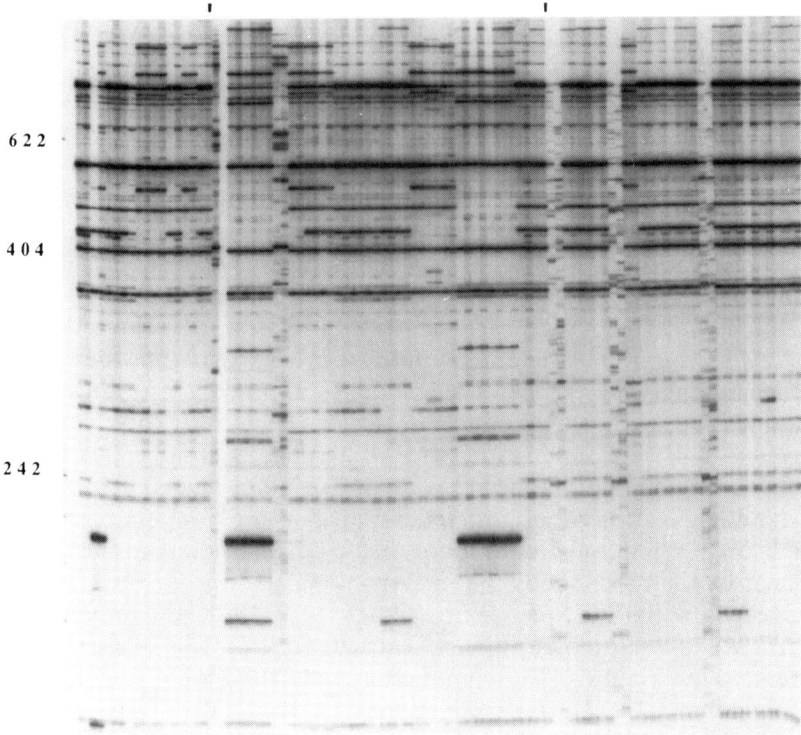

Figure 1 Arbitrarily primed PCR on *Staphylococcus aureus* genomes. Forty-eight strains of methicilin-resistant *S. aureus* were grown in a 96-well microtiter plate. DNA was prepared and fingerprinted using the M13 sequencing primer. Ten and 40 ng of genomic DNA were fingerprinted for each strain. The fingerprints were resolved on a sequencing gel and visualized by autoradiography. A few strains did not grow well under selection, and these do not show reproducible fingerprints (example indicated by arrow). Molecular mass markers on the left are in bases.

Protocol for DNA Fingerprinting by Arbitrarily Primed PCR with 18-mer Primer

Equipment and Reagents

Thermal cycler and related hardware (Perkin–Elmer 9600 model)
PCR tubes (Microamp, Perkin–Elmer, Norwalk, CT)
Micropipette, multichannel

2× arbitrarily primed PCR reaction mixture [20 mM Tris (pH 8.3), 100 mM KCl, 8 mM MgCl$_2$, 0.2 mM each deoxynucleoside triphosphate (dNTP), 0.1 U/1 μl AmpliTaq polymerase (Perkin–Elmer), 0.1 μCi/1 μl α[^{32}P]-dCTP and 1–5 μM of an arbitrary primer]

Method

Reactions are prepared at two DNA concentrations. It is wise to initially titrate the DNA over two orders of magnitude to find the concentrations that give the most robust fingerprints. The best fingerprints are usually obtained in the range of 5–50 ng per 20 μl reaction volume. The optimal concentration of the primer must be determined empirically.

1. DNA is prepared at 2× final concentration (e.g., 1 ng/μl and 40 ng/μl) and 10 μl are distributed to tubes.
2. To this are added 10 μl of 2× arbitrarily primed PCR reaction mixture. The proportions of DNA to reaction mixture can, of course, be changed, as can the total final reaction volume as long as the final concentrations of reaction components are kept the same. Many laboratories run 10-μl reactions, saving half of the cost of the components (particularly the expensive polymerase).
3. The thermocycling profile depends on the length of the primer. For 20-mers, we use a two-step protocol with 2 to 5 low-stringency steps, followed by 35 high-stringency steps as follows: 94°C for 1 min, 40°C for 5 min, 72°C for 5 min, for 2 to 5 cycles, followed by 94°C for 1 min, 60°C for 1 min, 72°C for 2 min for 35 cycles. After two low-stringency cycles, exact copies of the primer sequence flank a handful of anonymous sequences. Thus, when longer primers are used, the annealing temperature can be raised after a few cycles and the reaction allowed to continue under standard, high-stringency PCR conditions. The two-step low–high stringency protocol was designed to avoid internal priming within a larger amplifying product to maximize character independence.
4. The products are diluted 1 : 4 in 80% formamide containing 10 mM EDTA and tracking dye, heated to 65°C for 15 min, and 2 μl electrophoresed through a denaturing sequencing-type polyacrylamide gel. The gel may be dried and exposed to X-ray

film without hindering the ability to clone fragments from the gel, if desired. Alternatively, the radioactive label can be omitted entirely and native agarose gels followed by ethidium bromide staining can be used.

Note: Primers can be used in pairwise combinations to generate largely unique patterns, greatly reducing the cost of generating many fingerprints (Welsh and McClelland, 1991). However, for taxonomic studies, no primer should be used in more than one pairwise combination to ensure independence of characters between fingerprinting experiments.

Protocol for DNA Fingerprinting by Arbitrarily Primed PCR with a 10-mer Primer

Equipment and Reagents

Thermal cycler and related hardware (Perkin–Elmer 9600 model)
PCR tubes (Microamp, Perkin–Elmer, Norwalk, CT)
2× arbitrarily primed PCR reaction mixture [20 mM Tris (pH 8.3), 20 mM KCl, 10 mM MgCl$_2$, 0.2 mM each dNTP, 0.1 U/μl *Taq* polymerase Stoffel fragment (Perkin–Elmer), 0.1 μCi/1 μl α[^{32}]P-dCTP, and 1 μM of an arbitrary 10-mer primer]

Method

1. Reactions are prepared at two DNA concentrations. The best fingerprints are usually obtained in the range of 5–50 ng per 20-μl reaction volume. DNA is prepared at 2× final concentration and 10 μl are distributed to tubes.
2. To this are added 10 μl of 2× arbitrarily primed PCR reaction mixture.
3. The thermocycling profile depends on the length of the primer. For 10-mers, a one-step, high-stringency protocol is used: 94°C for 1 min, 35°C for 1 min, 72°C for 2 min for 40 cycles.
4. The products are diluted 1:4 in 80% formamide containing 10 mM EDTA and tracking dye, heated to 65°C for 15 min, and

2 μl electrophoresed through a denaturing sequencing-type polyacrylamide gel. The gel may be dried and exposed to X-ray film without hindering the ability to clone fragments from the gel, if desired. Alternatively, the radioactive label can be omitted entirely and native agarose gels followed by ethidium bromide staining can be used. We have also been able to resolve such fingerprints on single-stranded conformational polymorphism (SSCP) gels (McClelland et al., 1994a). This strategy increases the number of polymorphisms detected and is suitable for genetic mapping. However, arbitrarily primed PCR–SSCP is not suitable for taxonomic measurements because SSCP gels routinely yield two resolved strands for most of the PCR products, confounding the issue of independence of characters. The issue of independence of characters also means that when using primers in pairwise combinations, each primer should be used in only *one* pairwise combination.

Note: We have succeeded in using 10-mer primers in pairwise combinations to generate largely unique patterns from complex genomes (McClelland et al., 1994a). However, in bacterial genomes, only pairs of longer primers have succeeded in generating finger- of longer primers have succeeded in generating fingerprints. We do not know why 10-mers do not seem to work well in pairs on simple genomes.

Interpretation of Fingerprinting Data

Interpretation is, of course, the most subtle aspect of fingerprinting, and depends on the nature of the experiment being performed. However, there are several general issues that frequently arise. First, as predicted from an approximate kinetic model for arbitrarily primed PCR, the intensity of an amplified band will depend on the efficiency of the interaction of the genomic sequence with the primer during the initial steps, the efficiency of amplification once the primer has been incorporated into the sequence, and the starting relative concentration of the template for that band. Even though some repetitive elements in the genome are at a very high relative concentration, their low sequence complexity prevents them from interacting well with most primers. Differences in the efficiency of amplification of two bands can be seen when different numbers of effective cycles of exponential amplification are employed. The number of effective cycles of exponential amplification can be varied deliberately, or can result from template concentration effects. A higher input template concentration will result in fewer effective cycles before a reaction component becomes limiting. Side-by-side fingerprinting of each template at more than one concentration controls for this problem. In the special case where allelic losses or gains are compared between two genomes, Perucho and colleagues have shown that the intensities of invariant bands in the fingerprint serve as internal controls for quantitative comparisons of those few bands that change (Peinado et al., 1992).

Many laboratories use arbitrarily primed PCR to explore taxonomic and phylogenetic issues. As a taxonomic indicator of family relatedness or of strain distinction, arbitrarily primed PCR is a very powerful tool (Welsh and McClelland, 1993; Lynch and Milligan, 1994; Lamboy, 1994a,b). There is, however, some controversy over the appropriateness of arbitrarily primed PCR characters in phylogenetic analysis. Several problems have been discussed, including the independence of characters and the rates of forward and reverse character transitions, as well as misscoring from gels. These are substantial problems in parsimony analysis. Nevertheless, in general, arbitrarily primed PCR data seem to accord well with other types of phylogenetically useful data.

Some failures in PCR-based fingerprinting have been published and may discourage others from using the method. One concern is unreliability within the same experiment on the same day. The other concern is that the patterns may vary from day to day or from lab to lab.

The problem of intra-experiment variability can usually be attributed to inadequately prepared DNA. High-quality fingerprints that are reproducible at DNA concentrations differing by 2-fold or more have been obtained from genomes in every kingdom and over a wide spectrum of G+C contents. Thus, a failure of this kind must be attributed to inadequate DNA or reagents. A way to find out if the DNA is of sufficient quality is to perform 2-fold serial dilutions of the DNA over a wide range from about 200 ng to 200 pg. If the DNA does not produce reliable fingerprints over a number of 2-fold dilutions with a number of different primers, then the DNA quality is suspect. Second, primers that give moderately complex patterns should be used because simple patterns tend to vary more than complex patterns. We have reached the somewhat surprising conclusion that the reliability of the fingerprints seems to derive primarily from their complexity rather than the quality of the match of the primer with the template.

Day-to-day variation and interlab variation is a more genuine concern. Such variation occurs because all PCR-based fingerprinting varieties are sensitive to the buffer conditions, enzyme quality, and the primer preparation. While this is easy to control in a particular experiment, it is harder to control between experiments. The ratio of intensities among products within a single fingerprint lane may vary from day to day. However, as long as the fingerprint pattern is complex, the variation does not extend to variability in the presence or absence of bands from day to day and the ratio of intensities for a particular product does not change between RNA samples. These differences between experiments and between experimenters are generally rather subtle so they can be accommodated by the simple expedient of fingerprinting DNA from reference strains on each gel.

RNA Fingerprinting by Arbitrarily Primed PCR

Fingerprints can be generated for RNA through a cDNA intermediate. Such fingerprints reflect relative abundances of individual messenger RNA species, and can be used to detect and isolate differentially expressed genes. This method has many applications, including the identification of developmentally regulated genes, the detection of genes that respond to various hormones, growth factors or ectopic gene expression; and the identification of the tissue specificity of gene expression, to name a few. RNA fingerprinting by arbitrarily primed PCR (RAP–PCR) can be viewed as a molecular phenotype that reflects coordinate regulation of multiple genes in response to experimental manipulation. A molecular phenotype of this type is potentially very useful. For example, we can easily recognize molecular differences between tissue culture cells that have stopped growth as a result of contact inhibition and those that have stopped growing because of treatment with transforming growth factor-β (TGF-β) (Ralph et al., 1993; McClelland et al., 1994b).

Genetic and reverse genetic methods of cloning interesting genes derive their power from the ability of the investigator to discern and follow a phenotype. Methods that rely on a biological assay of gene function have been very useful, particularly in the area of cancer research. However, appropriate bioassays for most genes do not exist. When no phenotypic assay can be found, or when genetics is inconvenient or impossible, other methods must be employed. Methods for detecting and cloning differentially expressed genes that do not rely on a biological assay of phenotype include subtractive hybridization and differential screening strategies, as well as RAP–PCR fingerprinting. Besides these methods, there is the brute force yet powerful method of cloning and characterizing everything, one at a time (Adams et al., 1991).

Each of these strategies has strengths and weaknesses. Subtractive hybridization is technically challenging. The difficulties with subtractive hybridization methods derive mainly from the fact that rare messages have unfavorable hybridization kinetics, and many interesting genes are of this class. Abundant genes hybridize faster and more completely than low abundance genes, making them more amenable to subtractive methods. A second problem is that subtraction requires exhaustive hybridization of driver and target to avoid sequences that are not differentially expressed. This has the undesirable effect of obscuring significant differences in gene expression that do not fall into the "all-or-nothing" category. Another shortcoming of the method is that it is most conveniently used only for pairwise comparisons.

Differential screening suffers from similar drawbacks. In a typical differential screening experiment, radioactive probe is made from cDNA from two cell types, for example, and used to screen a cDNA library prepared from one of the two. Occasionally, clones from the library hybridize to one or the other but not both probes. Unfortunately, low abundance messages do not yield sufficient probe mass to allow favorable hybridization kinetics and detection levels. Differential screening is also tailormade for pairwise comparisons. RNA arbitrarily primed PCR fingerprinting (Welsh et al., 1992a; Wong et al., 1993; Ralph et al., 1993) avoids some of these problems, as we shall discuss later. A similar method has been developed by Liang and Pardee (1992) and their term "differential display" could be applied to any RNA fingerprinting strategy.

RNA Arbitrarily Primed PCR

RNA arbitrarily primed PCR is essentially the same as the genomic fingerprinting method described above, except that the RNA sequences must first be reverse transcribed into cDNA. This can be accomplished using either arbitrary priming or oligo(dT) priming from poly(A) tails. In one variant of RAP–PCR, first-strand cDNA synthesis by reverse transcriptase is initiated from an arbitrarily chosen primer at those sites in the RNA that best match the primer. The 3' seven or eight nucleotides in the primer usually match well with the template, but the nucleotides toward the 5' end of the primer also influence which sequences amplify. Second-strand synthesis is achieved by adding a thermostable DNA polymerase and the appropriate buffer to the reaction mixture, with priming at sites where the primer finds the best matches. Poorer matches at one end of the amplified sequence can be compensated for by very good matches at the other end. These two steps result in a collection of molecules that are flanked at their 3' and 5' ends by the exact sequence (and complement) of the arbitrary primer. These serve as templates for high-stringency amplification, resulting in fingerprints similar to those generated from genomic DNA. Alternatively, first-strand synthesis can be primed using oligo(dT). The first-strand cDNA can then be fingerprinted in two different ways, using a single arbitrary primer and two initial low-stringency cycles, incorporating the arbitrary primer at both ends for subsequent PCR amplification, or a single arbitrary primer for priming in both directions after converting the RNA to cDNA using oligo(dT). Liang and Pardee (1992) use a clamped oligo(dT)-CA primer to sample the mRNA.

Each of these approaches has strengths and weaknesses. The use of oligo(dT) priming favors the fingerprinting of bona fide messages rather than the abundant structural RNAs. However structural RNAs do not generally vary between treatment groups and would not be targets for cloning of differentially amplified products, so sampling them does not constitute a serious problem. Using methods that bias toward the 3' untranslated portion of the message can be a disadvantage because the 3' ends of messages tend to be noncoding and their sequences are poorly conserved between species and gene families, making comparisons with the databases less informative. When arbitrary priming is used for both first- and second-strand synthesis, the products are not biased toward the 3' end of the message. Furthermore, RNA that does not naturally possess a poly(A) tail, such as some bacterial RNAs, can be fingerprinted. The disadvantage of arbitrary priming in both directions is that introns in hnRNA might contribute more often to the fingerprint because removal of introns sometimes takes place before poly(A) addition to mRNAs. To avoid the 3' bias, poly(A)-selected RNA can be fingerprinted. However, bleeding of oligo(dT) from the column could result in some subsequent unwanted priming on poly(A).

Size of a RAP–PCR Experiment

Given two or more RNA populations, differences in the fingerprints will result when corresponding templates are represented in different amounts. However, while these fingerprints generally contain anywhere from 40 to 150 prominent bands, typical RNA populations for eukaryotic cells have complexities in the tens of thousands of molecules. Therefore, it would usually be unwise to search for a *particular* differentially expressed gene. Rather, the method is more appropriate for problems where many differentially expressed genes are anticipated. This is not as great a limitation as it might initially appear. First, many developmental and pathological phenomena are accompanied by changes in more than one in a thousand mRNAs, representing many dozens or even hundreds of alterations in gene expression. Second, much of the technological development associated with sequencing, such as fluorescent-tagged primers and automated gel reading, and capillary electrophoresis, is readily adaptable to RAP–PCR. Therefore, a large fraction of the genes expressed in many situations can be surveyed with existing technology, and this capability will be greatly enhanced by further technological developments.

The scale of a RAP–PCR experiment is determined by two factors: (1) the number of tissues, cell types, or more generally, RNA populations to be compared; and (2) the number of products that one wishes to compare. Each RNA is fingerprinted at two concentrations, providing information on reproducibility. Typically, 40 to 150 prominent bands might be observed in a single lane, depending on the choice of primer and quality of materials. Thus, using a 96-well format PCR machine and 96-well gels, it is possible to examine about 2400 messages in *pairwise* comparison in a single experiment (100 bands per lane × 96 wells/2 concentrations/2 RNA types, not corrected for double hits). It is wise to screen a large number of primers against one or a few RNA preparations to maximize the likelihood of choosing useful primers. The scale of the experiment depends, then, on the thoroughness of the search desired by the investigator. RAP–PCR is ideally suited to situations where the number of genes that are differentially expressed is fairly high (i.e., >1 per 1000). RAP–PCR would be less appropriate in experiments where, for example, only a single gene or very few genes are expected to vary.

There are other intrinsic limitations that should be considered in designing an experiment based on RNA arbitrarily primed PCR.

Primarily, the influence of the abundance of messenger RNA on the fingerprint must be considered. This will be discussed in detail later.

Elsewhere we argue that RAP–PCR should not be used primarily as a tool to compete with differential screening or subtractive hybridization (McClelland et al., 1994b). Instead, the fact that many RNAs can be compared in parallel means that the method can be used to compare the effect of many modulators of transcript abundance simultaneously. Also, instead of concentrating on the nature of the genes that are regulated, these genes can be thought of as anonymous markers to measure the interactions among different modulators of transcription. If two modulators overlap in the genes they regulate, then they must share at least some of the signal transduction pathways that regulate these shared genes. If many modulators are examined together, then the number of potentially observed regulatory categories in which a gene is up- or downregulated is enormous (McClelland et al., 1994b). With only eight different treatments, there are 6561 (3^8) such categories.

RNA Purification

The guanidinium thiocyanate-cesium gradient method of RNA purification described by Chirgwin et al. (1979) has been used successfully, as well as the acid guanidinium thiocyanate-phenol-chloroform extraction method (Chomczynski and Sacchi, 1987), but other methods may also work. Most purification methods yield RNA that is not entirely free of genomic DNA. This can be a serious problem because the genome is generally much more complex than the RNA population, resulting in better matches with the primer. RNA should be treated with RNase-free DNase 1 before fingerprinting. The RNA concentration is checked spectrophotometrically and equal aliquots are electrophoresed on a 1% agarose gel and ethidium bromide stained to compare large and small ribosomal RNAs qualitatively and confirm the spectroscopically determined yield.

Choice of Primers

Primers are chosen with several criteria in mind. First, the primers should not have a stable secondary structure. Second, the sequence should be chosen such that the 3' end is not complementary to any other sequence in the primer. In particular, palindromes should be avoided. Third, primers 10 to 20 nucleotides in length can be used with the appropriate modification of cycling parameters. Longer primers may have some small advantages. In particular, after the initial low-stringency step, 20-mers can be used at fairly high stringency so that any DNA contamination that persists will not contribute significantly to the fingerprint (Welsh et al., 1992a). Also, as the reaction proceeds, some internal priming on amplified PCR products at the low stringency might be anticipated. Aside from these features, longer primers can contain more sequence information designed to aid in subsequent steps in the experiment, such as cloning and sequencing. When primers are used in pairwise combinations, those prducts that contain a different primer at each end can be sequenced directly with one of the primers. Primers 10 bases in length can be obtained in kits from Genosys, Woodlands, TX. Finally, primers can be used in pairwise combinations, with minimal chances of double scoring the same product.

Protocol for RNA Arbitrarily Primed PCR with an 18-mer Primer

Equipment and Reagents

Thermal cycler and related hardware (Perkin–Elmer 9600 model)
PCR tubes (Microamp, Perkin–Elmer)
Micropipette, multichannel
Arbitrary primer, e.g., M13 sequencing or reverse sequencing primer
  2× DNase 1 treatment mixture [20 mM Tris–HCl (pH 8.0), 20 mM MgCl$_2$, 40 U/ml RNase-free DNase 1 (Boehringer Mannheim Biochemicals, Indianapolis, IN)]
2× First-strand reaction mixture [100 mM Tris–HCl (pH 8.3), 100 mM KCl, 8 mM MgCl$_2$, 20 mM DTT, 0.2 mM each dNTP, 1–10 μM primer and 2 U/μl murine leukemia virus reverse transcriptase (MuLVRT)]
2× Second-strand reaction mixture [10 mM Tris–HCl (pH 8.3), 25 mM KCl, 2 mM MgCl$_2$, 0.1 μCi/μl α[^{32}]P-dCTP and 0.1 U/μl *Taq* AmpliTaq DNA polymerase (Perkin–Elmer)]

Method

1. Prepare total RNA by guanidinium thiocyanate-cesium chloride centrifugation (Chirgwin et al., 1979) or guanidinium thiocyanate-acid phenol-chloroform extraction (Chomczynski and Sacchi, 1987). Dissolve the final pellet in 100 µl of water.
2. Add 100 µl of 2× DNase 1 treatment mixture and incubate at 37°C for 30 min. Phenol-chloroform extract and ethanol precipitate.
3. Prepare treated RNA at two concentrations of about 20 ng/µl and 10 ng/µl by dilution in water. It is often wise to use more than two different amounts of RNA in the first set of experiments with a new RNA preparation so as to determine the quality of the RNA (i.e., reproducibility of the fingerprint over a wide range of concentrations).
4. Add 10 µl first-strand reaction mixture to 10 µl RNA at each concentration. The reaction is ramped to 37°C over 5 min and held at that temperature for an additional 10 min, followed by 94°C for 2 min to inactivate the polymerase, and finally cooled to 4°C.
5. Add 20 µl of the second-strand reaction mixture to each first-strand synthesis reaction. Cycle through one low-stringency step (94°C, 1 min; 40°C, 5 min; 72°C, 5 min) followed by 35 high-stringency steps (94°C, 1 min; 60°C, 1 min; 72°C, 1 min).
6. Add 2 µl of each reaction to 10 µl of denaturing loading buffer, and electrophorese on a 4% or 6% polyacrylamide sequencing-type gel containing 40 to 50% urea in 0.5 × Tris–borate–EDTA buffer.

Protocol for RNA Arbitrarily Primed PCR with Sequential Application of Two 10-mer Primers

Equipment and Reagents

Thermal cycler and related hardware (Perkin–Elmer 9600 model)
PCR tubes (Microamp, Perkin–Elmer)

Micropipette, multichannel

Arbitrary primer, e.g. M13 sequencing or reverse sequencing primer

2× DNase 1 treatment mixture [20 mM Tris–HCl (pH 8.0), 20 mM MgCl$_2$, 40 U/ml RNase-free DNase1 (Boehringer Mannheim Biochemicals, Indianapolis, IN)]

2× First-strand reaction mixture [100 mM Tris (pH 8.3), 100 mM KCl, 8 mM MgCl$_2$, 20 mM DTT, 0.2 mM each dNTP, 4 μM of the first 10-mer primer and 3.75 U/μl MuLVRT (Stratagene Inc, La Jolla, CA)]

2× Second-strand reaction mixture [10 mM Tris (pH 8.3), 10 mM KCl, 3 mM MgCl$_2$, 0.4 mM each dNTP, 8 μM second 10-mer primer, 0.1 μCi/μl of α-[^{32}P]dCTP, 0.4 U/μl *Taq* polymerase Stoffel fragment (Perkin–Elmer, Branchburg, NJ)]

Method

1. Prepare total RNA as described earlier. Dissolve the final pellet in 100 μl of water.
2. Add 100 μl of 2× DNase 1 treatment mixture and incubate at 37°C for 30 min. Phenol-chloroform extract and ethanol precipitate.
3. Prepare treated RNA at two (or more) concentrations of about 20 ng/μl and 10 ng/μl by dilution in water.
4. Add 10 μl first-strand reaction mixture to 10 μl RNA at each concentration. The reaction is ramped to 37°C over 5 min and held at that temperature for an additional 1 hr, followed by 94°C for 2 min to inactivate the polymerase, and then cooled to 4°C. The reaction is then diluted 4-fold with 60 μl of water.
5. Add 10 μl of the second-strand reaction mixture to 10 μl of the diluted first-strand synthesis reaction. Cycle through 35 steps (94°C, 1 min; 35°C, 1 min; 72°C, 1 min).
6. Add 2 μl of each reaction to 10 μl of denaturing loading buffer, and electrophorese on a 4% or 6% polyacrylamide sequencing-type gel containing 40 to 50% urea in 0.5× Tris–borate–EDTA buffer.

Note: It is wise to perform a mock reverse transcription reaction to determine if the RNA samples are free of DNA. If a reproducible pattern is generated at various concentrations of template even without reverse transcriptase, then the sample is contaminated with significant amounts of DNA. DNase treatment will need to be performed again. If this control gives a series of irreproducible sporadic products, then the preparation is essentially DNA free. Do not expect this control to be blank! Arbitrarily primed PCR will occur sporadically on even the tiniest amount of DNA but if the pattern is not reproducible at different concentrations, then the DNA is of such low concentration that it will not compete with the RNA in the experimental lanes.

Note: Because the primers are used sequentially, the second primer is unlikely to reside at the 3' end relative to the sense strand of the transcript. Thus the sense strand is known in those products that reveal different primers at each end upon cloning (McClelland and Welsh, 1994).

Some primers work better than others. It is usually a good idea to screen several primers and use those that give the most qualitatively robust patterns for further work. Note that *Thermus thermophilus* (*Tth*) polymerase has reverse transcriptase activity and can substitute for both the RT and the DNA polymerase when a buffer containing some Mn^{2+} is used.

An example of a RAP–PCR fingerprint generated using a pairwise sequential protocol and 10-mers is presented in Fig. 2. The use of at least two concentrations of RNA for each experimental condition is absolutely vital. The most common reason that a differentially amplified product is not actually differentially expressed is that the RNAs compared were of nonidentical quality or quantity and therefore generated quality- or concentration-dependent differences. Both of these problems can be eliminated almost completely by comparing two or more concentrations of RNA for each sample. If the pattern is not reproduced, then the product is rejected. If the two concentrations differ substantially, then the RNA is rejected or further purified.

Nested RAP–PCR

Nested RAP–PCR is a method designed to partially normalize the fingerprint with respect to mRNA abundance (McClelland *et al.*, 1993; Ralph *et al.*, 1993). The strategy is very similar to standard nested PCR methods, except that we do not know a priori the internal sequences of the amplified products. In this method, the fingerprinting protocol is applied to the RNA, as described above. Then, a small aliquot of the first reaction is further amplified using a second nested primer having one, two, or three additional *arbitrarily chosen* nucle-

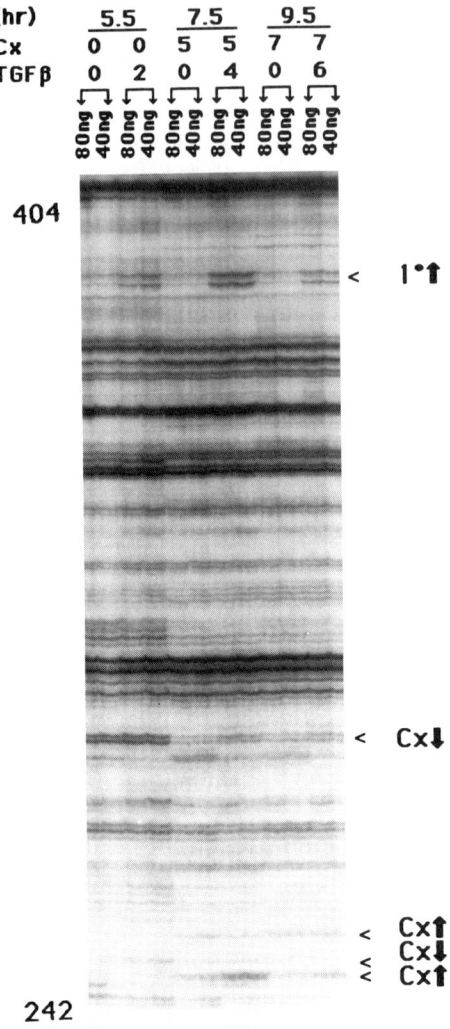

Figure 2 RNA-arbitrarily primed PCR. Lung epithelial cells were released from contact inhibition. After 2.5 hr, some plates of cells were treated with cycloheximide (CX). One hour later some plates were treated with transforming growth factor-β1 (TGF-β) which halts progression into the S-phase. Total RNA was prepared from treated and untreated cells at time points after TGF-β treatment, as shown. The RNAs were sampled by RAP–PCR using OPN24 (5'-AGGGGCACCA) in the first-strand synthesis followed by addition of DD2 (5'-GATGAGGCTGA) before the second-strand synthesis. The fingerprints were resolved on a sequencing gel and visualized by autoradiography. Arrows indicate some of the genes that are differentially expressed in this experiment. Molecular mass markers on the left are in bases.

otides at the 3'-end of the first primer sequence. This method is designed to improve upon the abundance normalization of sampling, as will be further explained in the protocol. While the protocol shown is for 18-mer primers, it could be adapted to nested sets of 10-mers.

Protocol for RNA Arbitrarily Primed PCR Using Nested 18-mer Primers

Equipment and Reagents

96-well format thermal cycler
0.5-ml Reaction tubes
A nested series of primers. We have used the series
   ZF38 CCACACAGAAACCCACCA
   ZF39 CACACAGAAACCCACCAG
   ZF40 ACACAGAAACCCACCAGA
   ZF41   CACAGAAACCCACCAGAG
Micropipette, multichannel
First-strand reaction mixture [50 mM Tris (pH 8.3), 50 mM KCl, 4 mM MgCl$_2$, 20 mM DTT, 0.2 mM each dNTP, 1–10 μM primer, and 2 U/μl MuLVRT].
Second-strand reaction mixture [10 mM Tris (pH 8.3), 50 mM KCl, 2 mM MgCl$_2$, and 0.1 U Taq polymerase (AmpliTaq, Cetus)].
Nested primer reaction mixture [10 mM Tris (pH 8.3), 50 mM KCl, 2 mM MgCl$_2$, 1–10 μM nested primer and 0.1 U/μl *Taq* polymerase (AmpliTaq, Cetus)].
Kodak AR-5 X-Omat X-ray film.

Method

1. Perform steps 1 to 5 of the previous 18-mer protocol for standard RAP–PCR fingerprinting. It is convenient to perform this step using a number of different primers from a nested series.
2. Transfer 3.5 μl of each reaction to 36.5 μl of nested primer reaction mixture, using the primers that are nested relative to the initial primer by 0, 1, 2, and 3 nucleotides. The nested primers

are at about 10 times the carryover concentration of the original primer.
3. Perform 40 high-stringency PCR cycles (94°C, 1 min; 40°C, 1 min; 72°C, 1 min).
4. Add 2 µl of each reaction to 10 µl of denaturing loading buffer, and electrophorese 2 µl on a 4 to 6% polyacrylamide sequencing-type gel containing 50% urea in 0.5× Tris–borate–EDTA buffer.
5. Wrap the gel in plastic wrap or dry the gel and autoradiograph.

This nesting strategy partially normalizes abundance in the sampling during RNA fingerprinting. If two messages have equally good matches and equally good amplification efficiency but differ by 100-fold in abundance, then the products derived from them will differ by 100-fold after RAP–PCR. Thus RAP–PCR fingerprinting produces a background of products that are not visible on the gel. This background includes products derived from low abundance messages. A secondary round of amplification using a primer identical to the first except for an additional nucleotide at the 3' end of the molecule can be expected to selectively amplify those molecules in the background that, by chance, share this additional nucleotide. The additional nucleotide will occur in 1/16 of the background molecules, accounting for both ends. There are many more molecules of low abundance in the RNA population than of high abundance, so most products generated by the high-stringency nesting step should be sampled from the low complexity class. Each additional nucleotide at the 3' end of the initial primer will contribute, in principle, a factor of 1/16 to the selectivity. In practice, the selectivity is probably somewhat less than this, because, while *Taq* polymerase is severely biased against extending a mismatch at the last nucleotide, it is more tolerant of mismatches at the second or third positions. Nonetheless, our initial experiments are consistent with the interpretation that additional selectivity is achieved by this nested priming strategy. Heteronuclear RNA is thought to be about ten times as complex as mRNA. Complete normalization, therefore, can be expected to sample primarily hnRNA. If sampling of hnRNA is not desired, then poly(A) selection of the RNA population should be performed, or the strategy should be modified to include first-strand synthesis using oligo(dT) priming. Then, the cDNA product can be fingerprinted using a single primer [reducing the contribution of carryover oligo(dT) from the reaction by dilution], as in the protocol for DNA fingerprinting, followed by nested RAP–PCR. Priming with oligo(dT) will not completely eliminate sampling of hnRNA because poly(A) addition sometimes takes place before intron removal.

Cloning

After AP–PCR, bands can be excised from the gel with a razor blade, eluted, reamplified, and cloned. The band is cut from the gel (either wet or dry) and placed in 50 μl of TE (Tris–EDTA buffer) for several hours to elute. Reexposure of the gel confirms that the correct band has been excised. Five microliters are then reamplified with a secondary PCR reaction using the same primer that was used to generate the fingerprint, followed by blunt-end cloning into any one of a number of commercially available vectors. We have had good luck with Bluescript II and pCR-script and Epicurian Coli XL1-Blue cells (Stratagene, La Jolla, CA). *Eco*RV or *Srf*I are used during the ligation to linearize vectors that have closed on themselves, thereby improving the cloning efficiency. The main difficulty with cloning AP–PCR products is contamination of the clones with unwanted comigrating species. There are several strategies to minimize this problem. One procedure is to perform complementation tests (c-tests) on several independent clones. Alternatively, the clones are sequenced. The most abundant product of the secondary PCR amplification is more likely to c-test positively or occur more than once among the sequences than low-level background contamination. Southern transfer of the fingerprint gel and probing with these clones verifies that the correct clone has been amplified. Because the sequence complexity of the predominant bands in the fingerprint is low, Southern transfer is fairly simple, even from dried gels. A dried, arbitrarily primed PCR gel can be transferred to a blotting membrane. It is possible to demonstrate whether the probe hybridizes to the expected band by Southern hybridization using the cloned band as probe. This procedure, suggested to us by Perucho and colleagues, is described in Wong and McClelland (1994).

The clones represent a short segment of the regulated transcript. As such they are similar to expressed sequence tags (ESTs). Mapping of genes has proved useful in many genetic diseases. However, in addition to being of possible utility as ESTs, these RAP–PCR products are mapped in "gene regulation space." Mapping in "regulatory space" may ultimately be most interesting because it is in this space that most genes interact.

Motif Sequences Encoded in Primers

Completely arbitrary priming lies at one end of a spectrum of possible targeting strategies for fingerprinting. The other end of the spectrum uses primers derived from known perfect or near-perfect dispersed repeats. Examples of genomic DNA from PCR include Alu-PCR (Nelson et al., 1989), tDNA-intergenic length polymorphisms (Welsh and McClelland, 1991b, 1992), and REP–PCR (Versalovic et al., 1991). Purine-pyrimidine motifs have been successfully used to produce PCR fingerprints (Welsh et al., 1991). Primers directed toward microsatellites reveal more polymorphisms than seen in the average arbitrarily primed fingerprint, making them particularly useful for detecting polymorphisms between closely related individuals (Welsh et al., 1991; Wu et al., 1994).

One possibility for further adaptation of RAP–PCR is to encode conserved motif sequences in the primers. Then one could search for differentially expressed genes using motifs directed toward particular gene families of relevance to a particular biological phenomenon. Thus, one could attempt to bias in favor of zinc fingers, as we did with the primer ZF-1 (Ralph et al., 1993). Such efforts are an extension of the work with motifs that have been tried for fingerprinting total genomic DNA, such as promoter and amino acid motifs (Birkenmeier et al., 1992). However, protein coding motifs are much rarer than Alu or RY repeats that have been used successfully in DNA fingerprinting (e.g., Welsh et al., 1991) and thus in order to be represented in the final fingerprint, must compete more effectively with arbitrary priming events. Furthermore, the consequence of wobble bases in the translation of a conserved amino acid motif is that a primer that has no redundant positions will match perfectly with only a small fraction of the nucleic acid sequences encoding the motif. One strategy that has had some success is to encode relatively long motifs (Stone and Wharton, 1994). This success holds out the prospect for RNA fingerprinting that remains largely arbitrary but nevertheless biased toward particular gene families.

Acknowledgments

We thank Drs. Bruno Sobral, Rhonda Honeycutt, and K. K. Wong for their helpful comments, and Rita Cheng for technical assistance. This work was supported by

National Institutes of Health grants AI34829, NS33377, and HG00456 to MM and AI32644 to JW.

Literature Cited

Adams M. D., M. Dubnick, A. R. Kerlavage, R. F. Moreno, J. M. Kelley, T. R. Utterback, J. W. Nagle, C. Fields, and J. C. Venter. 1991. Sequence identification of 2,375 human brain genes. *Nature* **355**:632–634.

Al-Janabi, S. M., R. J. Honeycutt, M. McClelland, and B. W. S. Sobral. 1993. A genetic linkage map of *Saccharum spontaneum* (L.) 'SES 208'. *Genetics* **134**:1249–1260.

Bachmann, K. 1994. Molecular markers in plant ecology. *New Phytologist* **126**:403–418.

Birkenmeier, E. H., U. Schneider, and S. J. Thurston. 1992. Fingerprinting genomes by use of PCR primers that encode protein motifs or contain sequences that regulate gene expression. *Mamm. Genome* **3**:537–545.

Bowditch, B. M., D. G. Albright, J. G. Williams, and M. J. Braun. 1993. Use of randomly amplified polymorphic DNA markers in comparative genome studies. *Methods Enzymol.* **224**:294–309.

Chalmers, K. J., R. Waugh, J. I. Sprent, A. J. Simons, and W. Powell. 1992. Detection of genetic variation between and within populations of *Gliricidia spium* and *Gliricidia maculata* using RAPD markers. *Heredity* **69**:465–72.

Chirgwin, J., A. Prezybyla, R. MacDonald, and W. J. Rutter. 1979. Isolation of biologically active ribonucleic acid from sources enriched in ribonuclease. *Biochemistry* **18**:5294–5299.

Chomczynski, P., and N. Sacchi. 1987. Single-step method of RNA isolation by acid guanidinium thiocyanate-phenol-chloroform extraction. *Anal. Biochem.* **162**:156–159.

Clark, A. G., and C. M. S. Lanigan. 1993. Prospects for estimating nucleotide divergence with RAPDs. *Mol. Biol. Evol.* **10**:1096–1111.

Haymer, D. S., and D. O. McInnis. 1994. Resolution of populations of the Mediterranean fruit fly at the DNA level using random primers for the polymerase chain reaction. *Genome* **37**:244–248.

Ionov, Y., M. A. Peinado, S. Malkhosyan, D. Shibata, and M. Perucho. 1993. Ubiquitous somatic mutations in simple repeated sequences reveal a new mechanism for colonic carcinogenesis. *Nature* **363**:558–561.

Kesseli, R. V., I. Paran, R. W. Michelmore. 1994. Analysis of a detailed genetic linkage map of *Lactuca sativa* (lettuce) constructed from RFLP and RAPD markers. *Genetics* **136**:1435–1446.

Kubota, Y., A. Shimada, and A. Shima. 1992. Detection of g-ray-induced DNA damages in malformed dominant lethal embryos of the Japanese medaka (*Oryzias latipes*) using AP–PCR fingerprinting. *Mutation Res.* **283**:263–270.

Lamboy, W. F. 1994(a). Computing genetic similarity coefficients from RAPD data: Correcting for the effect of PCR artifacts caused by variation in experimental conditions. *PCR Methods and Appl.* **4**:31–37.

Lamboy W. F., 1994(b). Computing Genetic Similarity Coefficients from RAPD data: The effects of PCR artifacts. *PCR Meth. Appl.* **4**:38–43.

Liang, P., and A. Pardee. 1992. Differential display of eukaryotic messenger RNA by means of the polymerase chain reaction. *Science* **257**:967–971.

Lynch, M., and B. G. Milligan. 1994. Analysis of population genetic structure with RAPD markers. *Mol. Ecol.* **3**:91–99.

McClelland, M., C. Peterson, and J. Welsh. 1992. Length polymorphisms in tRNA intergenic spacers detected using the polymerase chain reaction can distinguish streptococcal strains and species. *J. Clin. Microbiol.* **30**:1499–1504.

McClelland, M., K. Chada, J. Welsh, and D. Ralph. 1993. Arbitrary primed PCR fingerprinting of RNA applied to mapping differentially expressed genes. In *DNA fingerprinting: State of the science* (eds. S. D. Pena, R. Charkraborty, J. T. Epplen, and A. J. Jeffereys), pp. 103–115. Birkhauser Verlag, Basel, Switzerland.

McClelland, M., H. Arensdorf, R. Cheng, and J. Welsh. 1994(a). Arbitrarily primed PCR fingerprints resolved on SSCP gels. *Nucleic Acids Res.* **22**:1770–1771.

McClelland, M., D. Ralph, R. Cheng, and J. Welsh. 1994(b). Interactions among regulators of RNA abundance characterized using RNA fingerprinting by arbitrarily primed PCR. *Nucleic Acids Res.* **22**:4419–4431.

McClelland, M., and J. Welsh. 1994. RNA fingerprinting by arbitrarily primed PCR. *PCR Meth. Appl.* **4**:S66–S81.

Mendel, Z., D. Nestel, and R. Gafny. 1994. Examination of the origin of the Israeli population of *Matsucoccus josephi* (Homoptera: Matsucoccidae) using random amplified polymorphic DNA-polymerase chain reaction method. *Ann. Entomol. Soc. Am.* **87**:165–169.

Michelmore, R. W., I. Paran, and R. V. Kesseli. 1991. Identification of markers linked to disease resistance genes by bulked segregant analysis: A rapid method to detect markers in specific genomic regions using segregating populations. *Proc. Natl. Acad. Sci. U.S.A.* **88**:9828–9832.

Nadeau, J. H., H. G. Bedigian, G. Bouchard, T. Denial, M. R. Kosowsky, R. Norberg, S. Pugh, E. Sargeant, R. Turner, and B. Paigen. 1992. Multilocus markers for mouse genome analysis: PCR amplification based on single primers of arbitrary nucleotide sequence. *Mamm. Genome* **3**:55–64.

Neale, D., and R. Sederoff. 1991. Genome mapping in pines takes shape. *Probe* **1**:1.

Nelson, D. L., S. A. Ledbetter, L. Corbo, M. F. Victoria, R. Ramirez-Solis, T. D. Webster, D. H. Ledbetter, and C. T. Caskey. 1989. Alu polymerase chain reaction: a method for rapid isolation of human-specific sequences from complex DNA sources. *Proc. Natl. Acad. Sci. U.S.A.* **86**:6686–6690.

Peinado, M. A., S. Malkhosyan, A. Velazquez, and M. Perucho. 1992. Isolation and characterization of allelic losses and gains in colorectal tumors by arbitrarily primed polymerase chain reaction. *Proc. Natl. Acad. Sci. U.S.A.* **89**:10065.

Ralph, D., M. McClelland, and J. Welsh. 1993. RNA fingerprinting using arbitrarily primed PCR identifies differentially expressed genes in Mink lung (Mu1Lv) cells growth arrested by transforming growth factor-beta1. *Proc. Natl. Acad. Sci. U.S.A.* **90**:10710–10714.

Reiter, R. S., J. G. Williams, K. A. Feldmann, J. A. Rafalski, S. V. Tingey, and P. A. Scolnik. 1992. Global and local genome mapping in *Arabidopsis thaliana* by using recombinant inbred lines and random amplified polymorphic DNAs. *Proc. Natl. Acad. Sci. U.S.A.* **89**:1477.

Serikawa, T., X. Montagutelli, D. Simon-Chazottes, and J.-L. Guenet. 1992. Polymorphisms revealed by PCR with single, short-sized, arbitrary primers are reliable markers for mouse and rat gene mapping. *Mamm. Genome* **3**:65–72.

Sobral, B. W. S., and R. Honeycutt. 1993. High output genetic mapping of polyploids using PCR-generated markers. *Theor. Appl. Genet.* **86**:105–112.

Stone, B., and W. Wharton. 1994. Targeted RNA fingerprinting: The cloning of

differentially-expressed cDNA fragments enriched for members of the zinc finger gene family. *Nucleic Acids Res.* **22**:2612–2618.
Thomas, E. P., D. W. Blinn, and P. Keim. 1994. A test of an allopatric speciation model for congeneric amphipods in an isolated aquatic ecosystem. *J. N. Amer. Benth. Soc.* **13**:100–109.
Versalovic, J., K. Thearith, and J. R. Lupski. 1991. Distribution of repetitive DNA sequences in eubacteria and application to fingerprinting of bacterial genomes. *Nucleic. Acids. Res.* **19**:6823–6831.
Welsh, J., and M. McClelland. 1990. Fingerprinting genomes using PCR with arbitrary primers. *Nucleic Acids Res.* **18**:7213–7218.
Welsh, J., and M. McClelland. 1991(a). Genomic fingerprinting with AP–PCR using pairwise combinations of primers: Application to genetic mapping of the mouse. *Nucleic Acids Res.* **19**:5275–5279.
Welsh, J., and M. McClelland. 1991(b). Genomic fingerprints produced by PCR with consensus tRNA gene primers. *Nucleic Acids Res.* **19**:861–866.
Welsh, J., and M. McClelland. 1992. PCR-amplified length polymorphisms in tRNA intergenic spacers for categorizing staphylococci. *Mol. Microbiol.* **6**:1673–1680.
Welsh, J., C. Petersen, and M. McClelland. 1991. Polymorphisms generated by arbitrarily primed PCR in the mouse: Application to strain identification and genetic mapping. *Nucleic Acids Res.* **19**:303–306.
Welsh, J., K. Chada, S. S. Dalal, D. Ralph, R. Chang, and M. McClelland. 1992(a). Arbitrarily primed PCR fingerprinting of RNA. *Nucleic Acids Res.* **20**:4965–4970.
Welsh, J., C. Pretzman, D. Postic, I. Saint Girons, G. Baranton, and M. McClelland. 1992(b). Genomic fingerprinting by arbitrarily primed PCR resolves *Borrelia burgdorferi* into three distinct groups. *Int. J. Systematic Bacteriol.* **42**:370–377.
Welsh, J., and M. McClelland. 1993. The characterization of pathogenic microorganisms by genomic fingerprinting using arbitrarily primed polymerase chain reaction (AP–PCR). In *Diagnostic molecular microbiology* (eds. D. H. Persing, T. F. Smith, F. C. Tenover, and T. J. White), pp. 595–602. ASM Press, Washington D.C.
Williams, C. L., S. L. Goldson, D. B. Baird, and D. W. Bullock. 1994. Geographical origin of an introduced insect pest, *Listronotus bonariensis* (Kuschel), determined by RAPD analysis. *Heredity.* **72**:412–419.
Williams, J. G., A. R. Kubelik, K. J. Livak, J. A. Rafalski, and S. V. Tingey. 1990. DNA polymorphisms amplified by arbitrary primers are useful as genetic markers. *Nucleic Acids Res.* **18**:6531–6535.
Williams, J. G., M. K. Hanafey, J. A. Rafalski, and S. V. Tingey. 1993. Genetic analysis using random amplified polymorphic DNA markers. *Methods Enzymol.* **218**:704–740.
Wong, K. K., S. C-H. Mok, J. Welsh, M. McClelland, S-W. Tsao, and R. S. Berkowitz. 1993. Identification of differentially expressed RNA in human ovarian carcinoma cells by arbitrarily primed PCR fingerprinting of total RNAs. *Int. J. Oncology* **3**:13–17.
Wong, K. K., and M. McClelland. 1994. Stress-inducible gene of *Salmonella typhimurium* identified by arbitrarily primed PCR of RNA. *Proc. Natl. Acad. Sci. U.S.A.* **91**:639–643.
Woods, J. P., D. Kersulyte, R. W. Tolan, Jr., C. M. Berg, and D. E. Berg. 1994. Use of arbitrarily primed polymerase chain reaction analysis to type disease and carrier strains of *Neisseria meningitidis* isolated during a university outbreak. *J. Infect. Dis.* **169**:1384–1389.

Woodward, S. R., J. Sudweeks, and C. Teuscher. 1993. Random sequence oligonucleotide primers detect polymorphic DNA products which segregate in inbred strains of mice. *Mamm. Genome* **3**:73–78.

Wu, K-S., R. Jones, L. Danneberger, and P. A. Scolnik. 1994. Detection of microsatellite polymorphisms without cloning. *Nucleic Acids Res.* **22**:3257–3258.

Yu, K. F., A. Van Denze, and K. P. Pauls. 1993. Random amplified polymorphic DNA (RAPD) analysis. In *Methods in plant molecular biology and technology* (eds. B. R. Glick and J. E. Thompson). CRC Press, FL.

21

PCR-BASED SCREENING OF YEAST ARTIFICIAL CHROMOSOME LIBRARIES

Eric D. Green

Yeast artificial chromosome (YAC) cloning provides the means to isolate very large (~100,000–1,000,000 bp) segments of DNA (Burke *et al.*, 1987; Hieter *et al.*, 1990) and represents a powerful tool for dissecting complex genomes. Established applications of YAC cloning range from searching for human disease genes to large-scale genome mapping. Associated with these applications is the need to screen YAC libraries for clones of interest in a routine, reliable, and efficient fashion.

While hybridization of yeast colonies has been used successfully to identify appropriate clones within YAC libraries (Brownstein *et al.*, 1989; Traver *et al.*, 1989; Anand *et al.*, 1990; Larin *et al.*, 1991), such methods are hindered by a number of inherent limitations. First, the density of target DNA immobilized on a hybridization filter is significantly less for a lysed yeast colony than for a lysed bacterial colony, since YACs are present in roughly single copy within a host genome of higher complexity (typically representing <3% of the total yeast DNA) and the number of cells in a yeast colony is at least 10-fold lower than in a bacterial colony. As a result, hybridization signals are significantly lower with YAC-containing yeast colonies than with plasmid-bearing bacterial colonies. Second, the lysis of yeast cells and subsequent immobilization of DNA is prone to problems, leading to further decreases in hybridization signal. Third, standard protocols for constructing YAC libraries require the embedding of transformed spheroplasts within soft agar to promote regeneration of the yeast cell wall. The resulting clones must be individually picked from beneath the agar surface and, most often, arrayed in microtiter plates. Thus, replicas made from primary transformation plates are not routinely prepared and used for hybridization analysis (at least not in cases where all the resulting clones are incorporated into the library). Fourth, colony hybridization of YAC clones is prone to difficulties relating to both cloned repetitive sequences present in the YAC insert (e.g., *Alu* sequences in YACs containing human DNA) and contaminating probe sequences (e.g., pBR322-derived) that are complementary to the YAC vector. Such problems often lead to ambiguous results, which require the selection, isolation, and analysis of numerous false positive clones in order to identify the true positive ones. Finally, and perhaps most relevant to the large-scale application of YAC cloning for genome mapping, hybridization-based protocols are logistically difficult to perform in large numbers or to automate fully. Collectively, these features hamper the ability to screen comprehensive YAC libraries by colony hybridization, especially on the scale required for mapping large genomes.

To provide an alternative and more efficient means for isolating clones from comprehensive YAC libraries, we developed a PCR-based screening strategy that is tailored for analyzing clones maintained in simple, ordered arrays (Green and Olson, 1990). This approach has been found to yield several important and interrelated benefits. To begin with, the PCR-based approach is both highly sensitive, allowing the detection of positive YACs within pools of thousands of clones, and specific, providing information about the "authenticity" of a candidate clone early in the screening protocol (e.g., by detecting a PCR product of the expected size) (Green and Olson, 1990; Heard *et al.*, 1989). In addition, this approach has proven less prone to problems relating to cloned repetitive sequences than hybridization-based protocols. Furthermore, PCR-based methods are becoming increasingly amenable to automation, which is an important characteristic of any technique being used extensively for large-scale genome mapping. Finally, the screening of YAC libraries by PCR represents a key component of our model for constructing physical maps of DNA by detecting landmarks, called sequence-tagged sites (STSs), within recombinant clones (Green and Green,

1991). The advantages of using STSs as physical DNA landmarks have been discussed (Olson et al., 1989; Green and Green, 1991).

General Protocol

The fundamental features of a typical PCR-based screening strategy are depicted in Fig. 1. In most YAC libraries, each clone is stored in an individual well of an 8 × 12 (96-position) microtiter plate. One approach is to inoculate a nylon filter with the clones from four such plates slightly offset to one another, creating an array of 384 YACs (Brownstein et al., 1989). The filters are placed on solid medium, and the clones are grown at 30°C for 4–6 days to a stage where the colonies are all roughly equal in size. The longer growth period provides a larger number of yeast cells and, hence, greater yields of DNA. Similarly, the addition of a higher (two-fold) amount of adenine in the media will promote the development of larger colonies. Following growth (and, if desired, storage of the filters at −80°C), the colonies on each filter are pooled by simple washing, and total DNA from the pooled yeast cells is purified (Green and Olson, 1990), yielding "single-filter pools" of DNA derived from 384 YAC clones. Equal aliquots of DNA from each single-filter pool are then mixed together in groups of five to yield "multi-filter pools," each representing the DNA from 1920 YAC clones.

In the first stage of the screening protocol, each multi-filter pool is tested for the presence of a DNA segment of interest by using a corresponding PCR assay that specifically amplifies all or part of that segment. Products are separated by polyacrylamide or agarose gel electrophoresis and detected by staining with ethidium bromide. Any multi-filter pool of DNA that supports the amplification of a product of the appropriate size is scored as positive. In the second stage of the screening protocol, each of the five single-filter pools that comprises a positive multi-filter pool is analyzed with the same PCR assay.

The results of two sequential rounds of PCR analysis limit the location of positive YACs to individual filters, each containing 384 clones. For the final stage of the screening protocol, two different approaches can be employed to identify the precise address of each positive clone. In one approach, a replica filter containing the lysed

280 Part Three. Research Applications

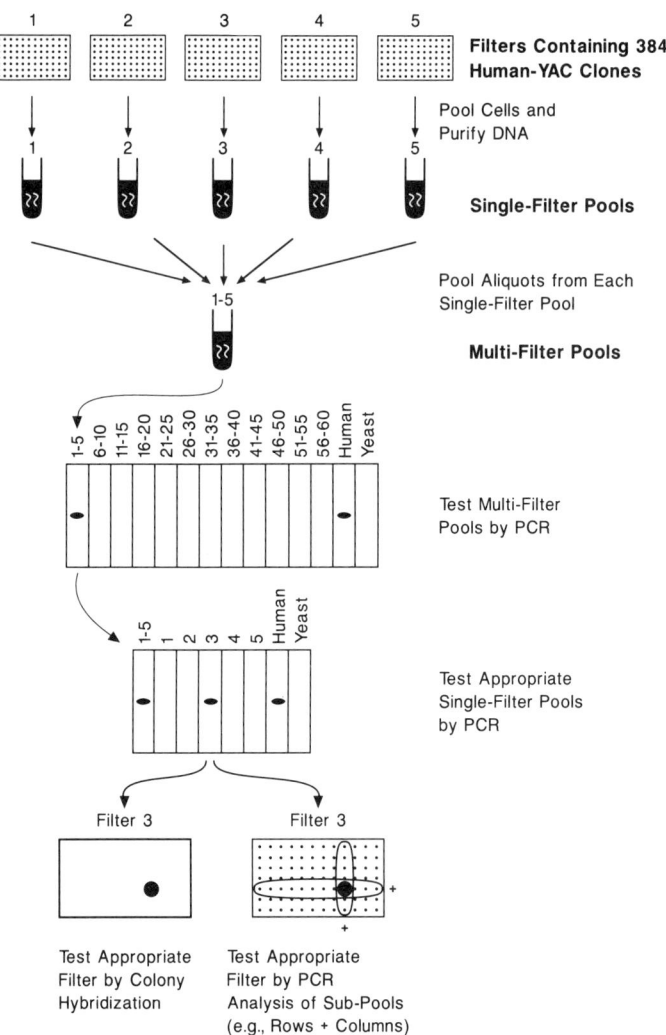

Figure 1 Schematic representation of the PCR-based strategy for screening YAC libraries. Details of the illustrated approach are provided in the text.

colonies of the 384 clones (corresponding to one single-filter pool) is hybridized with either the radiolabeled PCR product or another corresponding probe, and the location of the positive clone(s) within the 16 × 24 array is identified by autoradiography (Green and Olson, 1990). In our experience, PCR products >200 bp in size provide reliable hybridization probes when labeled by standard methods (Feinberg and Vogelstein, 1983), while the results with PCR products <100 bp in size are more inconsistent. While other hybridization probes (e.g., genomic or cDNA clones) often yield excellent results, they must be free of contaminating pBR322 sequences and cannot hybridize to repetitive elements (e.g., *Alu* sequences) present in the YAC insert. The latter problem can be overcome by the presence of large amounts of unlabeled source DNA (e.g., human placental DNA) in the hybridization cocktail. In the second approach, additional rounds of PCR analysis are used to determine the location of the positive YAC(s) within each 384-clone pool. In one strategy (Kwiatkowski et al., 1990), clones positioned in a given 16 × 24 array are pooled by rows and columns, and a crude cell lysate is prepared from each pool. The 40 resulting DNA pools are then individually tested by PCR, and the position(s) of the positive clone(s) corresponds to the intersection of a positive row and a positive column (see Fig. 1). Of course, other more complicated pooling schemes can be envisioned to track the location of a positive clone within a 384-clone array, and these can vary both in the number of sequential rounds of PCR analysis required and the overall complexity of the pooling strategy (discussed later and in Green and Olson, 1990). It is worth noting that for the smaller pools analyzed in the later stages (typically containing <50 clones), PCR can be performed by using DNA purified from individual YAC clones (MacMurray et al., 1991), crude cell lysates (Kwiatkowski et al., 1990), or nontreated yeast cells (Huxley et al., 1990; E. D. Green, unpublished data).

Following the localization of a positive YAC clone by any of the routes described above, a final confirmatory test of individual yeast colonies should always be performed with the same PCR assay as that employed to screen the YAC library. This is particularly important for libraries in which the clones were not colony-purified prior to storage, since two or more clones are often inadvertently picked together (i.e., mixed) during the transfer of primary yeast transformants into microtiter plates. For this final stage of the screen, PCR can be performed directly with cells removed from a yeast colony (Huxley et al., 1990) rather than with purified DNA. Typically, multiple independent yeast colonies should be tested, since mixed populations of clones are often encountered.

The PCR-based screening approach depicted in Fig. 1 has now been used by many laboratories to perform numerous YAC library screens; however, it represents only one scheme for constructing hierarchical pools of DNA from collections of clones. With the increasing demand to provide investigators access to comprehensive YAC libraries, various arrangements and alternative strategies for PCR-based screening have been established. The most straightforward of these has been the distribution of higher level (e.g., multi- and single-filter) pools of DNA for the initial rounds of PCR testing and, once determined, the appropriate filter(s) for the hybridization analysis. The major limiting feature of such an arrangement is the inherent inefficiency and inconsistency of the hybridization-based step (see later discussion). As a result, many centralized YAC screening facilities have implemented fully PCR-based strategies, thereby allowing them to distribute exclusively DNA pools (using any one of a number of established pooling schemes). There are now commercial vendors offering PCR-based YAC screening services, either by providing the DNA pools to investigators for PCR analysis or by screening the YAC library with PCR assays provided by investigators. The overall efficiency of PCR-based YAC library screening is thus allowing a number of different screening services to be established and, in the long run, will provide an increasing number of scientists greater access to large YAC libraries.

Additional Protocol-Related Issues

Yeast DNA Template

The quality and amount of purified yeast DNA used for PCR analysis must be carefully monitored. For consistent detection of positive clones within the larger (e.g., 1920- and 384-clone) pools, we have found it necessary to utilize more rigorous purification methods, involving either phenol extractions (Green and Olson, 1990) or immobilization in agarose and treatment with lithium dodecyl sulfate (MacMurray et al., 1991; Anand et al., 1989; Southern et al., 1987). More simple and rapid approaches for preparing yeast DNA have been found to yield templates of poorer quality that are often unsuitable for PCR amplification. For example, the crude methods for template preparation mentioned above (e.g., cell lysates) are not adequate for analyzing the larger pools of clones.

Careful monitoring of the final DNA concentrations is also important. We have often encountered complete inhibition of PCR in the presence of $>300-500$ ng of purified yeast DNA in a 5-μl reaction. The nature of the inhibition has not been established, although prior boiling, additional phenol-chloroform extractions, or gel filtration fails to eliminate the inhibitory effect. An optimal amount of yeast DNA appears to be 2–20 ng/μl of final reaction volume. As a result, we often measure the concentration of purified yeast DNA preparations by using a fluorometric assay (Labarca and Paigen, 1980) and dilute samples to roughly the same DNA concentration. To verify the quality of the DNA, we also test purified DNA samples with two control PCR assays (Green and Olson, 1990).

Control DNA Templates

During each stage of a PCR-based library screen, a reaction containing the same source of DNA as that used to construct the YAC library (e.g., human) should always be included as a positive control. For a negative control, we routinely use DNA from a YAC clone containing an anonymous segment of yeast DNA as the cloned insert (Green and Olson, 1990). The use of this negative-control DNA has proven more effective than simply not adding exogenous DNA. In several instances, the contamination of reagents (e.g., oligonucleotides) with target DNA was only detected when this DNA sample was added, with no amplification occurring in the absence of any exogenous yeast DNA. Presumably, the yeast DNA serves as a carrier for the trace amounts of contaminating template, thereby increasing the likelihood that amplification occurs. Various PCR products will often be generated with the negative-control template. When such products are the same size as the product of interest, the corresponding PCR assay cannot easily be used to screen a YAC library.

Optimization of PCR Assays

The robust behavior of the PCR assay is critical to the success of a PCR-based YAC library screen. Standard approaches for optimizing PCR conditions (e.g., alterations in thermal cycling, changes in magnesium concentration) should be used to maximize an assay's performance. However, in our experience, an occasional PCR assay that yields a convincing amplification product with positive-control templates will fail to detect positive clones within the larger pools (e.g., 1920 clones). For screening a human-YAC library, we have found that a good way to "mimic" a library screen is to use DNA purified from a pool of 384 YAC clones (at ~33 ng/μl) that has been "spiked" with ~5 ng/μl of total human DNA (Green et al., 1991). Successful amplification of the correct product from such a DNA mixture has proven to correlate well with a PCR assay's performance during a YAC library screen. As a result, we only use a PCR assay to screen a YAC library if it has successfully amplified the appropriate product from such a human DNA/YAC pool DNA mixture.

Analysis of PCR Products by Gel Electrophoresis

The amplification of "inappropriate" DNA fragments of different size(s) than the product of interest is often observed with DNA from YAC pools. Some of these products appear to derive from yeast DNA, while others are specific to the cloned source DNA (often amplifying with some YAC pools but not the positive-control template). The generation of PCR products with YAC pools but not with positive-control DNA suggests that those sequences can only be amplified when they are present at relatively high concentrations. As a practical matter, the use of high-resolution native polyacrylamide gels (Green and Olson, 1990) most often allows discrimination between appropriate and inappropriate PCR products, thereby providing a critical element of specificity to the PCR assay. In general, this system is preferable to agarose gels in terms of resolution and sensitivity. However, we have recently found that certain preparations of agarose designed for the separation of small DNA fragments (e.g., MetaPhor agarose from FMC) provide resolution and sensitivity that rival (but do not surpass) polyacrylamide gels. Of course, horizontal agarose gels are generally more convenient to use than vertical polyacrylamide gels, especially in terms of the ease of preparation and suitability for analyzing large numbers of samples.

Schemes for Constructing Pools of YAC Clones

The scheme shown in Fig. 1 is just one example of an approach for constructing pools of YAC clones. Numerous other strategies can be envisioned (Green and Olson, 1990). Important issues that must be considered include the number of sequential rounds of PCR analysis required to perform a screen and the total number of PCR assays required to identify each positive clone. Highly sophisticated pooling schemes can be designed that require a fewer number of PCR assays to be performed (Green and Olson, 1990); however, these are accompanied by the effort required to construct the more complicated pools and, in some cases, to perform a larger number of sequential rounds of PCR analysis en route to identifying positive clones.

Another issue to consider in designing YAC pooling schemes is the maximum size of the highest-level pool. In principle, pools containing more than 1920 YAC clones can be analyzed in the first stage of the screening protocol (Green and Olson, 1990). While some PCR assays can readily detect a single positive YAC within a pool of ~60,000 clones, many assays do not reliably provide such a high level of sensitivity. The sensitivity of the latter PCR assays could be enhanced by using a number of established methods (e.g., reamplification of products using nested primers, multiple additions of DNA polymerase, detection of products by hybridization or primer labeling). In general, we have chosen not to employ such methods, largely because they render the PCR assays excessively sensitive to trace quantities of contaminating DNA. As a result, we have found the use of 1920-clone pools for the first stage of screening to be highly reliable with a wide range of PCR assays.

Future Improvements

Additional refinements in the PCR-based strategy described for screening YAC libraries will most likely be directed at several interrelated problems that currently limit the efficiency of the method. Most important, approaches that employ PCR but not hybridization analysis are essential. Any hybridization-based assay rapidly becomes the rate-limiting step in the overall screening process and severely hampers efforts to increase overall throughput. Strategies that use PCR exclusively require a significant increase in the total number of PCR assays that must be performed to identify each positive clone. Improvements in the ability to automate the PCR setup and analysis will result in the increased utilization of such fully PCR-based schemes. Semi-automated preparation of PCR assays can, for the most part, already be performed with current robotic instrumentation. The construction of more complex pools that accompany fully PCR-based schemes can also be readily automated with DNA prepared from individual YAC clones (MacMurray *et al.*, 1991).

The implementation of more automated methods for analyzing PCR products will likely be more challenging. Ideally, this would include the complete elimination of gel electrophoresis as the fundamental technique for analyzing PCR products, since such a labor-intensive step cannot keep pace with automated PCR assay preparation. It is hoped that approaches will be developed that can be automated and that can offer a level of specificity comparable to that provided by gel electrophoresis. The latter feature is especially critical for screening YAC libraries, since numerous irrelevant PCR products of various sizes are routinely amplified from YAC pools that do not contain positive clones.

A technique that provides an efficient, highly specific alternative to gel electrophoresis is the oligonucleotide ligation assay (OLA). Using OLA, an amplified PCR product can be analyzed for the presence of a precise DNA sequence that will support the ligation of two oligonucleotides that anneal immediately adjacent to one another within the amplified DNA segment (Nickerson et al., 1990; Landegren et al., 1988). One oligonucleotide typically contains a biotin moiety that serves as an "anchor" group, while the other bears a molecule (e.g., digoxigenin) that serves as a "reporter" group. Successful ligation of the two oligonucleotides allows their co-capture and subsequent detection by using an enzyme-linked immunosorbent assay (ELISA) specific for the reporter group (Nickerson et al., 1990). High specificity is provided by the use of at least one oligonucleotide corresponding to the DNA sequence between the two primers used for PCR amplification. In addition, the use of an ELISA-based assay is advantageous in that it can be readily automated. Kwok et al. (1992) have, in fact, developed and implemented a PCR-based method for screening YAC libraries that uses OLA to detect the PCR products. While there are limitations associated with OLA (e.g., preparation of oligonucleotides with appropriate modifying groups), such a system provides a means to automate the analysis of large numbers of PCR products while retaining specificity and eliminating gel electrophoresis. Ultimately, the ability to extend the capacity for screening YAC libraries on the scale required for many genome mapping projects will depend upon the application of such nonelectrophoretic systems for detecting PCR products.

Acknowledgments

I thank Drs. Maynard Olson, Pui Kwok, and Gabriela Adelt Green for helpful comments. E. D. G. is a Lucille P. Markey Scholar, and this work was supported in part by a grant from the Lucille P. Markey Charitable Trust and in part from National Institutes of Health Grant P50-HG00201.

Literature Cited

Anand, R., A. Villasante, and C. Tyler-Smith. 1989. Construction of yeast artificial chromosome libraries with large inserts using fractionation by pulsed-field gel electrophoresis. *Nucleic Acids Res.* **17**:3425–3433.

Anand, R., J. H. Riley, R. Butler, J. C. Smith, and A. F. Markham. 1990. A 3.5 genome equivalent multi access YAC library: construction, characterisation, screening, and storage. *Nucleic Acids Res.* **18**:1951–1956.

Brownstein, B. H., G. A. Silverman, R. D. Little, D. T. Burke, S. J. Korsmeyer, D. Schlessinger, and M. V. Olson. 1989. Isolation of single-copy human genes from a library of yeast artificial chromosome clones. *Science* **244**:1348–1351.

Burke, D. T., G. F. Carle, and M. V. Olson. 1987. Cloning of large segments of exogenous DNA into yeast by means of artificial chromosome vectors. *Science* **236**:806–812.

Feinberg, A. P., and B. Vogelstein. 1983. A technique for radiolabeling DNA restriction endonuclease fragments to high specific activity. *Anal. Biochem.* **132**:6–13.

Green, E. D., and M. V. Olson. 1990. Systematic screening of yeast artificial-chromosome libraries by use of the polymerase chain reaction. *Proc. Natl. Acad. Sci. U.S.A.* **87**:1213–1217.

Green, E. D., and P. Green. 1991. Sequence-tagged site (STS) content mapping of human chromosomes: theoretical considerations and early experiences. *PCR Methods Appl.*, **1**:77–90.

Green, E. D., R. M. Mohr, J. R. Idol, M. Jones, J. M. Buckingham, L. L. Deaven, R. K. Moyzis, and M. V. Olson. 1991. Systematic generation of sequence-tagged sites for physical mapping of human chromosomes: application to the mapping of human chromosome 7 using yeast artificial chromosomes. *Genomics* **11**:548–564.

Heard, E., B. Davies, S. Feo, and M. Fried. 1989. An improved method for the screening of YAC libraries. *Nucleic Acids Res.* **17**:5861.

Hieter, P., C. Connelly, J. Shero, M. K. McCormick, S. Antonarakis, W. Pavan, and R. Reeves. 1990. Yeast artificial chromosomes: promises kept and pending. In: *Genome Analysis* (eds. K. E. Davies and S. M. Tilghman), pp. 83–120, Cold Spring Harbor Laboratory Press, Cold Spring Harbor, NY.

Huxley, C., E. D. Green, and I. Dunham. 1990. Rapid assessment of *S. cerevisiae* mating type by PCR. *Trends Genet.* **6**:236.

Kwiatkowski Jr., T. J., H. Y. Zoghbi, S. A. Ledbetter, K. A. Ellison, and A. C. Chinault. 1990. Rapid identification of yeast artificial chromosome clones by matrix pooling and crude lysate PCR. *Nucleic Acids Res.* **18**:7191–7192.

Kwok, P.-Y., M. F. Gremaud, D. A. Nickerson, L. Hood, and M. V. Olson. 1992. Automatable screening of yeast artificial-chromosome libraries based on the oligonucleotide-ligation assay. *Genomics* **13**:935–941.

Labarca, C., and K. Paigen. 1980. A simple, rapid, and sensitive DNA assay procedure. *Anal. Biochem.* **102**:344–352.

Landegren, U., R. Kaiser, J. Sanders, and L. Hood. 1988. A ligase-mediated gene detection technique. *Science* **241**:1077–1080.

Larin, Z., A. P. Monaco, and H. Lehrach. 1991. Yeast artificial chromosome libraries containing large inserts from mouse and human DNA. *Proc. Natl. Acad. Sci. U.S.A.* **88**:4123–4127.

MacMurray, A. J., A. Weaver, H.-S. Shin, and E. S. Lander. 1991. An automated method for DNA preparation from thousands of YAC clones. *Nucleic Acids Res.* **19:**385–390.

Nickerson, D. A., R. Kaiser, S. Lappin, J. Stewart, L. Hood, and U. Landegren. 1990. Automated DNA diagnostics using an ELISA-based oligonucleotide ligation assay. *Proc. Natl. Acad. Sci. U.S.A.* **87:**8923–8927.

Olson, M. V., L. Hood, C. Cantor, and D. Botstein. 1989. A common language for physical mapping of the human genome. *Science* **245:**1434–1435.

Southern, E. M., R. Anand, W. R. A. Brown, and D. S. Fletcher. 1987. A model for the separation of large DNA molecules by crossed field gel electrophoresis. *Nucleic Acids Res.* **15:**5925–5943.

Traver, C. N., S. Klapholz, R. W. Hyman, and R. W. Davis. 1989. Rapid screening of a human genomic library in yeast artificial chromosomes for single-copy sequences. *Proc. Natl. Acad. Sci. U.S.A.* **86:**5898–5902.

22

OLIGONUCLEOTIDE LIGANDS THAT DISCRIMINATE BETWEEN THEOPHYLLINE AND CAFFEINE

Rob Jenison, Stanley Gill, Jr., and Barry Polisky

The alkaloid theophylline is a bronchodilator commonly used to treat asthma. Because elevated levels of theophylline can have serious side effects, serum levels of theophylline in asthma patients must be carefully monitored. Several diagnostic assays are currently available that use antibodies that recognize theophylline (Hendeles and Weinberger, 1983). An important feature of such antibodies is their ability to discriminate between theophylline and closely related xanthine derivatives in the diet, particularly caffeine. Theophylline and caffeine differ only by a methyl group at the N-7 position, and thus provide a stringent test of specificity for binding (Fig. 1). The SELEX (*S*ystematic *E*volution of *L*igands by *EX*ponential Enrichment) procedure (Tuerk and Gold, 1990) uses large repertoire libraries of single-stranded oligonucleotides (either RNA or DNA) of random sequence to serve as a source of conformational complexity. The ensemble of random sequence oligonucleotides is allowed to interact with the target, and the subset of sequences with affinity for the target is separated from the sequences that have little or no affinity for the target. The bound sequences are separated from the target,

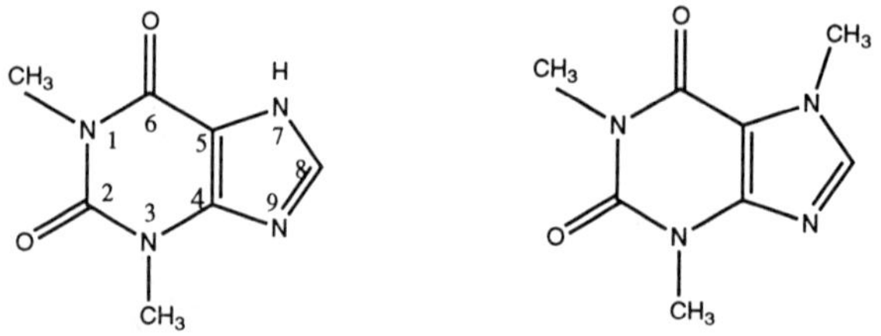

THEOPHYLLINE **CAFFEINE**

Figure 1 Structures of theophylline and caffeine.

converted to cDNA, and amplified to double-stranded DNA by a PCR. The DNA sequences contain a promoter for T7 RNA polymerase that permits transcription of the DNA population. Cycles of interaction of the RNA transcripts with target are repeated, imposing selection for those transcripts with high affinity for the target. SELEX has been used to isolate oligonucleotides with high affinity for a variety of protein targets such as bacteriophage T4 DNA polymerase (Tuerk and Gold, 1990), HIV reverse transcriptase (Tuerk et al., 1992), HIV Rev protein (Bartel et al., 1991), and basic fibroblast growth factor (Jellinek et al., 1993). In general, the affinity (K_D) of oligonucleotide ligands for protein targets has been observed to be in the 1–10 nM range. In addition to protein targets, techniques similar to SELEX have been used to isolate oligonucleotides with affinity for low-molecular-weight targets, such as tryptophan (Famulok and Szostak, 1992), ATP (Sassanfar and Szostak, 1993), and arginine (Connell et al., 1993). Taken together, these experiments have demonstrated the ability of single-stranded oligonucleotides to assume a high degree of higher order structure in solution. Of particular interest in exploring the capabilities of SELEX is the question of molecular discrimination of oligonucleotides between chemically similar low-molecular-weight targets. Theophylline and caffeine represent an interesting experimental challenge (Fig. 1). We describe here some properties of oligonucleotides selected to bind with high affinity to theophylline and which have low affinity for caffeine. A full description of the procedures used and results obtained in these experiments has appeared elsewhere (Jenison et al., 1994).

Materials and Methods

Template for SELEX

To initiate the SELEX procedure, a pool of single-stranded DNA was synthesized which, upon conversion to double-stranded DNA by a PCR, served as the template for selection and amplification. This double-stranded DNA, termed 40N1, contains a region 40 nucleotides in length which is random in sequence. Flanking this region are fixed regions containing the sequences necessary for PCR amplification, *in vitro* transcription by T7 RNA polymerase, and for cDNA synthesis. A primer, 5G1, contains the promoter sequences for *in vitro* transcription of RNA by T7 RNA polymerase and sufficient sequence complementarity to cDNAs of the transcripts to serve as a primer in PCR amplification. The 5G1 sequence is 5'CCGAAGC-TTAATACGACTCACTATAGGGAGCTCAGAATAAACGCTCAA3'. The primer 3G1 contains the sequence for reverse transcription of transcripts and also functions as a primer in PCR amplification. Its sequence is 5'-GCCGGATCCGGGCCTCATGTCGAA-3'.

PCR Amplification

To generate double-stranded DNA templates for transcription, PCR amplification of the template DNA or cDNAs from selection was necessary. In a 100-μl volume, 5–10 pmoles of DNA were combined with 200 pmoles each of the primers 5G1 and 3G1 and 1mM each of deoxyribonucleoside triphosphates (dNTPs) in a buffer containing 10 mM Tris–HCl (pH 8.4), 50 mM KCl, 7.5 mM MgCl2, and 50 μg/ml bovine serum albumin (BSA). To this, 20 units of *Taq* polymerase were added and the samples denatured at 93°C for 3 min. This was immediately followed by 15 cycles of PCR as follows: 93°C for 0.5 min, 53°C for 0.1 min, and 72°C for 1 min. The DNA was then phenol-chloroform (50 : 50) extracted and precipitated in 0.1 volumes of 3 M sodium acetate (pH 5.2) and 2.5 volumes of ethanol.

In Vitro Transcription

DNA transcription was performed with T7 RNA polymerase. In a 500-μl volume, 100 pmoles of DNA were mixed with 1 mM each of NTP and 100 μCi [α-^{32}P]ATP in 40 mM Tris–HCl (pH 8), 12 mM MgCl$_2$, 5 mM dithio threitol (DTT), 1 mM spermidine, 0.01% Triton X-100, and 4% PEG-8000 (w/v). To this, 2500 U of T7 RNA polymerase were added, and the reaction was allowed to run 2 hr at 37°C. The reaction was terminated by adding an equal volume of 10 M urea. The pool of random RNAs was purified by electrophoresis on an 8% denaturing polyacrylamide gel. The full-length product was cut and eluted from the gel. This gave an approximate yield of 2 nmoles of random RNA.

Column Selection

Affinity chromatography techniques were employed to isolate oligonucleotides with affinity for theophylline. A 1-/carboxypropyl derivative of theophylline was activated with N-/hydroxysuccinimide and 1-ethyl-3, 3-dimethylaminopropyl carbodimide (EDC) and then reacted with EAH Sepharose (Pharmacia Labs, Piscataway, NJ) to form a stable amide linkage. The concentration of theophylline on the column was about 0.5 mM. A control column was also prepared in which EAH Sepharose was acetylated with acetic anhydride. This column permitted removal of RNAs that have affinity for the Sepharose matrix.

For selection, the RNA pool was placed over 100 μl of precolumn resin packed into a Bio-Spin column (BioRad, Hercules, CA) in 100 μl of buffer A [0.1 M HEPES (pH 7.3), 0.5 M NaCl, 5 mM MgCl$_2$]. This column was incubated 15 min at room temperature and then 2× 400 μl fractions were collected and ethanol precipitated. This collected RNA pool was then resuspended in 80 μl of water. To this was added 20 μl of 5X buffer A, and after a 10-min incubation at room temperature, it was applied to a 100-μl theophylline column. The RNA was allowed to interact with the column for 20 min with mixing at room temperature. The column was then washed with 25 column volumes of buffer A to remove unbound RNAs. The remaining RNAs were then exposed to 400 μl of buffer B [0.1 M HEPES (pH 7.3), 0.5 M NaCl, 5 mM MgCl$_2$, 0.1 M theophylline]. The column was incubated for 30 min at room temperature. The bound RNAs were eluted with buffer B until background radioactivity levels were reached. Eluted RNA was pooled and reverse transcribed as will be described.

An alternative elution procedure called counter-SELEX was also used to increase the rate at which specific RNAs were selected. In this procedure, the RNA pool was loaded and the column washed with buffer A as described above. Next, 400 μl of buffer C [0.1 M HEPES (ph 7.3), 0.5 M NaCl, 5 mM MgCl$_2$, 0.1 M caffeine] were added and the column was incubated 15 min at room temperature. This step was designed to increase specificity by eliminating theophylline-binding RNAs that had substantial cross-reactivity for caffeine. After washing with 15 column volumes of buffer C, the remaining bound RNAs were eluted with theophylline as described earlier.

Reverse Transcription

The bound RNAs eluted from the column were ethanol precipitated and resuspended in water. They were then added to 200 pmoles of the 3G1 primer and 200 μM each of dNTP in 50 mM Tris–HCl (pH 8.3), 60 mM NaCl, 6 mM Mg(OAc)$_2$, and 10 mM DTT. To this were added 20 units of avian myeloblastosis virus (AMV) reverse transcriptase (Life Sciences, Denver, CO) and the reaction incubated at 48°C for 60 min. This reaction was then carried directly into PCR as described.

Preparation of Clones

Cloning was performed by cleaving PCR DNA pools with *Bam*H1 and *Hind*III endonucleases and ligation with similarly cleaved plasmid pUC18. DNA was transformed by heat shock into competent MC1061 cells. Individual colonies were picked, heated to 95°C in 20 μl of water for 5 min, and the supernatant PCR amplified in the presence of biotinylated 5g1. By incorporating biotin in one of the amplification primers, it was possible to physically separate the two strands produced by PCR (Mitchell and Merril, 1989). This material was exposed to immobilized streptavidin (Pierce, Rockford, IL); then the nonbiotinylated strand was eluted with 0.2 N NaOH. This single-stranded DNA was sequenced by the dideoxy method.

Equilibrium Filtration Binding Assay

A rapid procedure for assessing the affinity of RNAs for theophylline was developed. Called "equilibrium filtration" (Jenison et al., 1994), this assay was performed by adding [^{14}C]theophylline and RNA at indicated concentrations in a 150 μl reaction mixture containing 0.1 M HEPES (pH 7.3), 5 mM $MgCl_2$, and 50 mM NaCl. Each binding mixture was incubated 5 min at 25°C. The mixture was then placed in a Microcon 10 filtration device (Amicon, Beverly, MA) and centrifuged for 4 min at 13,000 g, allowing 40 μl to flow through the membrane. A 25-μl sample was removed from each side of the membrane and the radioactivity determined by scintillation counting. Bound theophylline was determined by the difference between the theophylline concentration of the filtrate and the theophylline concentration of the retentate. Data were fit by a least-squares analysis to a standard quadratic binding equation (Gill et al., 1991) using the observed 1 : 1 stoichiometry between RNA and theophylline (Jenison et al., 1994).

Results and Discussion

Radiolabeled RNA of random sequence was produced by transcription by T7 RNA polymerase and applied to the theophylline column. Bound RNA was eluted with excess free theophylline. The progress of the SELEX experiment was monitored by following the percent RNA eluted as rounds proceeded (Table 1). In the first round of SELEX, the theophylline column bound only 0.05% of the input radiolabeled RNA, but by round eight, termed TR8, 62% of the input RNA was bound. This represented an increase in affinity of at least 1240-fold relative to the starting random RNA population during the course of the experiment.

Elution with caffeine prior to theophylline was used to increase the rate at which oligonucleotides were selected that could discriminate between theophylline and caffeine. This protocol was termed "counter-SELEX." Counter-SELEX was initiated after five rounds of theophylline-elution SELEX. The effects of counter-SELEX are illustrated in Table 1. In the first counter-SELEX round, called TCTR1, 53% of the input radiolabeled RNA was eluted with caffeine

Table 1
RNA Binding vs SELEX Round

Round	% Eluted	Counter-SELEX % Eluted
1	0.05	—
2	0.05	—
3	0.58	—
4	0.07	—
5	7.80	—
6	49.2	0.3
7	61.3	32.8
8	62.8	53.0

Note: The numbers in the left column are the percent input radioactivity eluted by theophylline after the column was washed with loading buffer during the SELEX rounds. Counter-SELEX radioactivity refers to the percent radioactivity eluted with theophylline following an elution with caffeine.

while 0.3% was subsequently eluted with theophylline. In the second and third counter-SELEX rounds, TCTR2, 32.8% and 53% of total input radioactivity, respectively was eluted, after the initial caffeine challenge. In comparison, 13% and 37.5% of input radioactivity was specifically eluted with theophylline in similar rounds without prior caffeine challenge. The relative amount of specific theophylline binders increased dramatically during rounds 6–8 of the SELEX experiment. Further improvement was not observed in round 9 and SELEX was terminated.

Sequence analysis indicated that the RNA selected after the third round of counter-SELEX was clearly nonrandom in sequence, compared with the starting RNA population. Consequently, this population was cloned into the plasmid pUC18. The sequence of 11 clones is shown in Fig. 2A. Of these, eight unique sequences were obtained. All clones contained a 15-base consensus sequence in two separate regions of the RNA. Region 1 was 5'AUACCA3'; region 2 was 5'CCUUGG(C/A)AG3'. Similar results were observed for 6 clones derived from the RNA population selected without caffeine challenge. The spacing between the two regions varied from 8 to 20 residues in the different clones.

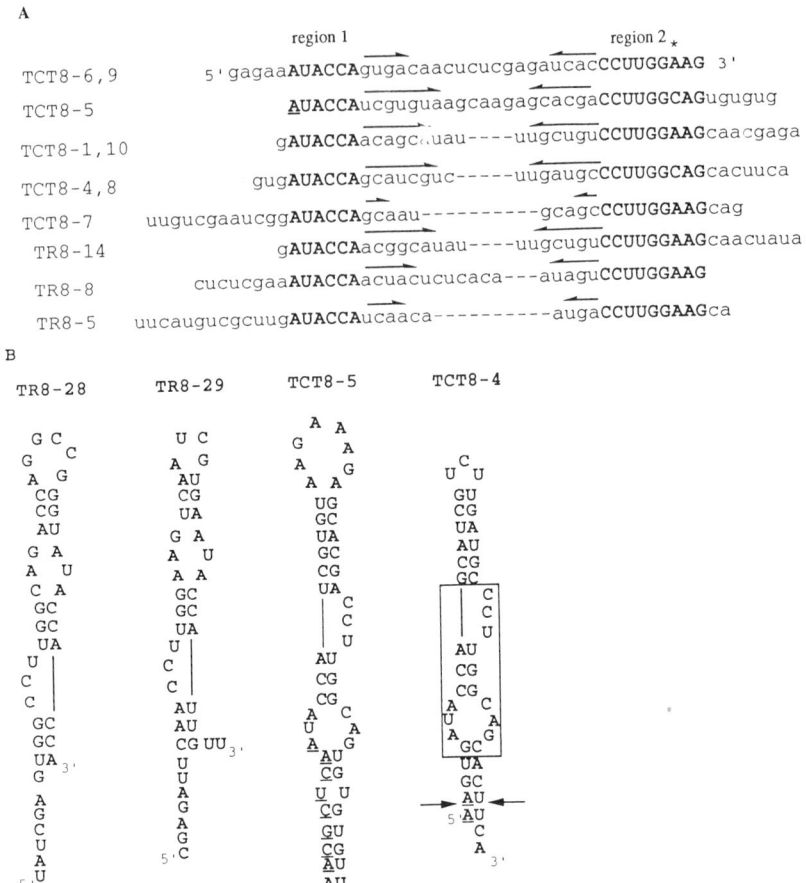

Figure 2 (A) Aligned sequences for eleven selected RNA molecules with affinity for theophylline. The clone number from which the sequence was derived is shown at the left of the sequence. In some cases, multiple isolates were obtained. Sequences shown comprised the 40-/nucleotide sequence that was initially random at the start of the SELEX process. The conserved sequences are highlighted in upper case and provide the basis for the alignment. The arrows overlay regions of potential base complementarity. The asterisk marks the single position in region 2 that shows variability. Dashes represent the absence of a nucleotide. (B) Potential secondary structures for theophylline-binding RNA species. Underlined bases were present in either the fixed 5' or 3' regions which flanked the random region. Note that either fixed region can contribute to the proposed structure. The arrows in the TCT8-4 ligand represent the terminals of the truncated mTCT8-4 except that the AU base pair above the arrows was changed to a GC base pair in mTCT8-4. Reprinted with permission from SCIENCE. Jenison, R. D., S. C. Gill, A. Pardi, and B. Polisky. 1994. High resolution molecular discrimination by RNA. *Science* 263:1425–1428, March 11, 1994. ©AAAS.

The dramatic conservation for the 15 bases in the 11 cloned sequences led us to postulate that these sequences comprised the binding site for theophylline in the selected RNAs. Inspection of the sequences revealed that it was possible to fold all 17 RNAs into a similar secondary structure in which the conserved sequences assumed a similar conformation (Fig. 2B). A key feature permitting us to postulate this common secondary structure was the observation that sequences between the conserved regions showed potential Watson–Crick complementarity despite varying in sequence from clone to clone. The postulated theophylline binding region consisted of a conserved asymmetric CCU bulge flanking a conserved three-base pair stem, and a six-nucleotide internal loop on the other side of the stem conserved at five of six positions (Fig. 2B). The sequence variability outside of the conserved regions suggested that other nucleotides do not play a direct role in theophylline binding. To test this idea, a 38-nucleotide truncated RNA was designed which consisted of the conserved regions and the flanking sequences of a particular RNA ligand called TCT8-4 (Fig. 2B). The truncated RNA lacked the fixed sequences present on the 87-nucleotide, full-length RNA. The truncate, designated mTCT4 RNA, was synthesized by *in vitro* transcription to determine the minimum requirements for high-affinity binding to theophylline (see later discussion).

The affinity of various RNAs derived from the SELEX experiment for theophylline was determined by equilibrium filtration analysis with [^{14}C]theophylline as described in the methods section. The affinity of the RNA population after round 8 of SELEX for theophylline was 2.5 μM. Individual RNA species after cloning showed K_Ds in the 0.5–3 μM range. These affinities are comparable to those described for monoclonal antibodies against theophylline (Poncelet et al., 1990).

Competition binding analysis was used to determine the ability of caffeine to occupy the theophylline binding site. The K_D of TCT8-4 RNA for caffeine was about 3.5 mM, approximately 7000-fold greater than the observed K_D for theophylline (Fig. 3). The K_D of the truncated version of TCT8-4 RNA for theophylline was 100 nM, about five fold lower than the full-length RNA, suggesting that the nonselected regions of the RNA present in the full-length RNA may actually interfere with binding. These data confirm that all the structural determinants required for high-affinity theophylline binding are contained in the conserved regions of the RNA.

These results indicate that oligonucleotides can recognize small-molecular-weight targets with high affinity and discriminate against

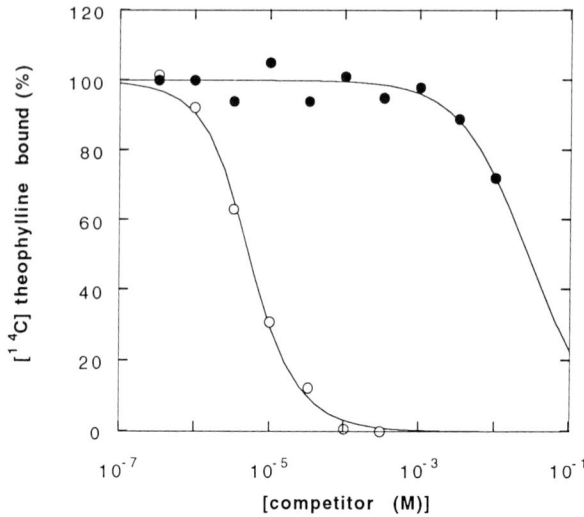

Figure 3 Competition binding studies, for ^{14}C-labeled theophylline (1 μM) bound to TCT8-4 RNA (3.3 μM) competed with either unlabeled theophylline (open circles) or caffeine (closed circles). The corresponding lines represent the nonlinear least-squares best fit. The calculated K_Ds were 0.5 μM for theophylline and 3500 μM for caffeine. Adapted with permission from SCIENCE. Jenison, R. D., S. C. Gill, A. Pardi, and B. Polisky. 1994. High resolution molecular discrimination by RNA. *Science* **263**:1425–1428, March 11, 1994. ©AAAS.

structurally similar targets. The application of these observations to the development of diagnostic reagents is a plausible extension for the future.

Literature Cited

Bartel, D. P., M. L. Zapp, M. R. Green, and J. Szostak. 1991. HIV-1 Rev regulation involves recognition of non-Watson-Crick base pairs in viral RNA. *Cell* **67**:529–536.

Connell, G. J., M. Illangesekare, and M. Yarus. 1993. These small ribooligonucleotides with specific arginine sites. *Biochemistry* **32**:5497–5502.

Famulok, M., and J. W. Szostak. 1992. Stereospecific recognition of trytophan agarose by *in vitro* selected RNA. *J. Am. Chem. Soc.* **114**:3990–3991.

Gill, S. C., S. E. Weitzel, and P. H. von Hippel. 1991. *E. coli* sigma 70 and NusA proteins. I. Binding interactions with core RNA polymerase in solutin and within the transcription complex. *J. Mol. Biol.* **220**:307–316.

Hendeles, L., and M. Weinberger. 1983. Theophylline: A "state of the art Review." *Pharmacotherapy.* **3**:2–43.

Jellinek, D., C. K. Lynott, D. B. Rifkin, and N. Janjic. 1993. High-affinity RNA ligands to basic fibroblast growth factor inhibit receptor binding. *Proc. Natl. Acad. Sci. U. S. A.* **90:**11227–11231.

Jenison R. D., S. C. Gill, A. Pardi, and B. Polisky. 1994. High resolution molecular discrimination by RNA. *Science* **263:**1425–1428.

Mitchell, L. G., and C. R. Merril. 1989. Affinity generation of single-stranded DNA for dideoxy sequencing following the polymerase chain reaction. *Anal. Biochem.* **178:**239–242.

Poncelet, S. M., J. N. Limet, J. P. Noel, M. C. Kayaert, L. Galanti, and D. Collet-Cassart. 1990. Immunoassay of theophylline by latex particle counting. *J. Immunoassay* **11:**77–88.

Sassanfar, M., and J. W. Szostak. 1993. An RNA motif that binds ATP. *Nature* **364:**550–553.

Tuerk, C., and L. Gold. 1990. Systematic evolution of ligands by exponential enrichment: RNA ligands to bacteriophage T4 DNA polymerase. *Science* **249:**505–510.

Tuerk, C., S. MacDougal, and L. Gold. 1992. RNA pseudoknots that inhibit HIV-1 reverse transcriptase. *Proc. Natl. Acad. Sci. U. S. A.* **89:**6988–6992.

23

GENERATION OF SINGLE-CHAIN ANTIBODY FRAGMENTS BY PCR

Jeffrey R. Stinson, Vaughan Wittman, and Hing C. Wong

The ability to generate monoclonal antibodies revolutionized the use of antibodies for scientific analysis, immunodiagnostics, and immunotheraputics. Monoclonal antibodies have traditionally been generated via hybridoma technology, which is expensive, time-consuming, and yields only a small number of clones producing antibodies of the desired specificity. Furthermore, after the proper cell clones are identified, extensive screening and passaging of the primary cell clones is then necessary to develop a stable cell line expressing the antibody. Recently, the production in bacteria of monoclonal antibodies generated through recombinant DNA technology (recombinant antibodies) has been demonstrated. This recombinant antibody technology is easier and less time-consuming than mammalian cell culture techniques. Recombinant antibody fragments assembled as fusion proteins on the surface of bacteriophages can be directly selected for antigen binding ability, circumventing the need for extensive screening procedures. The cloned antibody fragments can also be easily manipulated to improve their binding affinity through a number of techniques, including *in vitro* mutagenesis, a step which could not be carried out using the traditional hybridoma technology. For these reasons, recombinant antibody technology should soon replace traditional hybridoma techniques for creating antibody molecules for diagnostic and therapeutic use.

The use of *Escherichia coli* as a host for expression of recombinant antibodies has evolved since Boss and his colleagues first demonstrated that antibody fragments with variable regions could be coexpressed in bacteria. In 1988, Pluckthun, (Skerra and Pluckthun, 1988) and Better's (Better *et al.*, 1988) laboratories showed that *E. coli* could secrete active antibody fragments into the periplasm. Within a year, recombinant antibodies reactive against a transition state analog hapten were detected in a library constructed in *E. coli* by Lehrner's group (Huse *et al.*, 1989). Their library was generated from mRNA isolated from the spleen of immunized mice using immunoglobin-specific oligonucleotide primers and PCR amplification. Since then, several laboratories have produced recombinant antibody libraries in *E. coli* from both human and murine sources. For example, human banks have yielded antitetanus antibodies (Mullinax *et al.*, 1990; Persson *et al.*, 1991) as well as antibodies against foreign antigens for which the host had not been immunized (Marks *et al.*, 1991), and several "antiself" antibodies, such as a human antiidiotype antibody, an anticytokine (TNF) antibody, and antibodies against various tumor markers and the T-cell marker CD4 (Griffiths *et al.*, 1993).

Two forms of truncated immunoglobin fragments are commonly used when producing libraries or monoclonal antibodies in *E. coli*. The first form, the single-chain antibody or scFv, consists of the variable regions of the antibody connected by a flexible linker peptide which attaches the carboxy terminus of one variable region to the amino terminus of the other. It has been demonstrated that functional single-chain molecules can be produced independent of the order in which the variable regions occur and with a variety of different linkers (Whitlow and Filpula, 1991). The second form of immunoglobin fragment commonly expressed in *E. coli*, the Fab fragment, contains the variable and first constant region of the heavy chain and the entire light chain. Both of the forms of antibody fragment have inherent advantages and limitations when they are made in *E. coli* cells. The scFv fragment is produced as a single peptide by the cell and thus the production of light chain and heavy chain variable regions is coupled. The various Fab constructs depend on association of the antibody chains in the periplasm to produce a functional molecule. This can be a problem if both the chains are not produced with equal efficiency or if one of the chains is less stable or soluble in E. coli. The construction of independent heavy and light chain Fab fragment libraries may make some manipulations easier (for example, chain shuffling of the combinatorial library), since it would minimize the number of steps required in the process (Collet et al., 1992). Comparison of the binding affinities of the scFv and Fab forms of a small number of antibodies has shown that the Fab constructs had affinities that were equivalent to, or slightly higher than, those of the scFvs (Bird and Walker, 1991).

Regardless of the format of the antibody fragments or the vectors used to produce and/or display them, the generation of recombinant antibody libraries requires a set of specific primers and defined PCR conditions to amplify the appropriate gene fragments from the mRNA source. We have designed a series of primers for repertoire cloning single-chain antibody fragments from murine sources, and methods for mRNA isolation and PCR amplification. This chapter outlines the principles applied to primer design and describes the methods for isolating and amplifying recombinant antibody fragments.

Materials, Methods, and Results

We chose to clone our single-chain antibody fragments so that the heavy-chain variable region precedes the light-chain variable region and the two regions are linked through a flexible set of amino acids with the sequence (Gly-Gly-Gly-Gly-Ser)$_3$ as described by Genex (Bird et al., 1988). Several groups have shown this linker to be effective. Our version contains a *BspEI* restriction site, which is engineered for the convenience of chain shuffling. Our library and expression vectors both contain a *pelB* leader sequence in frame with the antibody sequence which is used to target the scFv protein to the periplasmic space of E. coli. The carboxy terminal amino acids of the leader are Ala-Met-Ala, and the cleavage by the signal peptidase occurs after the last alanine. The DNA sequence coding for these amino acids contains an *NcoI* restriction site (CCATGG) which we use as the 5' end restriction site for cloning the scFvs. This allows us to design primers that encode the native N-terminal amino acids of the heavy-chain variable region of the scFv rather than having to

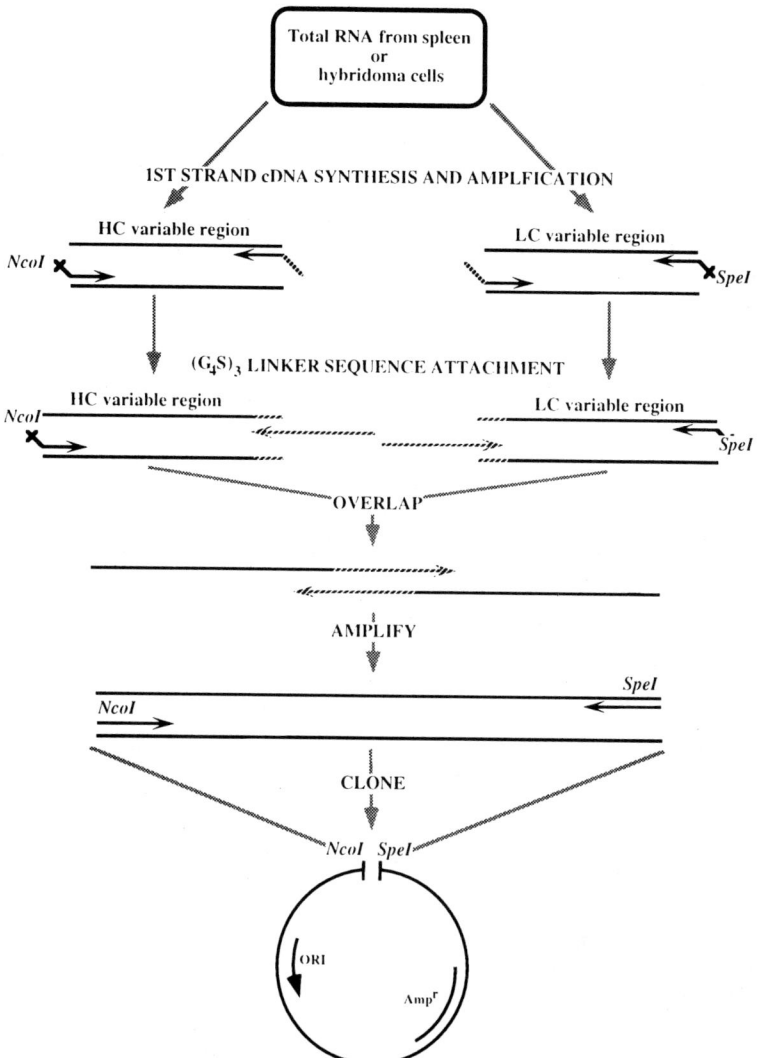

Figure 1 Single-chain antibody cDNA synthesis and cloning process. Details of the PCR-based cloning procedure are described in the materials and methods section.

fix a pair of amino acids necessary to encode a 5' end restriction site. At the 3' end of the cDNA inserts, we use an *spe*I restriction site that is present in the "back" primer set of our light-chain variable region. The overview of the cloning process is shown in Fig. 1.

Primer Design

The design of the primer sets was based on the DNA sequence information for the antibody variable regions in the murine kappa light and heavy chains as compiled by Kabat et al. (1991) (see Fig. 2).

In the case of the heavy-chain front primers, the first 12 nucleotides are the same since these encode the restriction site, a so-called "clamp" sequence 5' of the restriction site, and the last amino acid of the *pelB* leader. The remainder of the oligonucleotide codes for the first 6 $^{2}/_{3}$ amino acids of the variable region. The degeneracy of

Heavy Chain "front" primers

```
5'-GCCGGCCATGGCCCAGGTBCARCTKMARSARTC-3'(JS155)
5'-GCCGGCCATGGCCGARGTRMAGCTKSAKGAGWC-3'(JS156)
5'-GCCGGCCATGGCCGARGTYCARCTKCARCARYC-3'(JS157)
5'-GCCGGCCATGGCCCAGGTGAAGCTKSTSGARTC-3'(JS158)
5'-GCCGGCCATGGCCGAVGTGMWGCTKGTGGAGWC-3'(JS159)
```

Heavy chain "back" primers

```
5'-GCTGCCACCGCCACCTGMRGAGACDGTGASTGARG-3'(JS160)
5'-GCTGCCACCGCCACCTGMRGAGACDGTGASMGTRG-3'(JS161)
5'-GCTGCCACCGCCACCTGMRGAGACDGTGASCAGRG-3'(JS162)
```

Light Chain "front" primers

```
5'-GGAGGCGGCGGTTCTGACATTGTGMTGACCCAATC-3'(JS147)
5'-GGAGGCGGCGGTTCTGACATTGTGMTGWCACAGTC-3'(JS148)
5'-GGAGGCGGCGGTTCTGAYATTCAGATGACMCAGWC-3'(JS149)
5'-GGAGGCGGCGGTTCTSAAATTGTKCTSACCCAGTC-3'(JS150)
5'-GGAGGCGGCGGTTCTGATRTTKTGATGACCCARAC-3'(JS150A)
5'-GGAGGCGGCGGTTCTGATRTTGTGATGACKCARGC-3'(JS163)
5'-GGAGGCGGCGGTTCTGATRTTGTGATAACCCARGATG-3' (JS164)
```

Light Chain "back" primers

```
5'-TTCATAGGCGGCCGCACTAGTAGCMCGTTTCAGYTCCARC-3'(JS153)
5'-TTCATAGGCGGCCGCACTAGTAGCMCGTTTKATYTCCARC-3'(JS154)
```

(GGGGS)$_3$ Linker sequence

```
                                        (C)
5'-GGTGGCGGTGGCAGCGGCGGTGGTGGTTCCGGAGGCGGCGGTTCT-3'
3'-CCACCGCCACCGTCGCCGCCAGGAGGAAGGCCTCCGCCGCCAAGA-5'
                                        (G)
```

Figure 2 Oligonucleotide primer sets designed for use in cloning murine antibody cDNAs as single-chain antibody fragments in *E. coli* cells using the PCR. Restriction endonuclease sites used for cloning the fragments are underlined.

these primers ranges from 16- to 192-fold. The heavy-chain "back" primers each share a common sequence over the first 15 nucleotides since they represent the reverse complement of the nucleotides encoding the first 5 amino acids of the linker used to connect the variable regions.

The rest of the oligonucleotide represents the reverse complement of the nucleotides encoding the last 6 $^{2}/_{3}$ amino acids of the heavy chain variable region from amino acids 106 to 113 (as numbered by Kabat et al.) at the end of framework four. The degeneracy of these primers ranges from 48- to 96-fold. Although the degeneracy of these primer sets is quite high, we have consistently been able to amplify cDNA using them (see Fig. 3) The use of highly degenerate primers allows one to capture the diversity of the variable region sequences in heavy chains, while at the same time limiting the number of reactions during the PCR amplification and cloning process.

For the light-chain "front" primer sets, the first 15 nucleotides are common to all of the primers because these encode the last 5 amino acids of the linker region. The rest of the oligonucleotide corresponds to the first 6 $^{2}/_{3}$ (or in one case, 7 $^{1}/_{3}$) amino acids of the light-chain variable region. The degeneracy of these primers ranges from only two- to eight-fold. The two members of the light chain "back" primer set share the first 21 nucleotides since this region represents the reverse complement of the restriction site for cloning (SpeI) and the first 5 amino acids of a peptide tag sequence. The remainder of the oligonucleotide represents the reverse complement of the light-chain variable region from amino acids 102 to 109 at the end of framework four. The degeneracy of these primers ranges from 8- to 16-fold.

We have two pairs of oligonucleotides synthesized which encode the $(G_4S)_3$ linker sequence for the scFvs. One pair has a BspEI restriction site (TCCGGA, encoding Ser-Gly respectively) incorporated in the DNA sequence that encodes the tenth and eleventh amino acids of the linker. We have another pair without this restriction site because there is a strong bias against use of the GGA codon in E coli proteins that are expressed at high levels (Ernst, 1988). The primers were synthesized for us by Oligos Etc., Inc. (Oregon).

Total RNA Isolation

In general, we followed the isolation procedure described in Sambrook et al (1989), using the usual precautions to protect the RNA from degradation by RNase. An excised spleen (or 10^7–10^8 tissue

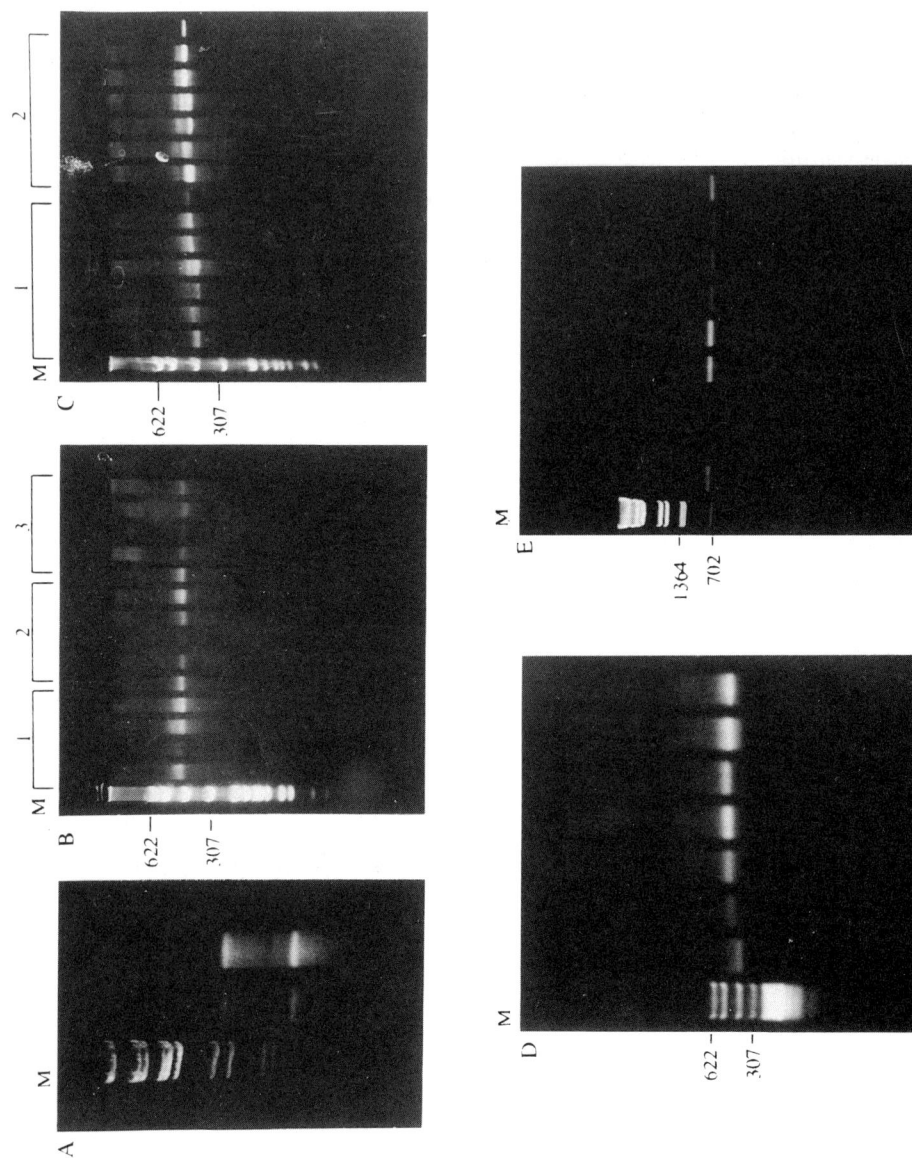

culture cells in the case of hybridoma cells) from a mouse inoculated and boosted with antigen was placed in 5 volumes of extraction buffer containing 4.0 M guanidinium thiocyanate, 0.1 M Tris–HCL (pH 7.5). The material was homogenized using a Tissue Tearer homogenizer for approximately 5 min to ensure cell breakage and complete shearing of the genomic DNA. One should note that a two-step homogenization occurs; initially the spleen is disrupted into its component cells and then the cells themselves are disrupted, releasing the intracellular material. Following homogenization, Sarcosyl was added to a final concentration of 0.5%. After thorough mixing, the solution was centrifuged at 5000 g for 10 min. The supernatant was then gently layered on top of a cesium chloride cushion consisting of 5.7 M $CsCl_2$, 0.01 M EDTA (pH 7.5) in a clear polycarbonate ultracentrifuge tube. The material was centrifuged in an SW41 rotor at 32,000 rpm for 24 hr at 20°C.

Following centrifugation, the supernatant was removed from above the $CsCl_2$ cushion with a Pasteur pipette, using care to avoid contaminating the RNA pellet on the bottom of the tube with any of the genomic DNA suspended at the interphase. Then as much of the CsCl solution as possible was removed from the tube without disturbing the RNA pellet. The pellet was washed once with approximately 500 μl of 70% ethanol. After the ethanol was removed, the pellet was air dried. The RNA was dissolved in 300 μl of diethyl pyrocarbonate (DEPC)-treated water containing 40 U of RNasin (Promega). The RNA was transferred to a microfuge tube. The bottom of the ultracentrifuge tube was rinsed with 50 μl of DEPC-water and this was added to the microfuge tube. One tenth volume of 3 M sodium acetate was added, followed by 2.5 vol of 100% ethanol to precipitate the RNA. We normally obtain 400–600 μg of total RNA from a spleen and 200–300 μg of total RNA from hybridoma cells by using this method (see Fig. 3A).

Figure 3 Agarose gel electrophoresis stained with ethidium bromide to show the products of the RNA isolation, cDNA synthesis, and amplification for cloning single-chain antibody fragments from mouse spleen cells. (A) Approximately 1.25 and 2.5 μg total RNA. (B) Heavy-chain variable region amplification products. Group 1 was amplified with the set of "front" primers and "back" primer JS160; group 2 was amplified with the set of "front" primers and "back" primer JS161; and group 3 was amplified with the set of "front" primers and "back" primer JS162. (C) Light-chain variable region amplification products. Group 1 was amplified with the set of "front" primers and "back" primer JS153; group 2 was amplified with the set of "front" primers and "back" primer JS154. (D) Linker sequence attachment reaction products. (E) Single-chain antibody cDNA products following overlap PCR, restriction digestion, and gel purification.

cDNA Synthesis

For synthesis of the first strand of the cDNA of the heavy- and light-chain variable regions, we used specific "back" primers to prime the reverse transcription reaction. These reactions were carried out under the following conditions:

Five micrograms of RNA were combined with 50 pmoles of the individual primer (three heavy-chain reactions and two light-chain reactions) in DEPC-treated water to a volume of 10 μl and heated to 65°C for 2 min, then snap cooled on ice. To each of these solutions was added a mixture of 10 μl of 5X reaction buffer, 5 μl of 10 mM deoxyriboncleaside triphosphates (dNTPs), 5 μl of 0.1 M dithiothreitol (DTT), 1μl RNasin (40 U, Promega), and 7 μl of DEPC-water. This was incubated at 42°C for 2 min; 2 μl of M-MLV (Maloney Murine Leukemia Virus) (400 units) reverse transcriptase (Life Technologies, Inc.) was added and the incubation was continued for 1 hr.

For the synthesis and amplification of the second strand of the cDNA, we performed separate reactions with each of the combinations of the heavy- or light-chain primer sets (for a total of 29 reactions) in a thermal cycler (Perkin–Elmer model 480) using DNA polymerase. The conditions for each reaction follow.

On ice, 5 μl of appropriate first-strand cDNA product was combined with 1 μl (10 pmoles) of each primer, 78.5 μl water, 10 μl 10× PCR reaction buffer (Perkin–Elmer, Norwalk, CT), 4 μl 5 mM dNTPs, and 0.5 μl AmpliTaq (Perkin–Elmer, Norwalk, CT) DNA polymerase. This mixture was kept on ice until the thermal cycler reached a temperature of at least 85°C, at which point the tubes were placed in the cycler and amplification was allowed to proceed.

We used the following amplification conditions. We began with a 5- min time delay at 96°C, followed by a thermocycle program of 1.5 min at 96°C, a rapid drop to 50°C, annealing at 50°C for 30 sec, a 30-sec ramp-up to 72°C, and an incubation at 72°C for 1 min. This thermocycle program was repeated 15 times. The program was linked to a step cycle program that alternated between 96°C for 1 min and 70°C for 1.5 min for 10–15 cycles. The products of the amplification were then analyzed on an agarose gel stained with ethidium bromide (see Fig. 3B and C). The number of cycles are adjusted to ensure that one does not overproduce the PCR products. This is based on the idea that the overproduction of amplified material during the PCR can lead to competition for reaction materials and thus increase the likelihood of misincorporation of nucleotides (D. Gelfand, personal communication).

Overlap PCR and Cloning

The first step in connecting the heavy- and light-chain variable regions was to attach the DNA encoding the linker sequence to each. The cDNA primers were removed with the use of a Centricon-100 concentrator (Amicon). The PCR products were applied to the membrane and one tenth TE [1 mM Tris (pH 8.0), 0.1 mM EDTA] was added to volume of 2.0 ml. This was centrifuged at 1000 g for 15 min. The retentate was then collected with a 2-min centrifugation at 200g. The retentate was typically 40 μl. At this point, amplification reactions using the appropriate linker oligonucleotide as a "back" primer (in the case of the heavy-chain variable regions) or a "front" primer (in the case of the light-chain variable regions) were performed. Thus, a total of seven reactions are required, five for the heavy chains and two for the light chains. The appropriate cDNA products from the previous amplification reactions were pooled where possible. For example, all of the PCR products produced using the different light-chain "front" primers but with the same "back" primer were pooled and reamplified using a combination of the common "back" primer and the linker "front" primer.

The conditions for the PCR were essentially as described in the section on cDNA synthesis, with some important exceptions. Only 1–2 μl of the pooled variable region cDNA were used in each reaction. The amplification profile was changed so that there were only 2 thermocycles rather than 15 and these were performed using an annealing temperature of 55°C rather than 50°C. The number of step cycles was also increased to 15–20. Production of PCR products was checked as described (see Fig. 3D).

The heavy and light chain variable regions were joined by overlap PCR carried out after the residual primers were removed using a Centricon-100. The concentration of DNA in each of the pools was estimated using agarose gel electrophoresis and ethidium bromide staining. Ten separate reactions were carried out to complete the overlap and construct the completed scFvs with the five heavy-chain "front" primer pools and two light-chain "back" primer pools. The overlap was performed as a two-step process. The first step, a series of thermocycles without the addition of "outside" primers, created full-length, double-stranded template and was followed by the addition of the appropriate primer sets and amplification of the full-length scFv molecules. The reactions were set up as follows:

Reactions were prepared on ice using equal amounts of the linker containing fragments from the heavy- and light-chain variable regions (based on ethidium bromide fluorescence, typically 2–4 µl in volume), 9 µl 10× reaction buffer, 4 µl of 5 mM dNTPs, 0.5 µl AmpliTaq polymerase, and water to 90 µl. The tubes were then placed in a hot thermal cycler for 5 cycles using the following profile: 96°C for 1 min, a rapid drop to 60°C, annealing at 60°C for 30 sec, a ramp-up to 72°C over a 30-sec interval, and a 1- min incubation at 72°C. After 5 of these cycles, a 10-µl aliquot was added that contained 10 pmoles of each appropriate "front" and "back" primer in 1 × PCR reaction buffer. The thermocycle profile was continued for 5 more cycles. At this point, the program was linked to a step cycle program that alternated between 96° and 70°C as described for 10–15 cycles (see Fig. 3E).

The resulting products are rarely homogeneous in size. For this reason the PCR products of the appropriate size are purified from an agarose gel following restriction digestion with the enzymes *Nco*I and *Spe*I. After the purification procedure, the digested fragments are ligated into our screening and/or expression vectors.

Discussion

We have used these methods to amplify scFv cDNAs from both spleen and hybridoma cells, and have cloned these cDNAs into plasmid and phagemid expression vectors. However, it has been our experience that certain cloning attempts present more of a technical challenge than others. Amplification products can sometimes be absent when certain primer combinations are used and some amplifications result in products that are highly heterologous in size. It is not clear exactly what factors play a role in this inconsistency. Of course, the RNA should be intact and of the highest quality possible. The source of the RNA, whether from a spleen or hybridoma cell line, may also have an impact. The spleen cell preparations contain a wide variety of anitbody mRNA sequences to which the primers may potentially anneal, whereas the hybridoma cells only produce mRNAs for a single heavy-chain sequence and one or possibly two light-chain sequences. The described procedures should therefore be used with the understanding that, as in all experiments involving the PCR process, a certain amount of optimization may be required.

We do not mix numerous front and back primers in our reactions. Although this necessitates performing many individual PCRs during the first stages of the cDNA production process (a total of 29), we have experienced cases in which mixing two or more primers of the same type results in only one of the primer sets being used during the amplification process. Apparently, under the appropriate conditions, one preferred reaction can effectively monopolize the polymerase and nucleotides during DNA synthesis.

We do not use oligo(dT) during the first-strand synthesis of the cDNA in order to reduce the proportion of nonantibody first-strand cDNAs in the reaction. This maximizes the yield of desirable cDNA templates (full-length, antibody-variable regions) synthesized for the subsequent PCR amplification step.

Although the protocols described here cover only the isolation of single-chain antibodies from murine sources and construction of a bank using a geneIII-type vector, with minor modifications of the primers and vector, the techniques can be used to prepare Fab libraries as well. Human oligonucleotide primers can be designed using the same principles we applied to the murine primers, thus allowing the creation of human libraries using the techniques described here.

Literature Cited

Better, M., C. P. Chang, R. R. Robinson, and A. H. Horwitz. 1988. *Escherichia coli* secretion of an active chimeric antibody fragment. *Science* **240**:1041–1043.

Bird, R. E., and B. W. Walker. 1991. Single chain antibody variable regions. *Tibtech* **9**:132–137.

Bird, R. E., K. D. Hardman, J. W. Jacobson, S. Johnson, B. M. Kaufman, S-M. Lee, T. Lee, S. H. Pope, G. S. Riordan, and M. Whitlow. 1988. Single chain antigen binding proteins. *Science* **242**:423–426.

Collet, T. A., P. Roben, R. O'Kennedy, C. F. Barbas III, D. R. Burton, and R. Lerner. 1992. A binary plasmid system for shuffling combinatorial antibody libraries. *Proc. Natl. Acad. Sci. U.S.A.* **89**:10026–10030.

Ernst, J. 1988. Codon usage and gene expression. *Tibtech* **6**:196–199.

Griffiths, A. D, M. Malmqvist, J. D. Marks, J. M. Bye, M. J. Embleton, J. McCafferty, M. Baier, K. P. Holliger, B. D. Gorick, N. C. Hughes-Jones, H. R. Hoogenboom, and G. Winter. 1993. Human ant-self antibodies with high specificity from phage display libraries. *EMBO J.* **12(2)**:725–734.

Huse, W. D., L. Sastry, S. A. Iverson, A. S. Kang, M. Alting-Mees, D. R. Burton, S. J. Benkovic, and R. Lerner. 1989. Generation of a large combinatorial library of the immunoglobulin repetoire in phage lambda. *Science* **246**:1275–1281.

Kabat, E. A., T. T. Wu, H. M. Perry, K. S. Gottesman, and C. Foeller, eds. 1991. *Sequences of proteins of immunological interest.* 5th ed. U.S. Department of Health and Human Services, Wasington, DC.

Marks, J. D., H. R. Hoogenboom, T. P. Bonnert, J. McCafferty, A. D. Griffiths, and G.

Winter. 1991. By-passing immunization: Human antibodies from V-gene libraries displayed on phage. *J. Mol. Biol.* **222**:581–597.

Mullinax, R. L., E. A. Gross, J. R. Amberg, B. N. Hay, H. H. Hogref, M. M. Kubitz, A. Greener, M. Alting-Mees, D. Ardourel, J. M. Short, J. A. Sorge, and B. Shopes. 1990. Identification of human antibody fragment clones library. *Proc. Natl. Acad. Sci. U.S.A.* **87**:8095–8099.

Persson, M. A. A., R. H. Caothien, and D. R. Burton. 1991. Generation of diverse high-affinity human monoclonal antibodies by repetoire cloning. *Proc. Natl. Acad. Sci. U.S.A.* **88**:2432–2436.

Sambrook, J., E. F. Fritsch, and T. Maniatis. 1989. *Molecular cloning: A laboratory manual*, 2nd. ed. Cold Spring Harbor Press, Cold Spring Harbor, NY.

Skerra, A. and A. Plückthun. 1988. Assembly of a functional immunoglobulin F_v fragment in *Escherichia coli*. *Science* **240**:1038–1041.

Whitlow, M., and D. Filpula. 1991. Single chain Fv proteins and their fusion proteins. *Methods* **2**(2):97–105.

24

Longer PCR Amplifications

Suzanne Cheng

One limitation of standard protocols for the polymerase chain reaction (PCR) is that amplifications of DNA fragments longer than 5 kb are not routine. The "long PCR" protocols described here were initially used to amplify up to 22 kb of the β-globin gene cluster from human genomic DNA, human genomic inserts of 9 to 23 kb from recombinant phage lambda (λ) plaques, and up to 42 kb from high-copy phage λ DNA (Cheng et al., 1994a). These protocols have since been used to amplify viral genomes and identify a 17-kb intron within the human MSH2 gene (Cheng et al., 1994b), and to amplify 16.3 kb of the mitochondrial DNA (mtDNA) genome (Cheng et al., 1994c) and up to 30 kb of the human β-globin gene cluster (Cheng, manuscript in preparation).

Development of these protocols was guided by several principles: (1) Template strands must be completely denatured to prevent renaturation before primers can anneal and be extended. (2) Extension times must be sufficiently long to permit the polymerase to complete strand synthesis. (3) Nucleotide misincorporations must be removed to prevent premature termination of strand synthesis (Barnes, 1994). (4) The template DNA must be protected against damage, such as depurination (Lindahl and Nyberg, 1972) during thermal cycling. (5) The specificity needed for single-copy gene amplifications from complex genomic DNA must be retained.

The long PCR protocols consequently have the following features: First, cosolvents that effectively lower strand separation temperatures, short denaturation times at moderately high temperatures, and increased buffer pH are collectively used to ensure template denaturation while protecting the strands from damage. Second, long extension times, cosolvents, and a secondary thermostable DNA polymerase with a 3'-to-5' -exonuclease or "proofreading" activity (Barnes, 1994) are used to facilitate the completion of strand synthesis in each cycle. Finally, a "hot start" method (Chou et al., 1992), relatively high annealing temperatures, reduced enzyme levels, and optimized salt (both monovalent and divalent cation) and cosolvent concentrations are used to enhance specificity.

Notes on Primer Design

Primer sequences should have the fewest possible secondary sites with significant complementarity to the 3' ends of the primers. This is true for all PCR primers, but particularly so for primers used in long amplifications because the longer the target, the greater the chance of having internal (in addition to external) secondary priming sites. Secondary products that are shorter than the targeted sequence are likely to be more efficiently amplified than the target, and therefore accumulate preferentially. Genomic primers within interspersed repetitive elements such as *Alu* sequences (Schmid and Jelinek, 1982) should be avoided, and if possible, candidate primers should be screened against any available sequence databases, such as known stretches of the targeted area. If necessary, mismatches may be introduced to selectively destabilize priming at secondary sites with minimal effect at the target site (see Kwok et al., 1990, 1994). Higher melting temperatures (e.g., 62°–70°C) that enable the use of relatively high annealing temperatures are best for specificity; primers of 20–24 bases should suffice if the G + C content is sufficiently high (50–60%), but longer primers may be needed if the A + T content is high. Primers previously used in standard PCR may work at the higher annealing temperatures recommended for specificity in long PCR, since longer incubation times are used. A number of methods to estimate melting temperatures are available, including the programs Oligo 5.0 (National Biosciences, Plymouth,

MN) and Melt (from J. Wetmur, Mt. Sinai School of Medicine, New York, NY), and the algorithms of Wu *et al.* (1991). Table 1 lists selected primers that can be used for control reactions. The *J* and *cro* gene primers should be useful for amplifying inserts that have been cloned with most λ-based vectors.

Notes on Template Preparation

Although specific methods of preparing templates for long PCR have not been explored fully, several comments can be made. Amplifica-

Table 1

Primers for control templates described in Cheng *et al.*, 1994a. All sequences are given 5' to 3'.

Left-hand primers for phage λ (GenBank assession number M17233)		
CF1001	GGTGCTTTATGACTCTGCCGC	pos. 304–324
SC1011*	GCTGAAGTGGTGGAAACCGC	pos. 506–525
(Use the left-hand primer CF1001 with right-hand primer CF1013, but SC1011 for the other right-hand primers.)		
Right-hand primers for phage λ		
CF1013	GCCACCAGTCATCCTCACGA	complements pos. 14551–14570
SC1003	GGGTGACGATGTGATTTCGCC	complements pos. 23335–23355
SC1008	GGCATTCCTACGAGCAGATGGT	complements pos. 26893–26914
SC1009	GGTCTGCCTGATGCTCCACT	complements pos. 28536–28555
SC1016	GTCGGACTTGTGCAAGTTGCC	complements pos. 30436–30456
Lambda vector primers, from the *J* and *cro* genes		
CF1018	AGAAACAGGCGCTGGGCATC	pos. 18872–18891
CF1019	CGGGAAGGGCTTTACCTCTTC	complements pos. 38197–38217
Left-hand primers for human β-globin gene cluster (accession number J00179)		
RH1022*	CGAGTAAGAGACCATTGTGGCAG	pos. 48528–48550
RH1024†	TTGAGACGCATGAGACGTGCAG	pos. 44348–44369
RH1025*	CCTCAGCCTCAGAATTTGGCAC	pos. 42389–42410
RH1026	GAGGACTAACTGGGCTGAGACC	pos. 40051–40072
Right-hand primer for human β-globin gene cluster		
RH1053*	GCACTGGCTTAGGAGTTGGACT	complements pos. 61986–62007

* Previously published in Kolmodin *et al.*, 1995.
† Previously published in Cheng *et al.*, 1995.

tions of genomic targets longer than 15 kb are particularly sensitive to template damage (Cheng, manuscript in preparation), and care should be taken to minimize shearing and nicking of template samples. Higher copy-number templates may tolerate more handling. For example, extraction of DNA from freshly plucked hair by boiling with Chelex (Walsh et al., 1991) yielded sufficient quantities of intact 16.6-kb mtDNA genomes for amplification of 16.3 kb (Cheng et al., 1994c). Alkaline agarose gel electrophoresis (as in Sambrook et al., 1989) is one method to evaluate relative levels of single-strand damage in templates. Long amplifications may also be more sensitive to inhibitors of PCR than amplifications of less than 1 kb.

To amplify inserts from recombinant phage λ clones (as in Cheng et al., 1994a), plate the clones on agar with top agarose. Remove the plaque plugs using siliconized Pasteur pipets, and elute into 30 μl of 25 mM Tris–Cl (pH 8.3 at 25°C) and 10 mM $MgCl_2$; store at 4°C.

Protocols

The GeneAmp XL PCR Kit is available from Perkin-Elmer (Applied Biosystems Division, Foster City, CA); these reagents and protocols were developed using the principles described in Cheng et al. (1994a). The following starting procedure may also be used (summarized in Table 2):

PCR Reagents

r*Tth* buffer: 20–25 mM Tricine, from a 1 M (pH 8.7 at 25°C) stock solution
80–85 mM K-acetate, from a 1 or 2 M stock solution (pH 8.3–8.7 at 25°C)
5–10% (w/v) glycerol and 1–5% (v/v) dimethylsulfoxide (DMSO)
0.2 mM each deoxyadenosine triphosphate (dATP), deoxycytidine triphosphate (dCTR), deoxyguanosine triphosphate (dGTP), and deoxythymidine triphosphate (dTTP), from pH 7 stock solutions
r*Taq* buffer: 25 mM Tris–acetate, from a 1 M (pH 9 at 25°C) stock solution
40–45 mM K-acetate, from a 1 or 2 M stock solution (pH 8.3–8.7 at 25°C)

5–12% (w/v) glycerol and up to 2% (v/v) DMSO
0.2 mM each dATP, dCTP, dGTP, and dTTP, from pH 7 stock solutions

For a high-copy template (e.g., 10^7–10^8 copies; λ DNA can be used as a control), use 0.4–0.5 µM each primer and 4–5 U of r*Tth* (*Thermus thermophilus*) DNA polymerase (Perkin-Elmer) per 100 µl. For inserts from recombinant λ plaques, use 10^5—10^7 phage (e.g., 1–3 µl of the plaque suspension described above), 0.4 µM each primer and 1.7–2 U r*Tth* polymerase per 100µl. For genomic targets (human β-globin gene cluster can be used as a control), use 25–250 ng total genomic DNA, 0.1–0.2 µM each primer, and 1.7–2 U r*Tth* polymerase per 100 µl. Low-copy λ DNA in a background of genomic DNA can also be used as a control, although λ DNA targets are amplified more efficiently than genomic targets (Cheng et al., 1994a). Recombinant *Taq* (*Thermus aquaticus*) DNA polymerase may also be used.

As the secondary DNA polymerase, use 0.02–0.2 U Vent (*Thermococcus litoralis*) DNA polymerase (New England Biolabs, Beverly, MA) per 100-µl reaction. The optimal level of this proofreading enzyme is best determined by a titration series because the 3'-to-5'- exonuclease activity levels may vary among tubes and over time. Optimal levels of other proofreading polymerases, such as Deep Vent (New England Biolabs), *Pfu* (Stratagene, La Jolla, CA) or *UlTma* (Perkin-Elmer) DNA polymerase, should be determined independently.

Nonacetylated bovine serum albumin (10–500 ng/µl BSA; from New England Biolabs) can enhance the yields of longer products, as in Fig. 1, possibly by binding nonspecific inhibitors. Acetylated BSA (2–50 ng/µl; GIBCO BRL, Life Technologies, Gaithersburg, MD) was inhibitory or had no detectable effect (not shown).

Mg-Acetate and Hot Start:

For a manual hot start, add Mg-acetate to 1.0–1.5 mM (total) from a 25 mM stock solution once all samples have reached 75–80°C. The accumulation of nonspecific products tends to be greater at higher Mg^{2+} levels.

If using a wax barrier (e.g., AmpliWax PCR Gem from Perkin-Elmer), divide the total reaction volume into two layers (top:bottom = 60:40 or 75:25), each containing reaction buffer, but separating the deoxyribonucleoside triphosphates (dNTPs), primers, and Mg-acetate (bottom layer) from the template and polymerases (top layer).

Table 2
Long PCR Recommendations

1.	Buffer	for rTaq:	25 mM Tris–Cl or Tris–OAc, pH 8.9–9
		for rTth:	20–25 mM Tricine, pH 8.5–8.9
2.	K^+	for rTaq:	30–45 mM KCl or 40–45 mM KOAc
		for rTth:	60–90 mM KCl or 80–90 mM KOAc
3.	Cosolvents		5–12% (v/v) glycerol with or without 0.5–5% (v/v) DMSO
4.	dNTPs		0.2 mM each dATP, dCTP, dGTP, TTP; pH 7 stock solutions
5.	Primers		20–23-mers with 50–60% G + C-content, or longer sequences, to permit the use of relatively high annealing temperatures
			0.4–0.5 μM for high-copy (10^7–10^8 copies) template
			0.1–0.2 μM for low-copy ($\leq 10^4$ copies) template
6.	DNA polymerase		Optimal concentrations are Mg^{2+}-dependent, but generally;
			2.5–5 U per 100 μl for high-copy (10^7–10^8 copies) template
			1.7–2 U per 100 μl for low-copy ($\leq 10^4$ copies) template
			rTth may be more reliable than rTaq DNA polymerase for long PCR.
7.	3′-to-5′-exonuclease		Should be titrated, and optimal concentrations are Mg^{2+}-dependent, but generally:

8.	Mg^{2+}	for r*Tth*: Vent DNA polymerase, 0.02–0.2 U per 100 μl for r*Taq*: *Pfu* DNA polymerase, at least 0.04 U per 100 μl 1–1.5 mM total; 0.2 mM changes can significantly affect reaction specificity.
9.	Other additives	Gelatin (e.g., 0.001–0.02%), polyethylene glycol 8000 ($\leq$ 0.2%), Tween 20 or Thesit (0.1%), and formamide (0.5–2%) may be useful in specific systems. Acetylated BSA (2–50 ng/μl) was generally detrimental. Nonacetylated BSA (10–500 ng/μl) was helpful, although accumulation of nontarget products sometimes increased.
10.	Hot start	Withhold one ingredient (dNTPs, enzyme, or Mg^{2+}) until all samples have reached 75–80°C, or use a wax-mediated protocol.
11.	Thermal cycling profile (GeneAmp PCR System 9600)	An initial denaturation step of 15–60 sec enhances strand separation for the first cycle. Two-temperature cycles can suffice if the annealing temperature is above ~ 60°C: a short denaturation step at a moderate temperature, e.g., 94°C for 10–15 sec, and a combined annealing and extension step at a relatively high temperature, e.g., 68°C for 10–14 min, with autoextensions of 15–20 sec per cycle for at least 5–8 cycles. Extension times are typically 30–60 sec per kb of target; longer times during later cycles help to maintain reaction efficiency. Total cycle numbers are between 25 and 40, depending upon the template copy number. A final hold at 72°C for at least 10 min allows final completion of strand synthesis.

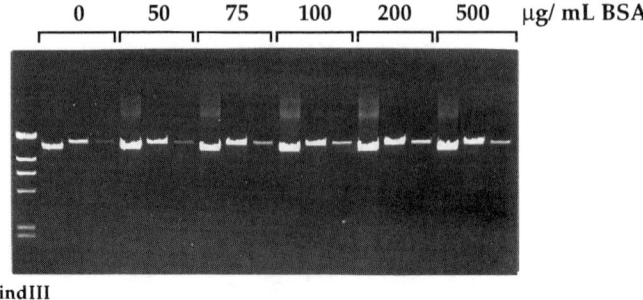

λ/HindIII

Product sizes: 13.5, 17.7, and 19.6 kb

Figure 1 Amplifications of 13.5-, 17.7-, and 19.6-kb sequences from the human β-globin gene cluster, showing the potential benefit of including up to 500 μg/ml nonacetylated BSA in the reaction mixture. Samples were analyzed on a standard 0.6% agarose gel with λ/HindIII as the molecular weight marker.

Thermal Cycling

These thermal cycling profiles were developed for a GeneAmp PCR System 9600 (Perkin-Elmer). Other types of thermal cyclers may require the adjustment of incubation or ramp times, particularly for complete template denaturation.

1. Hot Start (a) Manual: 75–80°C, e.g., 5 min for 20–30 tubes: Once the tubes have been incubated for 1.5 min, add the Mg-acetate (from a room temperature stock) to each tube. Allow 40–50 sec of additional incubation after closing the thermal cycler lid, before the first denaturation step. (b) Wax-mediated: 75–80°C for 2–5 min to melt the wax bead over the bottom reagent layer; 25°C for 2–5 min to solidify the wax barrier before adding the top reagent layer.
2. Initial mixing and denaturation with a 94°C hold for 10–30 sec if using a manual hot start; 30–60 sec if using wax.
3. Thermal cycling: Use 25–28 cycles for high-copy templates, 30–40 cycles for plaques and genomic targets. Too few cycles may yield little detectable product; too many cycles may lead to accumulation of nonspecific products or large networks of DNA.

   94°C denaturation for 10 sec (15 sec can also be used, if necessary). Templates with a high G + C-content may require higher denaturation temperatures. In general, moderately high denaturation temperatures and short incubation times preserve enzyme activity and minimize template damage.

60–70°C annealing and extension: The annealing temperature should be chosen based upon the melting temperatures of the primers. (The primers listed in Table 1 will work at 68°C.) In general, the optimal temperature will be determined by a balance between product yield and reaction specificity. If the optimal annealing temperature is less than 60–62°C, strand synthesis may be enhanced by adding a distinct extension step at 68–72°C, provided that the annealing step is sufficiently long for efficient priming and initial extension.

In general, allow 30–60 sec per kb of target. As a first protocol for high-copy targets up to 30 kb or genomic targets up to 15 kb long, incubate samples for 10–12 min per cycle. For longer targets, use an autoextension of 15–20 sec per cycle, either beginning with the first cycle or after up to 15–18 cycles for high-copy targets, 12–24 cycles for genomic targets. Yields of shorter targets may also increase with use of such autoextensions. Longer initial extension times may improve yields, although the accumulation of nonspecific products may be enhanced as well. Longer extension times during later cycles help to maintain reaction efficiency as the ratio of template-to-polymerase molecules increases, and incremental increases in time are more effective for improving yields with high reaction specificity than are constant, very long (e.g., >15 min) extension times.

4. 72°C final hold for at least 10 min for final completion of strand synthesis.
5. 4–15°C overnight hold, as necessary.

Agarose Gel Analysis of PCR Products

Samples can be loaded with a Ficoll-based 5–10× buffer: 12–25% (w/v) Ficoll 400 in 5× TAE (1× TAE is 40 mM Tris–acetate, 2 mM EDTA, pH 8–8.5) with bromophenol blue. The Ficoll percentage is critical if the PCR samples do not contain at least 5% (v/v) glycerol. For molecular weight markers, use λ/*Hin* d III and High Molecular Weight DNA Marker (8.3–49 kb; GIBCO BRL).

For quick, low-resolution analysis (as in Fig. 1), load 5–8 μl on a 0.6% (w/v) agarose gel (e.g., SeaKem GTG agarose, FMC BioProducts, Rockland, ME) in 1× TBE (89 mM Tris base, 89 mM boric acid, 1μM to 2mM EDTA) or 1× TAE. Run with 0.5 μg/ml ethidium bromide at 4–6 V/cm for 1.5–2 hr.

For higher resolution analysis, two alternatives are illustrated in Fig. 2.

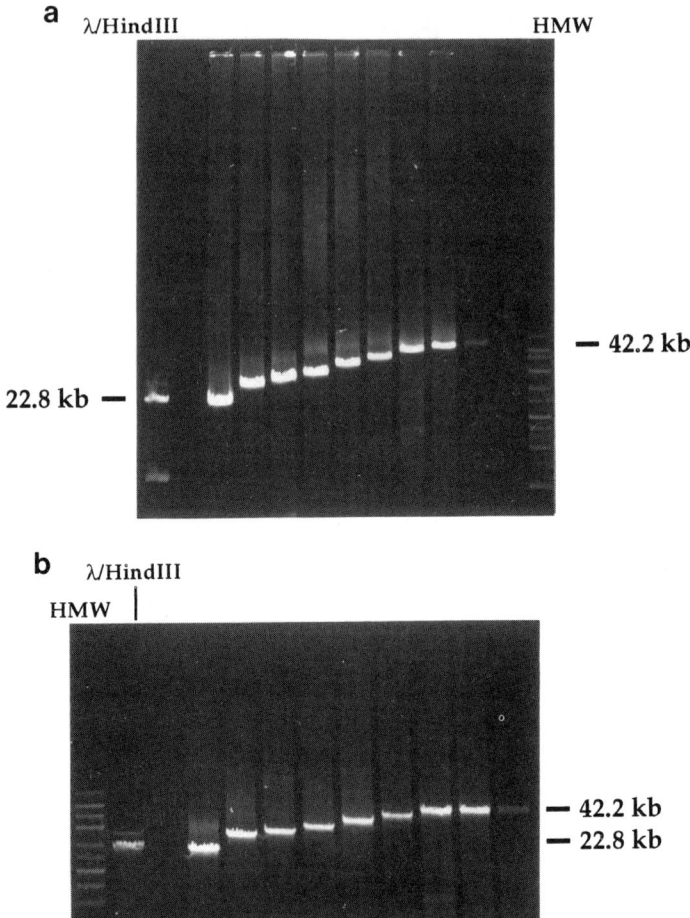

Figure 2 Higher resolution methods of gel electrophoresis, used for analysis of λ DNA PCR products ranging from 22.8 to 42.2 kb in size. (a) A 0.95% SeaKem GTG agarose gel after FIGE in a Hoefer (San Francisco, CA) system using 15 min at 110 V, then 25 hr at 145 V with a pulse time of 0.75 to 2.0 sec (forward:reverse = 3:1), at 15°C. (b) A 0.3% SeaKem Gold agarose gel after 2 min at 7 v/cm, then 6 hr at 1.5 V/cm.

(a) Load 3–7 µl on a 0.95% agarose (e.g., SeaKem GTG) gel for field inversion gel electrophoresis (FIGE; Carle *et al.*, 1986) in 0.5× TBE. Sharper bands may be observed if PCR samples are supplemented with 1–3 mM MgCl$_2$ and incubated at 37°C for 10 min before being loaded.

(b) Load 1–5 µl on a 0.3% Chromosomal Grade (Bio-Rad, Hercules, CA) or SeaKem GTG or Gold (FMC BioProducts) agarose gel in 1× TAE; the last may be the easiest to handle. Cool the gel at 4°C before removing the comb. Run in 1× TAE with 0.5 µg/ml ethidium bromide at 7 V/cm for 2 min, then at either 1.5 V/cm for 6 hr or 0.8 V/cm for 16 hr.

Further Adjustments

The accumulation of nonspecific products is a problem that is often system-dependent. Reaction specificity may be improved by lowering the enzyme, Mg^{2+}, or primer concentrations, or by raising the annealing temperature or the glycerol or DMSO levels. Nonspecific products can also accumulate if excessive extension times or numbers of cycles are used. New primers may need to be chosen if sufficiently specific reaction conditions cannot be identified.

In the absence of secondary products, yields may be improved in a number of ways. Longer extension times or additional cycles may be useful. Lower annealing temperatures, or higher primer or enzyme concentrations may improve yields if nonspecific products do not result. Some templates may benefit from slightly higher denaturation temperatures, although the half-life of enzyme activity may be reduced. Finally, adjustment of glycerol and DMSO levels or the concentrations of K^+ or Mg^{2+} may enhance yields. Some systems may also benefit from added gelatin, polyethylene glycol, formamide, or nonionic detergents (Table 2).

Conclusion

The protocols described here have been used to amplify much longer targets than previously possible, and should be useful in areas such as genome mapping, sequencing, and medical genetics. Further study of the fidelity of DNA replication under these conditions, of methods to minimize the preferential amplification of shorter alleles over longer alleles, and of methods to minimize the potential interference from G + C-rich secondary structures will be important. Successful amplifications of even longer genomic targets are likely to require methods to further increase the reaction specificity. Ultimately, template damage and secondary priming sites that contribute to competing products may limit the lengths amplifiable by PCR.

Acknowledgments

I would like to thank Russ Higuchi and Carita Fockler for the low percentage gel protocol and their contributions to the long PCR project from its inception, and Joanna Jung-Wong and Lori Kolmodin for the wax-mediated hot start protocol. I also thank Russ Higuchi for critically reading early drafts of this chapter.

Literature Cited

Barnes, W. M. 1994. PCR amplification of up to 35 kb DNA with high fidelity and high yield from λ bacteriophage templates. *Proc. Natl. Acad. Sci. U.S.A.* **91**:2216–2220.

Carle, G. F., M. Frank, and M. V. Olson. 1986. Electrophoretic separations of large DNA molecules by periodic inversion of the electric field. *Science* **232**:65–68.

Cheng, S., C. Fockler, and R. Higuchi. 1994(a). Efficient amplification of long targets from human genomic DNA and cloned inserts. *Proc. Natl. Acad. Sci. U.S.A.* **91**:5695–5699.

Cheng, S., S. -Y. Chang, P. Gravitt, and R. Respess. 1994(b). Long PCR. *Nature* **369**:684–685.

Cheng, S., R. Higuchi, and M. Stoneking. 1994(c). Complete mitochondrial genome amplification. *Nature Genet.* **7**:350–351.

Cheng, S., Y. Chen, J. A. Monforte, R. Higuchi, and B. Van Houten. 1995. Template integrity is essential for PCR amplification of 20- to 30-kb sequences from genomic DNA. *PCR Meth. Applic.* **4**:294–298.

Chou, Q., M. Russell, D. E. Birch, J. Raymond, and W. Bloch. 1992. Prevention of pre-PCR mis-priming and primer dimerization improves low-copy-number amplifications. *Nucl. Acids Res.* **20**:1717–1723.

Kolmodin, L., S. Cheng, and J. Akers. 1995. GeneAmp® XL PCR Kit. *Amplifications: A Forum for PCR Users*, Issue 13 (The Perkin–Elmer Corporation).

Kwok, S., D. E. Kellogg, N. McKinney, D. Spasic, L. Goda, C. Levenson, and J. J. Sninsky. 1990. Effects of primer-template mismatches on the polymerase chain reaction: Human immunodeficiency virus type 1 model studies. *Nucl. Acids Res.* **18**:999–1005.

Kwok, S., S. -Y. Chang, J. J. Sninsky, and A. Wang. 1994. A guide to the design and use of mismatched and degenerate primers. *PCR Meth. Applic.* **3**:S39–S47.

Lindahl, T. and B. Nyberg. 1972. Rate of depurination of native deoxyribonucleic acid. *Biochemistry* **11**:3610–3618.

Sambrook, J., E. F. Fritsch, and T. Maniatis. 1989. *Molecular Cloning: A Laboratory Manual*, 2nd ed. Cold Spring Harbor Press, Cold Spring Harbor, NY.

Schmid, C. W. and W. R. Jelinek. 1982. The Alu family of dispersed repetitive sequences. *Science* **216**:1065–1070.

Walsh, P. S., D. A. Metzger, and R. Higuchi. 1991. Chelex® 100 as a medium for simple extraction of DNA for PCR-based typing from forensic material. *BioTechniques* **10**:506–513.

Wu, D. Y., L. Ugozzoli, B. K. Pal, J. Qian, and R. B. Wallace. 1991. The effect of temperature and oligonucleotide primer length on the specificity and efficiency of amplification by the polymerase chain reaction. *DNA Cell Biol.* **10**:233–238.

25

DIRECT ANALYSIS OF SPECIFIC BONDS FROM ARBITRARILY PRIMED PCR REACTIONS

D. A. Carter, A. Burt, and J. W. Taylor

Arbitrarily primed PCR (AP-PCR) or Randomly amplified polymorphic DNA (RAPD) technology has revolutionized studies of population biology, ecology, and epidemiology by allowing the rapid generation of molecular genotypes (Welsh and McClelland, 1990; Williams et al., 1990; also see Welsh et al., Chapter 20 in this volume). In these techniques, polymorphism is sought among the genomic DNAs of a group of individuals by PCR amplification using a single primer of arbitrary sequence. When the amplified DNAs are electrophoresed, bands present in all individuals are assumed to be monomorphic, and those that are absent in some individuals or that have variable mobility are assumed to be polymorphic. Artifacts may arise, however, if the genomic DNAs are of different concentration or purity, as can be the case with organisms that are difficult or hazardous to culture. Even when artifacts are absent, there may be ambiguity in the basis of the polymorphism because a variety of different mutations may prevent a given band from amplifying, leading to a phenotypically indistinguishable result.

An extension of AP-PCR that eliminates this ambiguity is to search for hidden variation in the bands that are apparently monomorphic. The key to rapidly determining molecular genotypes from these monomorphic bands is to screen for sequence polymorphism using electrophoretic techniques that are sensitive to single-base changes. Single-strand polymorphism analysis (SSCP) (Orita et al., 1989), denaturing gradient gel electrophoresis (DGGE) (Myers et al., 1987), and heteroduplexing (White et al., 1992) are all capable of detecting variation in DNA fragments of equal length. Polymorphisms may then be further defined by DNA sequencing, and specific primers can be designed to amplify the allelic loci directly from additional DNA samples. The different alleles may be distinguished by any of the above techniques, or by restriction enzyme digestion (if an enzyme site has been created or destroyed), allele-specific hybridization (Saiki et al., 1989), or direct sequencing. This allows the rapid assembly of molecular multilocus genotypes in which each marker is reliable, informative, and easy to interpret.

When a single primer is used in the AP-PCR reaction, the resulting bands will be identical at both ends, and further analysis will require cloning prior to DNA sequencing. This is not only time-consuming but is also less reliable than direct sequencing, and it is usually necessary to sequence several clones for each band to ensure that sequence polymorphism is not the result of *Taq* DNA polymerase error. Here we describe two variations on the AP-PCR technique that allow the easy analysis of AP-PCR products. The first, sequencing with arbitrary primer pairs (Burt et al., 1994), uses two AP-PCR primers in a single AP-PCR reaction. One of these can then be used as a sequencing primer for direct double-stranded sequencing of the DNA band. In the second technique, constant plus arbitrary primer amplification, one primer (e.g., an M13 sequencing primer) is held uniform in every AP-PCR reaction and is paired with a variety of different arbitrary primers. The constant primer may then be modified, thereby allowing the PCR product to be modified in specific ways. Sequence variation can be analyzed by DGGE, for example, by incorporating a GC-clamp (a series of G and C nucleotides) into the specific primer (Abrams et al., 1990). Variation may also be screened for by fluorescent SSCP, in which the constant primer is attached to a fluorescent label, and the product analyzed on an SSCP gel in the ABI automated sequencing apparatus using Genescan software. Or, direct single-stranded sequencing of the PCR product can be performed by attaching a 5' biotin moiety to the constant primer and purifying the single strands by streptavidin–magnetic bead separation (Bowman and Palumbi, 1993). These modifications and their uses are further discussed in the section on polymorphism analysis of isolated AP-PCR bands.

Protocols

Band Generation and Isolation

AP-PCR for specific band analysis requires three stages for the generation of specific usable bands (Fig. 1):
1. AP-PCR using primers singly and in combination. Running these reactions in parallel on agarose gels ensures that the bands to be analyzed are the product of the two different PCR primers, and are not produced by either primer alone.
2. Agarose gel electrophoresis for identification and excision of appropriate bands.
3. Reamplification of DNA in the excised band using the same primer pair (with or without specific modifications to the constant primer)

1. AP–PCR

Reaction components (50-µl reaction volume)
Both primers arbitrary
  5 µl PCR buffer [100 mM Tris–HCl (pH 8.3), 500 mM KCl, 15 mM MgCl$_2$, 0.1% (w/v) gelatin]
  5 µl 50% (v/v) glycerol
  1.25 µl deoxyribonucleoside triphosphate (dNTP)mix (2.5 mM each dNTP)
  1 µl arbitrary primer 1 (10 µM)
  1 µl arbitrary primer 2 (10 µM)
  0.5 µl AmpliTaq (5 U/µl)
  x µl diluted DNA
  37.25 − x µl H$_2$O
Constant plus arbitrary primer
  As above but replacing arbitrary primer 2 with 1 µl of unmodified constant primer (10 µM).
Cycling parameters
  Initial denaturation: 94°C, 5 min
  2 cycles of 94°C, 5 min; 36°C, 5 min; 72°C, 5 min
  40 cycles of 94°C, 1 min; 36°C, 1 min; 72°C, 2 min
  Final elongation: 72°C, 7 min
Notes
  1. DNA should be first analyzed on an agarose gel and the concentration of each sample standardized as closely as possible. A

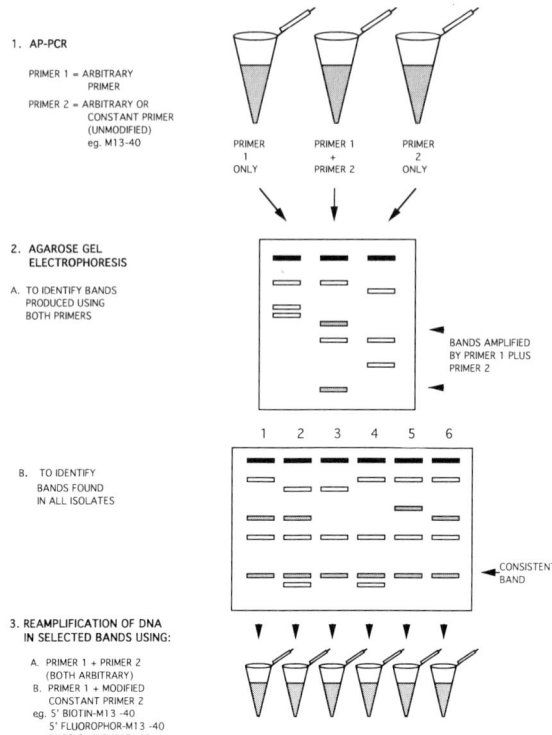

Figure 1 Generation and isolation of DNA bands by AP-PCR or RAPD amplification for analysis of sequence polymorphism.

range of 10-fold dilutions should be tried with a number of different arbitrary primers to find the lowest possible dilution that still results in clear amplification using most of these primers.
2. The choice of the constant primer to be used may depend in part on the organism under study. For example, in our work on human mycopathogens, the M13-40 sequencing primer, which only amplified very large fragments when used alone, and produced numerous nascent bands when used in combination with arbitrary primers, was much more useful than the M13-21 sequencing primer. The latter was able to amplify fragments in a range of sizes when used alone. This not only made identification and isolation of bands produced in combination with an arbitrary primer more difficult, but also had the effect of "swamping" amplification from the additional arbitrary primer, so that few primer combinations gave bands suitable for further analysis. It may be necessary, therefore, to try out a number of different

"constant" primers for each organism under study. A slightly longer (15–20 nt) primer than the 10-mers that are generally employed in AP-PCR is preferable if it is likely to be used later as a sequencing primer.

2. Agarose Gel Electrophoresis and Band Isolation

As with normal AP-PCR reactions, DNA should be fractionated in high-sieving gels at medium voltage. Low-melting agarose allows easy reamplification of isolated bands. We routinely use 2% NuSieve, 0.5% SeaKem (FMC) in 1× TBE (Tris–borate, boric acid EDTA). Following ethidium bromide staining, gels should be viewed on a long wavelength UV transilluminator to avoid DNA damage.

A simple and efficient way to excise DNA bands that avoids problems with cross-contamination is to use a micropipette tip (e.g., a Gilson p200 tip) that has 2–3 mm cut off from the fine end. The truncated tip is pushed through the DNA band in the gel so that a plug of agarose is removed, which is then excised into a microfuge tube. Fifty microliters of water are added, the tube is heated to 75°C and vortexed briefly to evenly disperse the molten agarose. One to 3 µl of this should be sufficient for the subsequent reamplification step.

Notes
1. The choice of the band to be excised will largely depend on its subsequent use. To avoid problems with cross-contamination, it should be clearly separated from all other bands. Sequencing and SSCP both require DNA fragments of 100–500 bp. For DGGE, fragments of up to 800 bp may be used. Sometimes larger bands may contain internal priming sites, so that some of the smaller AP-PCR bands are derivatives of larger bands. It is therefore preferable where possible to choose the smallest of the amplified fragments.

3. Reamplification of the Chosen Bands

Reaction components
Reamplification with the two arbitrary primers uses the same reaction components as the original amplification. When using constant plus arbitrary primers, modified versions of the constant primers may now be used (Fig. 1).

Cycling parameters
Reamplification requires much shorter cycling times and a fewer number of cycles than the original AP-PCR reaction.

Initial denaturation : 94°C, 5 min
20 cycles of 94°C, 30 sec; 36°C, 30 sec; 72°C, 1 min
Final elongation : 72°C, 7 min

Reamplification should result in a single-band species of high concentration that can be analyzed for polymorphism among the organisms under study.

Polymorphism Analysis of Isolated AP-PCR Bands

1. Fluorescent SSCP

SSCP allows sequence changes to be detected in small, single-stranded DNA fragments (100–500 nt) as altered mobility in nondenaturing polyacrylamide gels (Orita et al., 1989; also see Bailey, Chapter 9, this Volume). Attaching a fluorophor to one primer (here the constant primer in the reamplification step) enables SSCP to be performed in the ABI automated sequencer and analyzed with Genescan software. This eliminates the need for radioactive labeling, and since four different fluorophors exist, up to three differently labeled samples plus one labeled size standard can be run in a single lane and analyzed independently or together. Incorporation of size standards in each lane ensures that the gel has run correctly and allows altered mobilities to be clearly detected. This technique permits rapid screening for sequence polymorphism. The exact nature of the polymorphism can then be established by DNA sequencing.

2. DGGE

In DGGE, DNA fragments are electrophoresed through a polyacrylamide gel which contains an increasing gradient of chemical denaturants. As the DNA molecules denature, their mobility decreases; thus, since sequence polymorphisms may alter DNA melting properties, they can be detected as changes in electrophoretic mobility. Attaching a high-melting G+C-clamp via one of the PCR primers allows single-base changes to be detected throughout almost the entire length of the PCR product (Abrams et al., 1990). As with SSCP, the nature of the polymorphism may then be defined by DNA sequencing.

3. Direct Sequencing

Double-Stranded Sequencing with Arbitrary Primer Pairs
Double-stranded sequencing uses heat denaturation followed by snap cooling to generate single-stranded DNA template (Kusukawa et al., 1990). Primers as short as 10 bases have been used successfully to sequence AP-PCR products (Burt et al., 1994).

Single-Stranded Sequencing Following Magnetic Bead Purification
This requires that the constant primer in the reamplification step have an attached 5' biotin moiety, allowing adsorption of the PCR product to a magnetic bead via streptavidin. Alkaline denaturation results in efficient release and purification of the two single strands. Either strand may serve as template in single-stranded dideoxy sequencing (Bowman and Palumbi, 1993).

Design of Specific PCR Primers

Once sequence polymorphism has been found in the AP-PCR-generated bands, specific primers may be developed to allow direct amplification of the polymorphic locus. Completely new primers may be chosen from within the amplified sequence, or the AP-PCR primers may be extended at the 3' end to a size suitable for PCR amplification. If it is desirable to continue to incorporate the modifications introduced by the constant primer, this primer may be extended by three or more bases to allow specific amplification with its redesigned complementary primer. Amplification may then be initiated using limiting amounts of this primer, and the modified constant primer added subsequently. Amplification will continue using the initial amplicon as template, thus introducing the desired modification into all further PCR products.

These variations on the AP-PCR or RAPD techniques add reproducibility and higher resolution to the analysis of sequence variation in populations of organisms when it is difficult to obtain the same amounts of DNA from many isolates. In our initial work on populations of *Coccidioides immitis*, we are finding that approximately one in three AP-PCR-generated bands show sequence polymorphism between members of different geographic populations, and one in five bands show variation among members of a single population. Molecular genotypes based on discrete, characterized sequence polymorphisms can thus be constructed easily and rapidly, and can be used to answer questions of epidemiology, ecology, and population structure in organisms that have traditionally been difficult or hazardous to work with.

Literature Cited

Abrams, E. S., S. E. Murdaugh, and L. S. Lerman. 1990. Comprehensive detection of single-base changes in human genomic DNA using denaturing gradient gel electrophoresis and a GC-clamp. *Genomics* **7**:463–475.

Bowman, B. H., and S. R. Palumbi. 1993. Rapid production of single stranded sequencing template from amplified DNA using magnetic beads. *Methods Enzymol.* **224**:399–406.

Burt, A., D. A. Carter, T. J. White, and J. T. Taylor. 1994. DNA sequencing with arbitrary primer pairs (SWAPP). *Mol. Ecol.* **3**:523–525.

Kusukawa, N., T. Uemori, K. Asada, and I. Kato. 1990. Rapid reliable protocol for direct sequencing of material amplified by the Polymerase Chain Reaction. *Biotechniques* **9**:66–72.

Myers, R. M., T. Maniatis, and L. S. Lerman. 1987. Detection and localization of single base changes by denaturing gradient gel electrophoresis. *Methods Enzymol.* **155**:501–527.

Orita, M., M. Suzuki, T. Sekiya, and K. Hayashi. 1989. Rapid and sensitive detection of point mutations and DNA polymorphisms using the polymerase chain reaction. *Genomics* **5**:874–879.

Saiki, R. K., D. S. Walsh, and H. A. Erlich. 1989. Genetic analysis of amplified DNA with immobilized sequence-specific oligonucleotide probes. *Proc. Natl. Acad. Sci. U. S. A.* **86**:6230–6234.

Welsh, J., and M. McClelland. 1990. Fingerprinting genomes using PCR with arbitrary primers. *Nucleic Acids Res.* **18**:7213–7218.

Williams, J. G., A. R. Kubelik, K. J. Livak, J. A. Rafalski, and S. V. Tingey. 1990. DNA polymorphisms amplified by arbitrary primers are useful as genetic markers. *Nucleic Acids Res.* **18**:6531–6535.

White, M. B., M. Carvallo, D. Derse, S. J. O'Brien, and M. Dean. 1992. Detecting single base substitutions as heteroduplex polymorphisms. *Genomics* **12**:301–306.

Part Four

ALTERNATIVE AMPLIFICATION STRATEGIES

26

DETECTION OF LEBER'S HEREDITARY OPTIC NEUROPATHY BY NONRADIOACTIVE LIGASE CHAIN REACTION

John A. Zebala and Francis Barany

Leber's hereditary optic neuropathy (LHON) is a maternally inherited disease characterized by acute, bilateral, and synchronous optic nerve death that leads to complete blindness (Nikoskelainen et al., 1983, 1987; Nikoskelainen, 1984; Seedorf, 1985; Berninger et al., 1989). Investigations by Wallace et al. (1988) have shown that in 50% of cases there is a G→A mitochondrial mutation at position 11,778 in the gene encoding the fourth subunit of nicotinamide adenine dinucleotide dehydrogenase. The base change eliminates an SfaNI restriction site while creating a MaeIII restriction site. This restriction site polymorphism is the basis for genetic tests described by Stone et al. and Hotta et al. (1989). However, LHON is a relatively heterogeneous disease with only approximately 50% of cases containing the Wallace mutation (Holt et al., 1989; Vilkki et al., 1989). The other 50% of cases are most likely caused by mutations in other essential mitochondrial genes, and when found, may not cause alterations in coveniently assayable restriction sites.

In this chapter, we describe a ligase chain reaction (LCR) color assay for nonradioactive detection of the 11,778 mutation. The LCR uses thermostable ligase to discriminate and amplify single-base changes in genes of medical interest (Barany, 1991a,b). This enzyme specifically links two adjacent oligonucleotides when it is hybridized to a complementary target (see Fig. 1). Oligonucleotide products are exponentially amplified by thermal cycling of the ligation reaction in the presence of a second set of adjacent oligonucleotides that are complementary to the first set and the target. A single-base mismatch at the junction prevents ligation and amplification thus distinguishing a single-base mutation from the normal allele. The use of a thermostable ligase allows the enzyme to survive thermal cycling in a fashion analogous to *Taq* polymerase in PCR. Investigators have described other nonradioactive detection methods using T4 ligase (Landegren *et al.*, 1988; Nickerson *et al.*, 1990), and thermostable ligase as well (Bond *et al.*, 1990; Winn-Deen and Iovannisci, 1991; Wiedmann *et al.*, Chapter 27, this Volume). The development and refinement of such nonradioactive detection methods for DNA diagnostics is essential to the practical utilization of such tests in clinical laboratories.

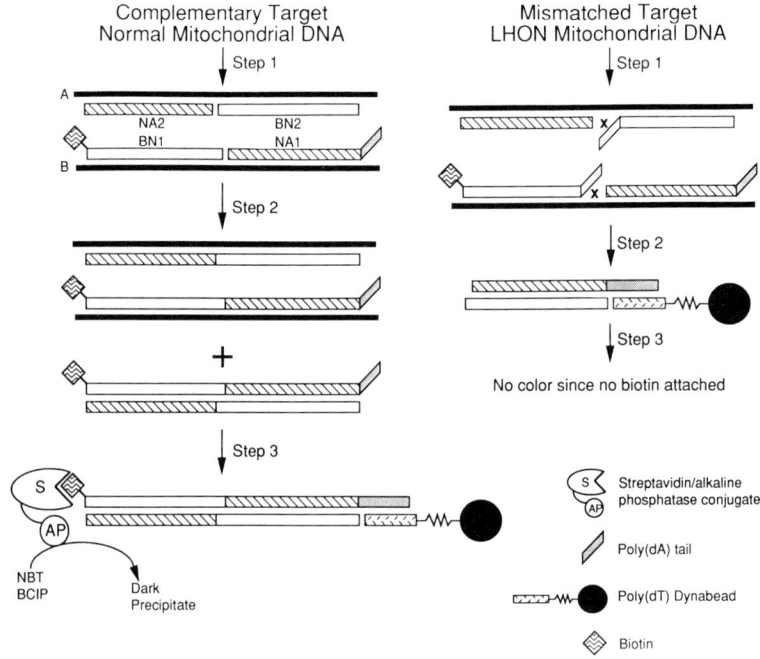

Figure 1 Schematic depicting detection and amplification of a single base change using the LCR color assay.

The method described here combines initial PCR amplification of a mitochondrial fragment containing the 11,778 mutation with subsequent LCR to detect the G→A transition. In the LCR part of the scheme, each reaction tube contains target DNA (PCR fragments containing normal or LHON alleles), four diagnostic oligonucleotides, and a thermostable ligase (Fig. 1). In step 1, the target DNA is heat denatured at 94°C, separating the two strands of the DNA duplex (lines A and B). Cooling to 60°C causes the four diagnostic oligonucleotides (NA1, NA2, BN1, and BN2 are complementary to the normal allele in this example) to hybridize to their complementary sequence on the target DNA (see Fig. 2 for sequence details). The four oligonucleotides are designed so that the base pair being assayed (the 11,778 mutation in this case) is at the junction between hybridized oligonucleotides. If there is perfect complementarity between the target and the hybridized oligonucleotides, as in the left portion of the figure using the normal allele as target, then the thermostable ligase will seal the nick between them by forming a covalent phosphodiester bond (step 2). As these two steps are re-

				T_m(°C)	Size
LHON	BL1	[B]-AAC TAC GAA CGC ACT CAC AGT C$\underline{A}$(OH)		66	23
Normal	BN1	[B]-AAC TAC GAA CGC ACT CAC AGT C$\underline{G}$(OH)		68	23
	NA1		(P)C ATC ATA ATC CTC TCT CAA GGA C-[A]$_{n=21}$	66	44
Normal target		AAC TAC GAA CGC ACT CAC AGT C$\underline{G}$C ATC ATA ATC CTC TCT CAA GGA C	Arg		46
		TTG ATG CTT GCG TGA GTG TCA G$\underline{C}$G TAG TAT TAG GAG AGA GTT CCT G	↓		
LHON target		--- --- --- --- --- --- --- C$\underline{A}$C --- --- --- --- --- --- --- -	His		46
		--- --- --- --- --- --- --- G$\underline{T}$G --- --- --- --- --- --- --- -			
	NA2	TG ATG CTT GCG TGA GTG TCA G(P)		64	21
Normal	BN2		(OH)CG TAG TAT TAG GAG AGA GTT CCT G 5'	70	24
LHON	BL2		(OH)TG TAG TAT TAG GAG AGA GTT CCT G 5'	68	24

Present in normal target: SfaNI GCATC(N)$_5$↓
Present in LHON target: MaeIII ↓GTNAC

Figure 2 Sequence of the diagnostic and target oligonucleotides used in detecting the LHON-associated mitochondrial mutation. Shown is a stretch of mitochondrial DNA sequence from normal and LHON alleles. The G→A change associated with LHON is underlined. The four oligonucleotides forming the normal diagnostic assay mix include BN1, BN2, NA1, and NA2. The four oligonucleotides forming the LHON diagnostic assay mix include BL1, BL2, NA1, and NA2. NA1 (44-mer) was ^{32}P-labeled on its 5' end to facilitate detection of ligated products (44 + 23 = 67-mer) via gel electrophoresis and autoradiography. NA2 was kinased with nonradioactive ATP. BL1 and BN1 were synthesized with 5' biotin molecules attached (indicated as [B]).

peated, the ligated products from one cycle go on to become targets for hybridization and ligation of more oligonucleotides in later cycles, amplifying the product in an exponential fashion. If there is no complementarity at the junction because of a base mismatch, as with the LHON allele (indicated by an X in the right part of the figure), no ligated products are formed, and there is no product amplification. After formation of sufficient products, the ligated molecules are captured from solution via hybridization of their poly(dA) tails with poly(dT) iron beads (Dynabeads) and subsequent magnetic separation (step 3). Only molecules that ligated efficiently carry a 5' attached biotin molecule (left portion of figure). Those which did not ligate do not carry the biotin (right side of figure). The presence of the biotin is detected by addition of a streptavidin–alkaline phosphatase conjugate and chromogenic substrate. In LCR reaction in Fig. 1, diagnostic oligonucleotides were used which are complementary for the normal allele and thus produce ligation products with a normal target but not a LHON target. Conversely, LCR reactions are also carried out using diagnostic oligonucleotides which are complementary for the LHON allele (NA1, NA2, BL1, and BL2; see Fig. 2) and thus produce ligation products with a LHON target but not a normal target.

Protocols

Extraction of Mitochondrial DNA from Peripheral Blood

1. Add 3 μl whole blood from normal or LHON individuals to 1 ml of sterile ddH$_2$O in a 1.5-ml tube.
2. Incubate 15 to 30 min at room temperature. Centrifuge for 3 min at 1000 g.
3. Carefully remove and discard supernatant, leaving a 30-μl pellet. Add 5% Chelex to total volume of 200 μl.
4. Incubate at 56°C for 15 to 30 min. Vortex 5 to 10 sec.
5. Incubate at 100°C for 8 min. Vortex 5 to 10 sec.
6. Centrifuge for 3 min at 1000 g. Use 2 to 10 μl for PCR. Store remainder at 4°C or frozen without resin.

PCR Amplification of Mitochondrial DNA

Reagents and Materials

10× PCR buffer 100 mM Tris–HCl (pH 8.3)
500 mM KCl
15 mM MgCl$_2$
0.1% (w/v) gelatin

Deoxyribonucleotide triphosphate (dNTP) mix [1.25 mM each of deoxyadenosine triphosphate (dATP), deoxycytidine triphosphate (dCTP), deoxyguanosine triphosphate (dGTP), and deoxythymidine triphosphate (dTTP)]

10 μM F (forward) primer - CCCACCTTGGCTATCATC (11,141-11,158)

10 μM R (reverse) primer - TAATAAAGGTGGATGCGACA (12,472-12,453)

AmpliTaq polymerase (5 U/μl)

Mineral oil

PCR Amplification

1. Prepare the following reaction mixture in a 0.5-ml microfuge tube: 2.0–10.0 μl Chelex DNA preparation from either normal or LHON patients
    3.0 μl F primer
    3.0 μl R primer
    10.0 μl 10× PCR buffer
    16.0 μl dNTP mix
    57.5 μl ddH$_2$O
    0.5 μl AmpliTaq polymerase
   100.0 μl
   Overlay with mineral oil
2. Place tubes in a thermal cycler as follows:
a. Denature for 1.5 min at 94°C
b. Run 35 cycles with: 45 sec denaturing at 94°C
                        30 sec annealing at 51°C
                        60 sec extensions at 72°C
c. The last extension step is for 5 min.
3. Heat denature the polymerase at 97°C for 25 min. Store at 4°C.

These reactions produce 1332-bp fragments containing the normal (N^{PCR}) and LHON (L^{PCR}) alleles.

Preparation of Oligonucleotides for LCR

Synthesis

Except for BN1 and BL1, oligonucleotides are assembled "trityl off" by the phosphoramidite method on an Applied Biosystems model 380A DNA synthesizer and gel purified. The oligonucleotide sequences are shown in Fig. 2. The oligonucleotides BN1 and BL1 are biotinylated as follows:

1. Dissolve 50 μmol of biotin phosphoramidite (DuPont Co., Wilmington, DE) in 0.5 ml anhydrous dichloromethane (Aldrich Chemical Co., Milwaukee, WI) and 0.5 ml anhydrous 0.5 M tetrazole in acetonitrile (Applied Biosystems, Foster City, CA). After the last step of synthesis is completed, a connected system is used to apply this mixture to the CPG column in 200-μl aliquots over a total of 15 min.
2. Wash the column with 2 ml of acetonitrile. Apply 2 ml of iodine–H_2O–pyridine–tetrahydrofuran for 30 sec to oxidize the phosphite, and wash the column again with 2 ml of acetonitrile.
3. After ammonium hydroxide treatment and lyophilization, the 5'-biotinylated (85%) and 5'-hydroxyl (15%) products are separated from each other by high-performance liquid-chromatography (HPLC) as follows: C-18 column, 0–15% acetonitrile over 15 min and 15% acetonitrile from 15 to 30 min in 0.1 M triethylammonium acetate (TEAA) (pH 7.0) at a flow rate of 1 ml/min. The 5'-biotinylated and 5'-hydroxyl products have retention times of 25 and 20 min, respectively.

Reagents and Materials

T4 polynucleotide kinase (10 U/μl)
10 × T4 polynucleotide kinase buffer: 500 mM Tris–HCl (pH 8.0)
100 mM MgCl$_2$
100 mM β-mercaptoethanol

[γ-³²P] ATP (6000 Ci/mmole) and 10 mM ATP
50% Trichloroacetic acid (TCA)
Sephadex G-25 equilibrated with 10 mM Tris–HCl (pH 8.0) and 1 mM EDTA

³²P-Labeling and Phosphorylation of Oligonucleotides NA1 and NA2

1. To a 1.5-ml tube containing 50 μl of lyophilized (83 pmoles) [γ-³²P]ATP add:
   4 μl of 50 μM (200 pmoles) NA1
   4 μl of 10× T4 polynucleotide kinase buffer
   3 μl of T4 polynucleotide kinase
   <u>29 μl H$_2$O</u>
   40 μl total
2. Incubate at 37°C for 1 hr.
3. Add 2 μl 10 mM ATP and continue the incubation for another 2 min at 37°C.
4. Add 1 μl of 0.5 M EDTA to terminate the reaction and then heat at 65°C for 10 min to heat inactivate the kinase.
5. Remove 1 μl from the labeling reaction and add to 9 μl H$_2$O. Remove 2 μl of this and add to 100 μl of 2 mg/ml salmon sperm DNA and mix. Add 10 μl 50% trichloroacetic acid. Spin at 1000 g for 3 min. Cherenkov count the pellet and supernatant. Calculate counts per mole of DNA by dividing the pellet counts by 0.93 pmol. This will allow easy and accurate calculation of concentrations after this step.
6. Add 10 μl of 20% glycerol–0.02% bromophenol blue to the remaining labeling reaction. Remove the unincorporated [γ-³²P]ATP label by chromatography of this mixture over Sephadex G-25. The final concentrations of the collected fractions are calculated by Cherenkov counting a given volume from each. Concentrations are adjusted to 800 nM by adding the appropriate amount of H$_2$O.

The specific activity is usually around 2×10^9 cpm/nmol (1.4×10^8 cpm/μg). The oligonucleotide NA2 is phosphorylated in the same manner except that 4 μl of 10 mM ATP are used in place of the [γ-³²P]ATP, and steps 3, 5, and 6 are omitted.

LCR

Reagents and Materials

N assay mix = 100 nM each of oligonucleotides BN1, BN2, [^{32}P]-NA1, and NA2

L assay mix = 100 nM each of oligonucleotides BL1, BL2, [^{32}P]-NA1, and NA2

10× LCR buffer   200 nM Tris–HCl (pH 7.6 at 25°C)
                     500 mM KCl
                     100 mM MgCl$_2$
                     10 mM EDTA

Taq DNA ligase (50 U/μl, prepared as described in Barany and Gelfand, 1991).

Note: addition of 0.1% or 1% Triton X-100 in the 10x LCR buffer has been shown to enhance ligations or enzyme stability (Wiedmann *et al.*, 1994; Winn-Deen and Iovannisci, 1991).

LCR Reactions

1. Prepare the LCR mixture in a 0.5-ml microfuge tube as follows:
   - 12 μl of N or L assay mix
   - 18 μl of N^{PCR} or L^{PCR} (10-fold dilution of stock)
   - 18 μl of 10x LCR buffer
   - 18 μl of 10 mM NAD$^+$
   - 18 μl of 100 mM dithiothreitol (DTT)
   - 22 μl of 10 mg/ml salmon sperm DNA
   - <u>74 μl</u> H$_2$O
   - 180 μl total

   Add 6 μl *Taq* DNA ligase (50 U/μl or 300 nick-closing units) last.

2. Overlay the mixture with a drop of mineral oil and incubate the reactions at 94°C for 1 min followed by 60°C for 4 min; repeat for a total of 10 cycles.

Analysis of LCR Products

Reagents and Materials

TBE buffer [100 mM Tris–borate (pH 8.9), 1 mM EDTA]

12% polyacrylamide gel

Kodak XAR-5 film

Stop solution = 95% formamide, 20 mM EDTA, 0.05% bromophenol blue, and 0.05% xylene cyanol

Gel Electrophoresis

1. Pour a 12% polyacrylamide sequencing gel containing 7 M urea and TBE.
2. Remove a 4-μl aliquot from the LCR mix and add it to 5 μl of the stop solution. Heat the sample to 65°C for 5 min and load 4 μl on the gel. Electrophorese for 2 hr at 60 W constant power or until the bromophenol blue is 2 inches from the bottom.
3. Remove the urea from the gels by soaking 10 min in 10% acetic acid and 15 min in H$_2$O. Dry the gel and autoradiograph on Kodak XAR-5 film.
4. Develop autoradiograph (see Fig. 3).

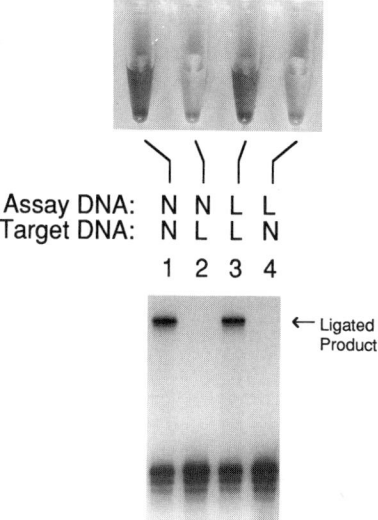

Figure 3 Autoradiograph showing the specificity of detecting the LHON-associated mutation using PCR targets from whole blood, LCR, and the corresponding color changes. The assay DNA is the normal (N) mix which includes BN1, BN2, NA1, and NA2 (1.2 pmol each); or the LHON (L) mix, which includes BL1, BL2, NA1, and NA2 (1.2 pmoles each). The target DNA is either the N^{PCR} (N) or L^{PCR} (L) 1332-bp fragments containing the normal and LHON alleles, respectively. Bottom panel, 10% polyacrylamide, 7 M urea sequencing gel of LCR reactions. Lanes 1 and 3: the assay and target DNA are of the same allelic sequences and show a positive ligation product (67-mer); 2 and 4: the assay and target DNA are of different allelic sequences. Top panel, the Me$_2$SO solubilized precipitate from the color reactions.

Reagents and Materials

poly(dT)$_{25}$ Dynabeads (Dynal, Inc. Lake Success, NY)
Chromogenic substrate   0.1 M borate (pH 9.5)
                        5 mM MgCl$_2$
                        1 mg/ml nitro blue tetrazolium (NBT)
                        0.5 mg/ml 5-bromo-4-chloro-3-indolyl phosphate (BCIP)
Wash buffer   20 mM Tris–HCl (pH 7.6)
              50 mM KCl
              10 mM MgCl$_2$
              1 mM EDTA
Streptavidin-alkaline phosphatase conjugate (0.13 μM in wash buffer)

Color Assay

1. Remove the buffer from 100 μl of Dynabeads by magnetic separation. Resuspend the beads with 155 μl of LCR mixture and incubate from 60° to 50°C over approximately 15 min. Slow cooling allows for maximum annealing between the oligo(dT)$_{25}$ on the bead and the poly(dA)$_{21}$ on NA1. Remove the supernatant by magnetic separation.
2. Resuspend the beads with 100 μl of streptavidin-alkaline phosphatase conjugate and incubate on ice for 25 min. Remove and discard the supernatant by magnetic separation and add 100 μl of wash buffer. Resuspend gently and discard the supernatant after magnetic separation.
3. Add 50 μl of chromogenic substrate and incubate 25 to 45 min at room temperature. A dense, black precipitate should form on the Dynabeads of those tubes which contain ligation product. Remove the supernatant by magnetic separation and rinse the beads once with 100 μl of H$_2$O. Solubilize the precipitate with 50 μl Me$_2$SO (see Fig. 3).

Comments

This chapter describes a ligase chain reaction color assay for detecting single-point mutations which we have applied to the mutation associated with Leber's hereditary optic neuropathy. The normal and LHON genes differ by a single G→A point mutation at position 11,778 in the mitochondrial genome. Figure 3 shows that is difference between the normal (N) and disease (L) alleles can be distinguished using the LCR test. Both nonradioactive and radioactive assays give a clear positive signal when the correct nucleotide is present, and a negative signal for the incorrect nucleotide (<2.5% background). This LCR color assay may also be useful in other applications involving point mutations.

Acknowledgments

The authors thank M. Lott and D. Wallace for the kind gifts of LHON DNA and for their DNA extraction protocol, and B. Rigas for the BCIP and NBT. F. B. is supported in part by a Hirschl/Monique Weill-Caulier Career Scientist Award. This study was supported by National Institutes of Health Grant GM-41337-03 and Applied Biosystems Inc. (F.B.).

Literature Cited

Barany, F. 1991a. Genetic disease detection and DNA amplification using cloned thermostabile ligase. *Proc. Natl. Acad. Sci. U.S.A.* **88**:189–193.

Barany, F. 1991b. The ligase chain reaction in a PCR world. *PCR Methods Appl.* **1**:5–16.

Barany, F., and D. Gelfand. 1991. Cloning, overexpression, and nucleotide sequence of a thermostable DNA ligase-encoding gene. *Gene* **109**:1–11.

Berninger, T. A., A. C. Bird, and G. B. Arden. 1989. Leber's hereditary optic atrophy. *Ophthalmic Paediatr. Genet.* **10**:211–227.

Bond, S., J. Carrino, H. Hampl, L. Hanley, L. Rinehardt, and T. Laffler. 1990. New methods of detection of HPV. In *Serono symposia* (ed. J. Monsonego), Vol. 78, p. 425. Raven Press, New York.

Holt, I. J., D. H. Miller, and A. E. Harding. 1989. Genetic heterogeneity and mitochondrial DNA heteroplasmy in Leber's hereditary optic neuropathy. *J. Med. Genet.* **26**:739–743.

Hotta, Y., M. Hayakawa, K. Saito, A. Kanai, A. Nakajima, and K. Fujiki. 1989. Diagnosis of Leber's optic neuropathy by means of polymerase chain reaction amplification. *Am. J. Ophthalmol.* **108**:601–602.

Landegren, U., R. Kaiser, J. Sanders, and L. Hood. 1988. A ligase-mediated gene detection technique. *Science* **241**:1077–1080.

Nickerson, D. A., R. Kaiser, S. Lappin, J. Stewart, L. Hood, and U. Landegren. 1990. Automated DNA diagnostics using an ELISA-based oligonucleotide ligation assay. *Proc. Natl. Acad. Sci. U. S. A.* **87**:8923–8927.

Nikoskelainen, E. 1984. New aspects of the genetic, etiologic, and clinical puzzle of Leber's disease. *Neurology* **34**:1482–1484.

Nikoskelainen, E., W. F. Hoyt, and K. Nummelin. 1983. Ophthalmoscopic findings in Leber's hereditary optic neuropathy, II: the fundus findings in the affected family members. *Arch. Ophthalmol.* **101**:1059–1068.

Nikoskelainen, E., M. L. Savontaus, O. P. Wanne, M. J. Katila, and K. U. Nummelin. 1987. Leber's hereditary optic neuroretinopathy: a maternally inherited disease. *Arch. Ophthalmol.* **105**:665–671.

Seedorf, T. 1985. The inheritance of Leber's disease. *Acta Ophthalmol.* **63**:135–145.

Stone, E. M., J. M. Coppinger, R. H. Kardon, and J. Donelson. 1990. *Mae*III positively detects the mitochondrial mutation associated with type I Leber's hereditary optic neuropathy. *Arch. Ophthalmol.* **108**:1417–1420.

Vilkki, J., M. L. Savontaus, and E. K. Nikoskelainen. 1989. Genetic heterogeneity in Leber's hereditary optic neuropathy revealed by mitochondrial DNA polymorphism. *Am. J. Hum. Genet.* **45**:206–211.

Wallace, D. C., G. Singh, M. T. Lott, J. A. Hodge, T. G. Schurr, A. M. Lezza, L. J. Elsas, and E. K. Nikoskelainen. 1988. Mitochondrial DNA mutation associated with Leber's hereditary optic neuropathy. *Science* **242**:1427–1430.

Wiedmann, M., J. Czajka, F. Barany, and C. A. Batt. 1992. Discrimination of *Listeria Monocytogenes* from non *Listeria monocytogenes* species by ligase chain reaction. *Appl. Environ. Microbiol.* **58**:3443–3447.

Winn-Deen, E. S., and D. M. Iovannisci. 1991. Sensitive fluorescence method for detecting DNA ligation amplification products. *Clin. Chem.* **37**:1522–1525.

27

DETECTION OF *LISTERIA MONOCYTOGENES* BY PCR-COUPLED LIGASE CHAIN REACTION

Martin Wiedmann, Francis Barany, and Carl A. Batt

Listeria is a Gram-positive bacillus, and while most species within the genus are nonpathogenic to humans, *L. monocytogenes* has the potential to cause listeriosis (Gellin and Broome, 1989). This species is primarily a livestock pathogen, but it has been seen sporadically in humans. One of the most important modes of transmission for humans is through the consumption of contaminated food. The selective detection of *L. monocytogenes* is of great importance because the ecology of *Listeria* is not well understood and the detection of *Listeria* spp. other than *L. monocytogenes* may not be an accurate indicator of a food's safety (Cox *et al.*, 1989).

The detection of *L. monocytogenes* using conventional culture methods is often very time-consuming and difficult; therefore, new approaches to detection have been developed over the past few years (Farber and Peterkin, 1991). These systems use either nucleic acid probes, monoclonal antibodies (see review by Gavalchin *et al.*, 1992), or the polymerase chain reaction (e.g., Bessesen *et al.*, 1990; Border *et al.*, 1990; Deneer and Boychuck, 1991; Golsteyn Thomas *et al.*, 1991; Niederhauser *et al.*, 1992; Wernars *et al.*, 1991).

The use of 16S rRNA as a distinct signature for bacteria has become the method of choice for identifying and differentiating microorganisms when no other obvious nucleic acid sequence uniquely defines the desired target (Woese, 1987). Differences in single base pairs between *L. monocytogenes* and *L. innocua* have been reported for the V2 and the V9 region of the 16S rRNA gene (16S rDNA) (Collins *et al.*, 1991; Czajka *et al.*, 1993; Wang *et al.*, 1991). The ligase chain reaction (LCR) has been shown to be a highly sensitive and specific method for discriminating between DNA sequences that differ in only a single base pair (Barany, 1991a; review by Barany, 1991b; Winn-Deen and Iovannisci, 1991). Single base pair differences can also be detected by allele-specific PCR, but LCR may be the more straightforward and universally applicable approach. Based on a single A to G difference in the V9 region of the 16S rRNA gene (16S rDNA), an LCR assay was developed to distinguish between *L. monocytogenes* and other *Listeria* species (Wiedmann *et al.*, 1992). Figure 1 shows a single base pair difference between *L. monocytogenes* and its most similar species, *L. innocua*, together with the LCR primers for its detection. All other *Listeria* spp. are also different than *L. monocytogenes* in this locus (Collins *et al.*, 1991).

Since even a single *L. monocytogenes* cell in food is a potential risk for human health, there is a need for a very sensitive detection system. A combination of the sensitivity of PCR with the specificity of LCR provides a fast and promising detection method. In the method described here, *Listeria*-encoded 16S rDNA is initially amplified via PCR, and *L. monocytogenes* species are differentiated from *L. innocua* and other *Listeria* spp. by LCR amplification (see

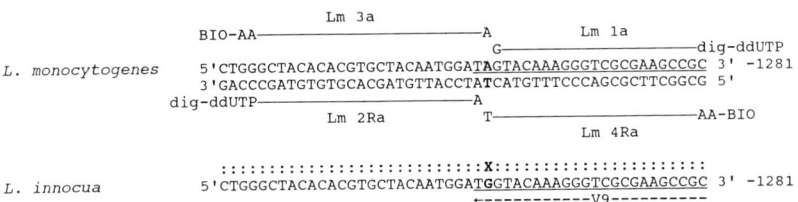

Figure 1 Nucleotide sequence of the V9 region of 16S rDNA in *L. monocytogenes* and in *L. innocua* used for LCR and the location of primers Lm 1a, Lm 2Ra, Lm 3a, Lm 4Ra. (Adapted from Wiedmann *et al.*, 1992 with permission from American Society for Microbiology, Journals Division.) Oligonucleotides Lm 3a and Lm 4Ra contained two nucleotide-length tails to prevent ligation at the wrong end. Size of primers are: Lm 1a, 21-mer; Lm 2Ra, 23-mer; Lm 3a, 26-mer; Lm 4Ra, 24-mer. X, Differences between *L. monocytogenes* and *L. innocua*. :, nucleotides identical for *L. monocytogenes* and *L. innocua*. Bold, target nucleotide for LCR. Primers are noted by a bar (—) where nucleotides are identical to *L. monocytogenes*; the base at the ligation junction is also depicted for clarity. The sequences for each primer (in the 5' → 3' orientation) are listed in Table 2.

Fig. 2). A single base pair difference, A – G, is distinguished (see Fig. 1). By using a subsequent LCR step, L. monocytogenes can be detected in some cases even when the signal from the initial PCR amplification is too weak to be detected in ethidium bromide-stained gels.

This chapter describes the procedures for DNA preparation, PCR, LCR, and two different detection methods for the LCR product. One way that the LCR product can be detected is by radioactively labeling one oligonucleotide of each of the two primer pairs. The ligated product is separated from the primers by denaturing polyacrylamide gel electrophoresis, then detected by autoradiography. As an alternative, a nonisotopic method which involves labeling one primer of a pair with biotin on the 5' end and labeling the second primer with digoxigenin on the 3' end was developed (see Fig. 1) (Landegreen et al., 1988; Nickerson et al., 1990; Wiedmann et al., 1993; Winn-Deen et al., 1993). After the ligation product is captured on streptavidin-coated wells, the digoxigenin can be detected with the help of antidigoxigenin antibodies, which are conjugated with alkaline phosphatase (AP). The oligonucleotide primers for PCR and LCR are shown in Table 1 and Table 2, respectively. A different approach to nonisotopic

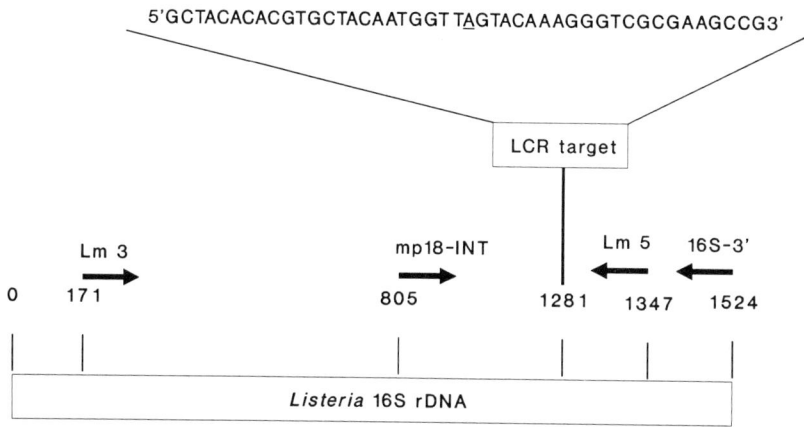

Figure 2 Diagram of the *Listeria* 16S rDNA gene. Shown are the location and orientation of the PCR primers and the location of the target region for the LCR. The numbers for the location are based on the numbering system used by Czajka et al. (1993). Primers are shown as → (in 5' to 3' orientation). The numbers for the location of these PCR primers give the location of the 5' end of the primers. The number for the LCR target region gives the location of the single base pair difference between *L. monocytogenes* and the other *Listeria* spp. In the nucleotide sequence of the LCR target for *L. monocytogenes*, the single base that differs between *L. monocytogenes* and *L. innocua* is underlined.

Table 1
Sequence of the PCR Primers for the 16S rDNA

Primer	Primer Sequence (5'-3')	Specificity
Lm 3	GGACCGGGGCTAATACCGGAATGATAA	*Listeria*
Lm 5	TTCATGGTAGGCGAGTTGCAGCCTA	*Listeria*
mp18-INT	ATTAGATACCCTGGTAGTCC	Universal
16S-3'	CCCGGGATCCAAGCTTTAACCTTGTTACGACTT	Universal

detection of LCR products is reported by Barany in Chapter 26 of this volume.

Protocols

DNA Preparation

A method adapted from Furrer et al. (1991) was used to prepare the DNA.

Table 2
Sequence and T_m of LCR Primers for V9 Region

Primer No.	Primer Sequence (5'-3'):	T_m
Lm 1a[a]	GTACAAAGGGTCGCGAAGCCG	68°C
Lm 2Ra[a]	ATCCATTGTAGCACGTGTGTAGC	68°C
Lm 3a[b]	AAGCTACACACGTGCTACAATGGATA	70°C
Lm 4Ra[b]	AACGGCTTCGCGACCCTTTGTACT	70°C

[a] Primers Lm 1a and Lm 2Ra are either radioactive or digoxigenin labeled.
[b] Primers 3a and 4Ra are synthesized with a biotin group at the 5' end when used for nonradioactive detection.

Reagents and Materials

Listeria enrichment broth (LEB) (BBL, Becton Dickinson Microbiology Systems, Cockeysville, MD)
Lysozyme
Proteinase K

Rapid Preparation of Genomic DNA for PCR Amplification

A. From Listeria enrichment broth
1. Transfer 500 µl of an 8-hr liquid culture in LEB (~10^9 cells/ml) to a 1.5-ml microfuge tube.
2. Pellet cells at 12,000 g for 10 min; discard supernatant.
3. Dissolve pellet in 100 µl 1× PCR buffer containing 2 mg/ml lysozyme and incubate at room temperature (RT) for 15 min.
4. Add 1 µl proteinase K (20 mg/ml) and incubate 1 hr at 55°C, then boil the lysate for 10 min to inactivate proteinase K.
5. Use a 1-µl aliquot of this lysate for the PCR amplification.

B. From milk samples
1. Incubate 1 ml of milk with 9 ml LEB at 37°C for 12 hr.
2. Transfer 1.5 ml of the broth to a 1.5-ml microfuge tube.
3. Follow steps 2–5 in part A.

PCR and Detection of the PCR Products

Reagents and Materials

10× PCR buffer [500 mM KCl, 100 mM Tris–HCl (pH 8.8 at 25°C), 15 mM MgCl$_2$, 1% Triton X-100]

1 mM deoxyribonucleoside triphosphates (dNTPs) [1 mM each of deoxyadenosine triphosphate (dATP), deoxycytidine triphosphate (dCTP), deoxyguanosine triphosphate (dGTP), and deoxythymidine triphosphate (dTTP)]
Primers Lm 3 and Lm 5 or mp18-INT and 16S-3' (100 pmol/µl stock)
Taq polymerase (5 U/µl)
Mineral oil
TBE buffer [89 mM Tris–borate, 2 mM EDTA (pH 8.2)]

PCR Amplification of 16S rRNA and Detection of PCR Products

For amplification of the V9 region, either two universal primers for the 16S rDNA 16S-3' and mp18-INT (Czajka et al., 1993; Weisburg et al., 1991) or the *Listeria*-specific primers Lm 3 and Lm 5 are used (see Table 1). The amplified DNA can be directly used in the subsequent LCR. The PCR products are analyzed for the presence of the expected 720-bp fragment (for primers 16S-3' and mp18-INT) or 1180-bp fragment (for primers Lm 3 and Lm 5) by agarose gel electrophoresis. Even if no PCR product is observed after ethidium bromide staining, conducting the subsequent LCR step may yield positive results for *L. monocytogenes*.

The PCR amplification procedure is as follows:
1. Prepare the following reaction mixture in a 0.5-ml microfuge tube:
   2.5 μl of 10× PCR buffer
   1 μl of target DNA (prepared as described)
   1.25 μl of 1 mM dNTPs
   0.5 μl of each primer (100 pmol/μl): either Lm 3 and Lm 5 or mp 18-INT and 16S-3'
   0.2 μl *Taq* polymerase
   19.05 μl of H$_2$O
   Overlay this mixture with 75 μl of mineral oil.
2. Place tubes in a thermal cycler and perform cycles as follows:
   a. Denature for 4 min at 94° C
   b. Run 40 cycles with: 1 min denaturing at 94° C
   1.5 min annealing at 55°C
   1.5 min extension at 72°C
   c. The extension step in the last PCR cycle should be for 8 min rather than 2 min as in the previous cycles.
   d. Run 1 cycle with 97°C for 25 min to inactivate the *Taq* polymerase.
3. Add 2 μl 5× bromophenol blue–xylene cyanole dye to 8 μl of the PCR reaction mix and run it on a 1.5% agarose gel in a TBE buffer.
4. Stain the gel with ethidium bromide and examine the DNA fragments using long-wave UV light.
5. Store the PCR reaction at 4°C until used in the LCR procedure.

Preparation of LCR Primers for Radioactive Detection

Reagents and Materials

T4 polynucleotide kinase (10 U/µl)
10× T4 polynucleotide kinase buffer (0.5 M Tris–HCl, 100 mM $MgCl_2$, 100 mM 2-mercaptoethanol)
[γ-^{32}P]ATP (6000 Ci/mM = 60 pmol ATP)
ATP (20 mM): must be stored at $-70°C$
TE buffer [10 mM Tris–HCl (pH 8.0 at 25°C), 1 mM EDTA]
Sephadex NAP 5 columns (Pharmacia Labs, Piscataway, NJ)

Radioactive Labeling of LCR Primers

For the LCR, one set of four different primers for the V9 region (Lm 1a, Lm 2Ra, Lm 3a, Lm 4Ra) is used. The primer sequences are shown in Table 2.

Primers Lm 1a and Lm 2Ra are 5' end-labeled with [γ-^{32}P]ATP as described by Barany (1991a).

1. Prepare labeling mix as follows:
    1.5 µl primer (Lm 1a or Lm 2Ra) (10 pmol/µl)
    2 µl 10× T4 polynucleotide kinase buffer
    1.5 µl T4 polynucleotide kinase (10 U/µl)
    4 µl [γ-^{32}P]ATP
    11 µl H_2O
2. Incubate at 37°C (water bath) for 45 min.
3. Add 1 µl unlabeled ATP (20 mM) and continue the incubation for another 2 min at 37°C (water bath).
4. Add 0.5 µl 0.5 M EDTA to terminate the reaction.
5. Heat to 65°C (water bath) for 10 min to heat inactivate the kinase.
6. Add 478.5 µl of TE buffer to bring the final volume to 500 µl.
7. Run the 500 µl of labeling mix through a NAP 5 column, equilibrated in TE buffer. Since the primers are eluted in 1 ml of TE buffer, the final concentration of the labeled primers will be 15 fmol/µl.

The radioactive labeled primers can be stored for up to 2 weeks after they have been kinased.

Preparation of LCR Primers for Nonradioactive Detection

The primers Lm 3a and Lm 4Ra are synthesized with a biotin group at the 5' end by the "Biotin-ON" (Clontech, Palo Alto, CA) phosphoramidite method using an ABI 392 DNA synthesizer (Applied Biosystems, Foster City, CA). Primers Lm 1a and Lm 2Ra are kinased at the 5' end and digoxigenin labeled at the 3' end using the following procedure.

Reagents and Materials

T4 polynucleotide kinase (10 U/μl)
10× T4 polynucleotide kinase buffer (0.5 M Tris–HCl, 100 mM MgCl$_2$, mM 2-mercaptoethanol)
ATP (20 mM): must be stored at $-70°$ C
Terminal deoxynucleotidyl transferase (17 U/μl)
5× terminal deoxynucleotidyl transferase buffer [500 mM sodium cacodylate (pH 7.2), 1 mM 2-mercaptoethanol, 10 mM CoCl$_2$]
Digoxigenin-11-dideoxy uracil triphosphate solution (1 mM) (Boehringer Mannheim Biochemica, Indianapolis, IN)

Kinasing of the Primers

1. Prepare kinasing reaction as follows:
   300 pmol primer (Lm 1a or Lm 2Ra)
   3 μl 10× T4 polynucleotide kinase buffer
   1 μl ATP (20 mM)
   6 U T4 polynucleotide kinase
   Make up with H$_2$O to 30 μl final volume
2. Incubate at 37°C (water bath) for 45 min, then heat at 65°C (water bath) for 10 min to inactivate the kinase.

Labeling of the Primers with Digoxigenin

1. Set up labeling reaction:
   50 pmol kinased primer (Lm 1a or Lm 2Ra)
   8 μl 5× terminal deoxynucleotidyl transferase buffer
   2 μl digoxigenin-11-ddUTP solution
   85 U terminal deoxynucleotidyl transferase
   Make up with H$_2$O to 40 μl final volume.

2. Incubate at 37°C (water bath) for 45 min, then heat at 65°C (water bath) for 10 min.
3. Dilute with H_2O to 100 fmol/μl and store at $-20°C$.

LCR

Reagents and Materials

10× LCR buffer [500 mM Tris–HCl (pH 8.2 at 25°C), 1 M KCl, 100 mM MgCl$_2$, 10 mM EDTA, 100 mM dithiothreitol, 0.1% Triton X-100]. This buffer has to be stored in small aliquots at $-20°C$ to avoid multiple freezing and thawing cycles. The inclusion of 0.1% Triton X-100 gives higher yields of ligation product in the LCR than a 10X buffer without Triton.

Primer Lm 3a and Lm 4Ra (100 fmol/μl)
NAD$^+$ (12.5 mM): prepare fresh every time used
Salmon sperm DNA (10 mg/ml): prepare by standard procedures (Sambrook et al., 1989)
Taq DNA ligase (37.5 U/μl) (Barany and Gelfand, 1991)
Termination mixture (10 mM EDTA, 0.2% bromophenol blue, 0.2% xylene cyanole in formamide)

LCR Reaction

1. Prepare LCR mixture in a 0.5-ml microfuge tube as follows for the radioactive labeled primers:
   1 μl of the PCR reaction
   2.5 μl 10× LCR buffer
   2 μl NAD$^+$
   1 μl salmon sperm DNA
   1 μl Taq DNA ligase (37.5 U/μl)
   0.5 μl of each unlabeled primer, i.e., Lm 3a and Lm 4Ra (=25 fmol in 25 μl)
   1.75 μl of each labeled primer, i.e., Lm 1a and Lm 2Ra (=25 fmol in 25μl)
   11.5 μl of H_2O
   Overlay this mixture with 75 μl of mineral oil.

If the primers are nonradioactively labeled, use 100 fmol each of Lm 1a, Lm 2Ra, Lm 3a, and 4Ra in the reaction mixture. Note that

the amount of water to be added has to be reduced, so that the final volume stays at 25 µl.
2. Place tubes in a thermal cycler and perform 25 cycles (for isotopic LCR) or 10 cycles (for nonisotopic LCR) as follows: 1 min at 94°C, followed by 4 min at 65°C.
3. Stop the reaction by adding 20 µl of termination mixture (only for radioactive detection).
4. The LCR reaction may be stored at −20°C until used for electrophoresis or for nonisotopic detection.

Analysis of LCR Products for Radioactive Detection

Reagents and Materials

TBE buffer [89 mM Tris–borate, 2 mM EDTA (pH 8.2)]
16% Polyacrylamide gel (Sequagel sequencing system, National Diagnostics, Manville, NJ)
Kodak X-OMAT AR film

Polyacrylamide Gel Electrophoresis and Autoradiography

1. Prepare a 16% polyacrylamide gel containing 7 M urea in a Mini-PROTEAN II electrophoresis cell (Bio-Rad, Richmond, CA) or a similar apparatus.
2. Allow gel to polymerize for 30 min and then place it in the electrophoresis apparatus with TBE buffer [89 mM Tris–borate, 2 mM EDTA (pH 8.2)]
3. Heat LCR samples at 90°C for 5 min and load 10-µl aliquots on the gel. It is important to flush the wells of the gel with buffer immediately before loading the sample to remove the urea.
4. Carry out electrophoresis at 175 V constant voltage for about 1 hr (until the front of the bromophenol blue reaches the bottom of the gel).
5. Remove gel from the apparatus and remove upper glass plate, then cover gel with plastic wrap (e.g., Saran Wrap).
6. Place gel on a Kodak X-OMAT AR film and autoradiograph at −20°C for 12 hr.
7. Develop autoradiogram.

Figure 3 shows an example of the results of the PCR-coupled LCR for different *Listeria* spp. as they are seen on the autoradiogram.

Analysis of LCR Products for Nonradioactive Detection

Reagents and Materials

High-binding EIA 8-well strips, Type I material (COSTAR, Cambridge, MA)

Streptavidin (Promega, Madison, WI)

Antidigoxigenin Fab fragment, conjugated with alkaline phosphatase (Boehringer Mannheim Biochemica, Indianapolis, IN)

Lumi-Phos 530 (Boehringer Mannheim Biochemica)

Buffer T [100 mM Tris–Cl (pH 7.5), 150 mM NaCl, 0.05% Tween 20]

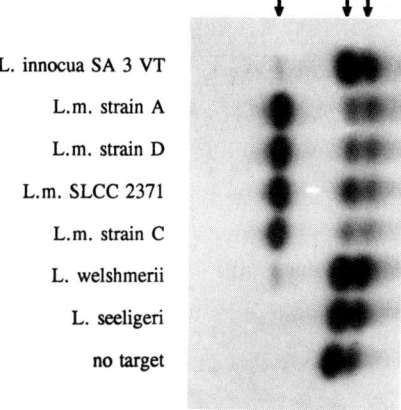

Figure 3 Autoradiogram showing LCR for a selected number of different *Listeria* species and strains. (Adapted from Wiemann *et al.*, 1992, with permission from American Society for Microbiology, Journals Division.) On the autoradiogram the two radioactive labeled primers are visible at the bottom of the gel; the lower band is primer Lm 1a (21 nucleotides), the higher one is primer Lm 2Ra (23 nucleotides). The ligated products for primer pairs Lm 1a and 3a and Lm 2Ra and 4Ra, are both 47 bases and therefore only one band appears on the autoradiogram. A strong band of ligated product is seen for *L. monocytogenes*, while the closely related *L. welshmerii* and *L. innocua* only show very weak bands, which is due to a very long exposure time (22 hr).

Buffer T without Tween [100 mM Tris–Cl (pH 7.5), 150 mM NaCl]
Carbonate buffer (20 mM Na$_2$CO$_3$, 30 mM NaHCO$_3$, 1 mM MgCl$_2$): pH should be approximately 9.6; store at 4°C
Plate binding buffer (1 M NaCl, 0.75 M NaOH)
Dry milk (Carnation Co., Los Angeles, CA)
Salmon sperm DNA (10 mg/ml): prepare by standard procedures (Sambrook et al., 1989).
CAMLIGHT (Camera Luminometer System, Analytical Luminescence Laboratory, San Diego, CA)
Polaroid instant image film type 612 (20,000 ISO)

Detection of LCR products

1. Add 60 µl streptavidin (100 µg/ml in carbonate buffer) to high-binding EIA well strips and incubate for 1 hr at 37°C. Discard the streptavidin solution.
2. Block for 20 min at RT with 200 µl buffer T + 0.5% dry milk + 100 µg/ml salmon sperm DNA. Discard the blocking solution.
3. Wash twice with 200 µl buffer T.
4. Add 5 µl LCR mixture diluted with 40 µl buffer T without Tween to one well and add 10 µl plate binding buffer. Incubate at RT for 30 min.
5. Wash twice with buffer T.
6. Wash twice with 200 µl 0.01 M NaOH, 0.05% Tween 20.
7. Wash three times with buffer T.
8. Add 50 µl of antidigoxigenin-AP, Fab fragment (diluted 1 : 1000 in buffer T + 0.5% dry milk) and incubate 30 min at RT.
9. Wash 6 times with 200 µl buffer T.
10. Add 50 µl Lumi-Phos 530 and hold 30 min at 37°C.
11. Expose to Polaroid-type 612 film in CAMLIGHT for 5 min.
12. Develop film.

Samples should always be run in duplicate (one suggestion regarding the LCR detection is that on rare occasions single wells might be contaminated with alkaline phosphatase so that even in the absence of ligation product, a false positive reaction is possible). A sample is positive only if both wells for one LCR reaction show color development.

Comments

It is recommended that the 16S universal primers (mp18-INT; 16S-3') be used for the PCR if *Listeria*-like colonies are grown in LEB and subsequently used for the PCR-coupled LCR. The rationale behind this suggestion is that the universal primers will amplify the 16S rDNA of all bacteria present in the sample and therefore act as a positive control for the PCR amplification. If no PCR product is observed, there might be inhibitors in the sample. Another reason for the lack of amplification could be an incomplete lysis of the bacteria.

On the other hand, the use of the *Listeria*-specific primer set (Lm 3 and Lm 5) is preferred to the 16S universal primer set in milk samples and other samples where a high concentration of other microorganisms and only a few *L. monocytogenes* are expected. This is because the universal primers could be titrated out during the amplification of the DNA from other microorganisms and therefore would be unavailable for the amplification of the *Listeria* DNA. Yet, in such cases, it is suggested that a parallel PCR using universal primers for the same sample be run to serve as an indicator for the presence of any PCR inhibitors.

In our experiments we never saw any inhibition of the LCR when there was no inhibition of the first PCR step, provided no more than 1 μl of the PCR reaction was used in the LCR. Therefore, we do not deem it necessary to include a positive control for each sample used in the LCR.

The results of the PCR-coupled LCR can only be interpreted as negative for *L. monocytogenes* if the presence of inhibiting factors for the PCR is ruled out.

Acknowledgments

This work was supported by the Northeast Dairy Foods Research Center (C.A.B.); a grant from the Cornell Center for Advanced Technology (CAT) in Biotechnology, which is sponsored by the New York State Science and Technology Foundation, a consortium of industries, and the National Science Foundation (C.A.B.); the National Institutes of Health (GM 41337-03) (F.B.) and Applied Biosystems Inc. (F.B.). M.W. was supported by a stipend from the Gottlieb Daimler- and Carl Benz-Stiftung (2.92.04).

Literature Cited

Barany, F. 1991a. Genetic disease detection and DNA amplification using cloned thermostable ligase. *Proc. Natl. Acad. Sci. U.S.A.* **88:**189–193.

Barany, F. 1991b. The ligase chain reaction in a PCR world. *PCR Methods Appl.* **1:**5–16.

Barany, F., and D. Gelfand. 1991. Cloning, overexpression and nucleotide sequence of a thermostable DNA ligase-encoding gene. *Gene* **109:**1–11.

Czajka, J., N. Bsat, M. Piani, W. Russ, K. Sultana, M. Wiedmann, R. Whitaker, and C. A. Batt. 1993. Differentiation of *Listeria monocytogenes* and *Listeria innocua* by 16S rDNA genes and intraspecies discrimination of *L. monocytogenes* by random amplified polymorphic DNA polymorphisms. *Appl. Environ. Microbiol.* **59:**304–308.

Bessesen, M. T., L. Qian, A. R. Harley, J. B. Martin, and R. T. Ellison III. 1990. Detection of *Listeria monocytogenes* by using polymerase chain reaction. *Appl. Environ. Microbiol.* **56:**2930–2932.

Border, P. M., J. J. Howard, G. S. Platsow, and K. W. Siggens. 1990. Detection of *Listeria* species and *Listeria monocytogenes* using polymerase chain reaction. *Lett. Appl. Microbiol.* **11:**158–162.

Collins, M. D., S. Wallbanks, D. J. Lane, J. Shah, R. Nietupski, J. Smida, M. Dorsch, and E. Stackbrandt. 1991. Phylogenetic analysis of the genus *Listeria* based on reverse transcriptase sequencing of 16S rRNA. *Int. J. Systematic Bacteriol.* **41:**240–246.

Cox, L. J., T. Kleiss, J. L. Cordier, C. Cordellana, T. Konkel, C. Pedrazzini, R. Beumer, and A. Siebenga. 1989. *Listeria* spp. in food processing, non-food and domestic environments. *Food Microbiol.* **6:**49–61.

Deneer, H. G., and I. Boychuck. 1991. Species-specific detection of *Listeria monocytogenes* by DNA amplification. *Appl. Environ. Microbiol.* **57:**606–609.

Farber, J. M., and P. I. Peterkin. 1991. *Listeria monocytogenes*, a food borne pathogen. *Microbiol. Rev.* **55:**476–511.

Furrer, B., U. Candrian, C. Hoefelein, and J. Luethy. 1991. Detection and identification of *Listeria monocytogenes* in cooked sausage products and in milk by *in vitro* amplification of haemolysin gene fragments. *J. Appl. Bacteriol.* **70:**372–379.

Gavalchin, J., K. Landy, and C. A. Batt. 1992. Rapid methods for the detection of *Listeria*. In: *Molecular approaches to improving food quality and safety* (ed. D. Bhatnagar and T. E. Cleveland), pp. 189–204. Van Nostrand Reinhold, New York.

Gellin, B. G., and C. V. Broome. 1989. Listeriosis. *J. Am. Med. Assoc.* **261:**1313–1320.

Golsteyn Thomas, E. J., R. K. King, J. Burchak, and V. P. J. Gannon. 1991. Sensitive and specific detection of *Listeria monocytogenes* in milk and ground beef with the polymerase chain reaction. *Appl. Environ. Microbiol.* **57:**2576–2580.

Landegren, U., R. Kaiser, J. Sanders, and L. Hood. 1988. A ligase-mediated gene detection method. *Science* **241:**1077–1080.

Nickerson, D. A., R. Kaiser, S. Lappin, J. Stewart, L. Hood, and U. Landegren. 1990. Automated DNA diagnostics using an ELISA-based oligonucleotide ligation assay. *Proc. Natl. Acad. Sci. U.S.A.* **87:**8923–8927.

Niederhauser, C., U. Candrian, C. Höflein, M. Jermini, H.-P. Bühler, and J. Lüthy. 1992. Use of polymerase chain reaction for detection of *Listeria monocytogenes* in food. *Appl. Environ. Microbiol.* **58:**1564–1568.

Sambrook, J. E., E. F. Fritsch, and T. Maniatis. 1989. *Molecular cloning: a laboratory manual*, 2nd ed. Cold Spring Harbor Laboratory, Cold Spring Harbor, NY.

Wang, R.-F., W.-W. Cao, and M. G. Johnson. 1991. Development of a 16S rRNA based oligomer probe specific for *Listeria monocytogenes*. *Appl. Environ. Microbiol.* **57**:3666–3670.

Weisburg, W. G., S. M. Barns, D. A. Pelletier, and D. J. Lane. 1991. 16S ribosomal DNA amplification for phylogenetic study. *J. Bacteriol.* **173**:697–703.

Wernars, K., K. J. Heuvelman, T. Chakraborty, and S. H. W. Notermans. 1991. Use of the polymerase chain reaction for direct detection of *Listeria monocytogenes* in soft cheese. *J. Appl. Bacteriol.* **70**:121–126.

Wiedmann, M., J. Czajka, F. Barany, and C. A. Batt. 1992. Discrimination of *Listeria monocytogenes* from other *Listeria* spp. by ligase chain reaction. *Appl. Environ. Microbiol.* **58**:3443–3447.

Wiedmann, M., F. Barany, and C. A. Batt. 1993. Detection of *Listeria monocytogenes* with a nonisotopic polymerase chain reaction coupled ligase chain reaction. *Appl. Environ. Microbiol.* **59**:2743–2745.

Winn-Deen, E. S., and D. M. Iovannisci. 1991. Sensitive fluorescence method for detecting DNA ligation amplification products. *Clin. Chem.* **37**:1522–1525.

Winn-Deen, E. S., C. A. Batt, and M. Wiedmann. 1993. Non-radioactive detection of *Mycobacterium tuberculosis* LCR products in a microtitre plate format. *Mol. Cell. Probes* **7**:179–186.

Woese, C. R. 1987. Bacterial evolution. *Microbiol. Rev.* **51**:221–271.

INDEX

2-Amino-9-(2'-deoxy-β-ribofuranosyl)-6-methoxyaminopurine, affinity in base pairs, 76
Antibody
  cloning in bacteria
    Fab fragment, 301–302
    mutagenesis, 300–301
    screening, 300
    single-chain antibody, 301–302
  generation by hybridoma technology, 300
  PCR generation of single-chain fragments
    cDNA synthesis, 308, 311
    cloning strategy, 302–303, 310
    linker sequence, 302, 305
    overlap PCR, 309
    primer design, 302–304
    RNA isolation, 305, 307, 310

AP-PCR, see Arbitrarily primed polymerase chain reaction
Arbitrarily primed polymerase chain reaction, see also DNA fingerprinting; RNA fingerprinting
  agarose gel electrophoresis of products, 329
  artifacts, 324
  bands
    generation, 327
    heterozygosity, 325–326
    isolation, 329
    polymorphism analysis, 330–331
    reamplification, 329–330
    sequencing, 331
  primer design, 328–329, 331–332
  reaction conditions, 327–329

Biotinylation, DNA, 75, 227–228, 234

C7dGTP, see 7-Deaza-2′-deoxyguanosine triphosphate
Caffeine, similarity to theophylline, 289, 297
CFTR gene
  identification of mutations
    MutS protein binding, 112–118
    reverse dot blot hybridization, 131–138
  mutations in cystic fibrosis, 111–112, 114–116
Chromatography, see High-performance liquid chromatography
CMV, see Cytomegalovirus
Cystic fibrosis
  carrier frequency, 115
  gene, see CFTR gene
Cytomegalovirus, quantification by PCR–TGGE, 190, 194–195

D1S80 allele
  allelic sizing ladder, 170–171
  forensic application, 163
  PCR amplification, 165–168
  sequence, 164
DARTT, see DNA amplification-restricted transcription translation
7-Deaza-2′-deoxyguanosine triphosphate
  destabilization of DNA duplex, 4, 71, 77
  resistance to endonucleases, 77–78
Denaturing gradient gel electrophoresis, polymorphism analysis of DNA, 330
Deoxyinosine, affinity in base pairs, 76

DGGE, see Denaturing gradient gel electrophoresis
Differential display, see RNA fingerprinting
Digoxigenin, labeling of primers, 354–355
Dimethyl sulfoxide
  amplification enrichment of G–C rich DNA, 4, 7
  effect on DNA melting temperature, 12
DMSO, see Dimethyl sulfoxide
DNA
  extraction from paraffin-embedded tissues
    digestion of tissue, 34–36
    PCR sensitivity, 32–33, 35, 37
    tissue section preparation, 32–34
  genomic, isolation, 227
  melting temperature
    calculation, 70, 78
    denaturation thermodynamics, 70–71
    effects of
      modified nucleotides, 77
      solvents, 9, 12–13, 15, 71
    profile determination, 6
  mitochondrial
    extraction from peripheral blood, 338
    PCR amplification, 339
  modification of nucleotides
    deoxyuridine substitution, 78
    endonuclease protection, 77
    fluorophores, 77
  secondary structure effect on primer hybridization, 74
DNA amplification-restricted transcription translation
  applications
    antisense RNA evaluation, 245
    epitope identification, 237, 239, 241, 245
    functional sites of polypeptides, 237

oligermization sites, 237
truncated polypeptide
generation, 237, 239
PCR
primers, 242–243
reaction, 241–242
protein products, functional
assays, 245, 247
tissue culture adaptation, 239,
246–247
transcription, 243–244
translation, 244
DNA fingerprinting, arbitrarily
primed PCR
applications, 250–251
cloning of products, 271
DNA
concentration-dependent
differences, 253, 255
preparation, 251–252
electrophoresis of products,
255–257
interpretation of data, 257–259
phylogenetic analysis, 258
polymerase selection, 253
primer
concentration, 252, 255
design, 249–250, 252–253
motif sequence encoding, 272
principle, 249
reaction mixture, 255
reproducibility, 258–259
sequence selectivity, 251, 257
Staphylococcus aureus genome,
253–254
thermocycling profile, 255–256
DNA polymerase, *see also*
Escherichia coli DNA
polymerase I
exonuclease activity
mechanism, 49
fidelity
error rate, 18-19, 22, 51
factors affecting, 18–20, 51
mechanisms, 18–19, 50
mismatch types, 18

kinetic models
extension efficiency, 23–26
extension fidelity, 21–23
nucleotide insertion, 21–23
processivity, 20, 25
thermostable polymerases, *see
also Pfu* polymerase; *Taq*
polymerase; *Tli* polymerase;
Tma polymerase; *Tth*
polymerase
exonuclease activity, 48
PCR applications, 52–53
properties, 40, 47
quaternary structure, 46
Dot blot, *see* Reverse dot blot
hybridization
Dropout, *see* Quantitative
polymerase chain reaction

Electroporation, DNA transfection,
246–247
Escherichia coli DNA polymerase
I, exonuclease activity
effect on fidelity, 50
mechanism, 48–49

Fetal cells
flow cytometry, 214–215
male sex diagnosis, 213–214
presence in maternal blood,
213–214
Y chromosome-specific PCR
amplification reaction, 216
contamination prevention, 215
primers, 215
product analysis, 216
Flow cytometry, fetal cells in
maternal blood, 214–215
5-Fluorodeoxyuridine, affinity in
base pairs, 76
Formamide
amplification enrichment of
G–C rich DNA, 6–7, 13–15

effects on
  DNA melting temperature, 4, 9, 12–13, 15
  Taq polymerase thermostability, 9

Gene, identification methods
  arbitrarily primed PCR, *see* RNA fingerprinting
  differential screening, 260
  subtractive hybridization, 260
Genomic subtraction
  adaptor
    capping of unbound DNA, 231–232, 234
    preparation, 229
  affinity matrix preparation, 229–230
  applications
    chromosome deletion identification, 222–223
    gene isolation, 224
    pathogen DNA identification, 224
    RFLP isolation, 235
  biotinylation of DNA, 227–228, 234
  controls, 226–227
  ethanol precipitation, 226
  exogenous DNA in subtraction sample, 225–226
  genetic prerequisites, 221–222
  genomic DNA isolation, 227
  PCR
    amplification, 233
    analysis of products, 233
  principle, 220–221
  reagents, 225
  restriction digestion of wild-type DNA, 229
  shearing of DNA, 227–228
  subtraction
    first round, 230–231
    subsequent rounds, 231
  time requirement, 225

Glycerol
  amplification enrichment of G–C rich DNA, 6–7, 13–15
  effects on
    DNA melting temperature, 9, 12–13, 15
  Taq polymerase thermostability, 9

Heteroduplex mobility analysis
  DNA isolation, 157
  gel electrophoresis, 158
  heteroduplex formation, 158
  PCR, 157
  phylogenetic analysis, 155–156, 158–159
  principle, 155–156
High-performance liquid chromatography
  column preservation, 151
  double-stranded DNA separation, 141
  maintenance of systems, 151
  optimization of PCR protocols, 142
  PCR product quantitation
    calibration of system, 146–148
    gel electrophoresis comparison, 141–142
    gradients, 143–144, 151
    internal standard assay, 149–150
    linearity, 149
    multiplex analysis, 142
    peak identification, 144–145
    precision, 141–142, 148–149
    preparative HPLC, 141–142
    product detection, 142, 144
    safety, 142
    standard curves, 145–146
    system configuration, 143
    time required, 141
HIV-1, *see* Human immunodeficiency virus

HMA, *see* Heteroduplex mobility analysis
HPLC, *see* High-performance liquid chromatography
Human immunodeficiency virus
 copy number in cells, 199–200
 heteroduplex mobility analysis, 155–159
 in situ PCR
  lymph node, 207
  primer design, 201, 203–204
 polymorphism analysis, 154
 quantitation of PCR products, 150

In situ polymerase chain reaction
 cell fixation, 200–201
  digestion of tissue, 34–36, 206
  paraffin, 205
  slides, 205
  suspensions, 204–205
 detection of products, 209–210
 primer
  concentration, 207
  design, 201, 203–204
  multiple primer set, 203–204, 208
  permeability of fixed cells, 201, 203
 reaction mixture, 206–207
 slides, 208–209
 suspensions, 208
 thermal cycling parameters, 208
Interleukin-2, mRNA
 quantification by PCR–TGGE, 190, 196

Klenow fragment, *see E. coli* DNA polymerase I

LCR, *see* Ligase chain reaction
Leber's hereditary optic neuropathy
 detection by LCR, 337–338, 344–345
 gene mutations, 335
 heredity, 335
LHON, *see* Leber's hereditary optic neuropathy
Ligase chain reaction
 detection of products
  agarose gel electrophoresis, 352
  nonradioactive detection methods, 336, 344, 349
 genomic DNA preparation from bacteria, 351
 LHON gene mutation, 337–338, 344–345
 Listeria monocytogenes, 347–350, 359
 PCR amplification of, 16S rRNA, 352
 primer
  design, 337, 349–350, 359
  digoxigenin labeling, 354–355
  radiolabeling, 341, 353
  synthesis, 340
 principle, 335–336
 product analysis
  autoradiography, 356–357
  color assay, 344, 349, 357–358
  gel electrophoresis, 343
 reaction conditions, 342, 355–356
Listeria monocytogenes
 culture, 347, 359
 detection by LCR, 347–350, 359
 PCR primers, 359
 ribosomal RNA marker, 347–348
 transmission of infection, 347–348

5-Methyldeoxycytidine, effect on DNA melting, 71, 77
1-Methyl-2-pyrolidone
 amplification enrichment of G–C rich DNA, 7
 effect on DNA melting temperature, 9, 12

N^6-Methoxy-2'-deoxyadenosine, affinity in base pairs, 76
Minisatellites, *see* Variable number tandem repeats
Mobility shift assay, *see also* Heteroduplex mobility analysis
  DNA isolation, 157
  gel electrophoresis, 158
  heteroduplex formation, 158
  PCR, 157
  phylogenetic analysis, 155–156, 158–159
  point mutation identification, 115, 118
  principle, 155–156
Mutagenesis, site-specific, *see* Site-specific mutagenesis
MutS
  immobiliation on solid support, 118
  mobility shift assay, 115, 118
  mutation identification in CFTR gene, 112, 115–118
  replication error correction, 111
  stability of complex with DNA, 118

Nested polymerase chain reaction, *see* RNA fingerprinting
NMP, *see* 1-Methyl-2-pyrolidone

Oligonucleotide ligation analysis, analysis of PCR products, 286

Paraffin-embedded tissue, extraction of nucleic acids
  digestion of tissue, 34–36, 206
  PCR sensitivity, 32–33, 35, 37
  tissue section preparation, 32–34, 205
PCR, *see* Polymerase chain reaction
Pfu polymerase, properties, 40, 45

Point mutation
  identification techniques, 121
  MutS protein identification
    amplification reaction, 113, 115
    CFTR gene, 116–117
    effect of polymerase fidelity, 117–118
    primer
      design, 112
      end-labeling, 113
    rationale, 112
    reannealing of products, 115
  reverse dot blot hybridization
    analysis of multiple mutations, 130
    color development, 134–135
    hybridization, 133–136
    membrane strip preparation, 132
    PCR amplification, 133
    probe
      binding to membrane, 132–133
      optimization, 135–136
    reagents, 131
    washing, 134
  SSCP analysis
    automation, 122
    clinical application, 121–122
    DNA denaturation, 124
    electrophoresis, 124–125
      cold gels, 126
      room temperature gels, 125–126
    PCR
      amplification reaction, 123–124
      product detection, 122, 126–127
    principle, 122–123
    sensitivity, 122, 124–126
Polymerase chain reaction, *see also* Arbitrarily primed polymerase chain reaction; *In situ* polymerase chain reaction;

Reverse transcription
polymerase chain reaction
allele-selective amplification, 17,
19, 23, 28
assay for modified nucleotide
incorporation, 74
efficiency
calculation, 85–87
effect on product
accumulation, 86
quantitation of products, 87
long extension amplification
electrophoresis of products,
321–323
extension time, 313
hot start PCR, 317
optimization, 323
primer design, 314–315
reagents, 316–319
specificity, 314
template
denaturation, 313–314
preparation, 315–316
protection during thermal
cycling, 313–314, 323
thermal cycling profiles,
320–321
optimization, 4
principle, 3
quantitation assay, *see*
Quantitative polymerase
chain reaction
reaction conditions, 6
size limitations for
amplification, 313
Primer
arbitrarily primed PCR, 328–329,
331–332
DNA fingerprinting, 249–250,
252–253, 255
RNA fingerprinting, 249–250,
261, 263–264, 267
DARTT, 242–243
degeneracy minimization, 75–76
DNA secondary structure effect
on primer hybridization, 74

effect of polymerase exonuclease
activity, 53
half-time for hybridization, 73
HIV, 201, 203–204
in situ PCR, 201, 203–204,
207–208
LCR, 337, 340–341, 349–350,
354–355, 359
Listeria monocytogenes, 16S
rRNA, 359
long extension PCR, 314–315
melting temperature of hybrid
calculation, 71–72, 78–80
effect of nucleotide
modification, 72, 80–81
modification
biotinylation, 75
digoxigenin labeling, 354–355
phosphorothiolation of 3' end,
53, 75, 187, 340
radiolabeling, 341, 353
single-chain antibody generation,
302–304
site-specific mutagenesis,
183–185
TGGE, 190–191
VNTRs, 166
Y chromosome-specific PCR, 215
Proteinase K, digestion of paraffin-
embedded tissue, 34–36, 206

Quantitative polymerase chain
reaction, *see also* High-
performance liquid
chromatography; Temperature
gradient gel electrophoresis
design of assay, 101–102, 104,
106
detection limit, overcoming in
DNA-probe assay
combined strategies, 97
internal control, 91, 96–97
PCR cycle limiting, 91, 93–95
product dilution, 91, 95–96
sample dilution, 91–93

dropouts
  causes, 104
  frequency, 104
dynamic range, 89, 93, 95
efficiency
  calculation, 85–87
  effect on product
    accumulation, 86
  quantitation of products, 87
plateau phase determination,
    87–88
Poisson distribution to correct
    for sampling error, 98–99
resolution, factors affecting
  amplification efficiency,
    98–100, 102
  detection assay variability, 98,
    100–101
  hybridization efficiency, 102
  inhibitor removal, 102, 104
  precision of assay, 102, 104,
    106
  variability in target molecule
    number, 98, 102
sensitivity minimum, 90–91,
    102
signal molecule incorporation in
  product, 89–90
target molecule range
  determination, 89, 93

Randomly amplified polymorphic
  DNA, see Arbitrarily primed
  polymerase chain reaction
Restriction fragment length
  polymorphisms, identification
  by genomic subtraction, 235
Reverse dot blot hybridization
  analysis of multiple mutations,
    130
  CFTR mutation identification,
    131–138
  clinical application, 138
  color development, 134–135

hybridization, 133–136
membrane strip preparation, 132
PCR amplification, 133
probe
  binding to membrane, 132–133
  optimization, 135–136
reagents, 131
washing, 134
Reverse transcription polymerase
  chain reaction
  cDNA synthesis, 58, 193, 308
  reverse transcriptase reaction, 59
  single-chain antibody fragments
    cDNA synthesis, 308, 311
    cloning strategy, 302–303, 310
    linker sequence, 302, 305
    overlap PCR, 309
    primer design, 302–304
    RNA isolation, 305, 307, 310
  Tth polymerase RT–PCR
    PCR, 62–64
    primer design, 64
    reagents, 60–61
    reverse transcription
      reaction, 61
    sensitivity, 65
    single buffer reaction,
      63–64, 66
RFLPs, see Restriction fragment
  length polymorphisms
RNA, see also Reverse
  transcription polymerase chain
  reaction
  extraction from paraffin-
    embedded tissues
    digestion of tissue, 34–36
    PCR sensitivity, 32–33, 35, 37
    tissue section preparation,
      32–34
  hydrolysis in high-temperature
    reactions, 65
RNA fingerprinting, arbitrarily
  primed PCR
  applications, 259–260
  bias towards 3' end, 261
  cloning of products, 271

concentration dependence, 262, 267
electrophoresis of products, 265–266, 270
nested PCR
 advantages, 267
 heteronuclear RNA interference, 270–271
 principle, 267
 reaction conditions, 269–270
 reagents, 269
primer
 design, 249–250, 261, 263–264, 267
 motif sequence encoding, 272
principle, 249
reagents, 264
reverse transcription, 260–261, 265–267
RNA purification, 263, 265, 270
scaling the size of experiments, 262–263
RT–PCR, see Reverse transcription polymerase chain reaction

SELEX, see Systematic evolution of ligands by exponential enrichment
Short tandem repeats
 frequency in genome, 173
 PCR amplification, 162
 size, 161, 173
Single-stranded conformational polymorphism analysis
 automation, 122
 clinical application, 121–122
 DNA
  denaturation, 124
  fingerprinting, 257
  electrophoresis, 124–125
   cold gels, 126
   room temperature gels, 125–126
 PCR
  amplification reaction, 123–124

product detection, 122, 126–127, 330
 principle, 122–123
 sensitivity, 122, 124–126
Site-specific mutagenesis
 PCR mutagenesis
  base changes within oligonucleotide sequence, 187
  FokI extensions, 180–181
  inverse PCR, 180
  megaprimer techniques, 183–185
  oligonucleotide mutation, 179–180
  overlap extension mutageneses, 181–183
  polymerase error rate, 185–187
  product screening, 186–187
  sequence insertion into single-stranded vector, 180–181
 VNTRs, 163, 168–169, 171
Southern blot, PCR products, 163, 168–169, 171
SSCP, see Single-stranded conformational polymorphism analysis
Stoffel fragment, see *Taq* polymerase
STRs, see Short tandem repeats
Systematic evolution of ligands by exponential enrichment
 oligonucleotide affinity for proteins, 290
 principle, 289–290
 quantitation of theophylline affinity chromatography of RNA, 292–293
 cloning, 293
 counter-SELEX, 294–295
 DNA transcription, 292, 297
 equilibrium filtration binding assay, 294, 297
 PCR amplification, 291
 reverse transcription, 293

RNA
    binding region, 297
    sequences with affinity, 295–296
    template synthesis, 291

T7 RNA polymerase, recombinant expression in mammalian cells, 246
Taq polymerase
    assay
        polymerase activity, 5
        thermal inactivation, 5–6
    effect of solvents
        DMSO, 4, 7
        formamide, 7
        glycerol, 7
        NMP, 7
        thermostability, 9
    exonuclease activity, 41, 49
    extension efficiency, 26–28
    fidelity, 18–19, 26–28, 41, 51, 65–66, 185
    inhibition by solvents, 4, 13
    processivity, 41
    specific activity, 40–41
    Stoffel fragment properties, 41–42, 52–53
    temperature optimum for DNA synthesis, 3, 41
    types, 45–46
Temperature gradient gel electrophoresis
    PCR
        internal control, 189, 197
        primer selection, 190–191
        synthesis of standards, 191
    quantification of PCR products
        accuracy, 189–190
        cytomegalovirus, 190, 194–195
        electrophoresis, 192–193
        interleukin-2, 190, 196
        plasmid copy number in *Escherichia coli*, 193–194
        principle, 189–190

Tetramethylammonium chloride, effect on PCR specificity, 4
TGGE, *see* Temperature gradient gel electrophoresis
Theophylline
    asthma therapy, 289
    caffeine similarity, 289, 297
    immunoassay, 289
    quantitation by SELEX
        affinity chromatography of RNA, 292–293
        cloning, 293
        counter-SELEX, 294–295
        DNA transcription, 292, 297
        equilibrium filtration binding assay, 294, 297
        PCR amplification, 291
        reverse transcription, 293*
        RNA
            binding region, 297
            sequences with affinity, 295–296
            template synthesis, 291
    structure, 290
Tli polymerase
    properties, 40, 44–45
    thermostability, 45
Tma polymerase
    fragment, 44
    properties, 40, 43–44
TMAC, *see* Tetramethylammonium chloride
Tth polymerase
    fidelity, 65–66
    metal dependence, 43, 59–60, 65
    properties, 40, 42–43
    reverse transcriptase activity, 59, 64–65
    RT–PCR
        PCR, 62–64
        primer design, 64
        reagents, 60–61
        reverse transcription reaction, 61
        sensitivity, 65
        single buffer reaction, 63–64, 66

thermostability, 43
types, 46

Uracil-*N*-glycosylase
  RT–PCR application, 59–60, 66
  thermolability, 59, 75

Variable number tandem repeats
  clinical applications of markers, 161–163
  PCR amplification
    cycle number, 166, 168
    efficiency, 162, 173
    optimization, 162–163
    polymerase concentration, 165
    primer concentration, 166
    product analysis
      acrylamide gel, 169
      allelic sizing ladder, 171
      Southern blotting, 163, 168–169, 171
    reaction conditions, 168
    specificity, 173–174
  size, 161
Visna-maedi virus
  copy number in cells, 199–200
  in situ PCR, 201, 203–204
VNTRs, *see* Variable number tandem repeats

VP4, epitope identification by DARTT, 239, 241, 245

Y chromosome-specific PCR, fetal cells
  amplification reaction, 216
  contamination prevention, 215
  primers, 215
  product analysis, 216
Yeast artificial chromosome
  hybridization of yeast colonies, difficulties, 277–278
  insert size, 277
  PCR screening
    advantages, 278
    amplification reaction, 279
    automation, 278, 285–286
    cell growth, 279
    commercial vendors, 282
    control templates, 283
    DNA concentration, 282–283
    optimization, 283–284
    pooling schemes, 284–285
    principle, 279–281
    probe labeling, 279–280
    product analysis
      electrophoresis, 284
      oligonucleotide ligation assay, 286
    template preparation, 282